AN INTRODUCTION TO ASTRONOMY

An Introduction to Astronomy

8th edition

Laurence W. Fredrick
Director, Leander McCormick Observatory
University of Virginia

Robert H. Baker

D. Van Nostrand Company
New York · Cincinnati · Toronto · London · Melbourne

D. Van Nostrand Company Regional Offices:
New York   Cincinnati   Millbrae

D. Van Nostrand Company International Offices:
London   Toronto   Melbourne

Library of Congress Catalog Card Number 73-18624
ISBN: 0-442-22436-2

Published by D. Van Nostrand Company
450 West 33rd Street, New York, N. Y. 10001

Published simultaneously in Canada by
Van Nostrand Reinhold Ltd.
10 9 8 7 6 5 4 3 2

Preface

A host of advances have been made in all areas of astronomy since the publication of the seventh edition of *An Introduction to Astronomy*. Man has walked on the moon, and we have had a series of close looks at Mars and its satellites. Now we are about to look at Jupiter and at least one of its satellites. If our ideas change as much about the cosmos in the next five years as they have changed in the past five, we can anticipate an extremely fascinating time indeed.

It is the intention of this new edition to provide the nonscience major with a basic astronomical knowledge of recent developments in astronomy and give some insight into the current frontiers of astronomy. To this end, the latest information and theories on such topics as pulsars, quasi-stellar objects, black holes, and the evolution of the universe have been presented. Where appropriate, explanations of opposing theories are offered, setting the stage for discussion and encouraging scientific thinking.

Past editions have stressed observational astronomy, and this orientation has received increased emphasis in the eighth edition. Because knowledge of the historical aspects of astronomy is essential for a complete understanding of modern developments and future progress in the field, an effort has been made to focus on the most significant past astronomical events. Two new chapters, Cosmology (Chapter 18) and Cosmogony (Chapter 19), have been added in the new edition to place all of the preceding information in the proper perspective.

The eighth edition of *An Introduction to Astronomy* has been written with the interests and backgrounds of liberal-arts students in mind. The nontechnical style has been maintained, and the use of mathematics has been kept to a minimum. An understanding of basic algebra is all that is required to perform even the most complex calculations. This edition follows the established organization of its predecessors and contains ample material to allow an instructor some flexibility in structuring a one-semester course.

The list of persons who supplied preprints, pictures, notes, etc., is too long to reproduce here, but the response to my requests leaves me in the debt of my world-wide colleagues. It is a pleasure to acknowledge the Austrian-American Educational Exchange Commission (Fulbright-Hays Program) and the hospitality of the Universitäts-Sternwarte Wien, under whose auspices almost all of this revision was carried out. Rebecca Myers Berg helped tremendously in smoothing the material presented and especially with the chapter-ending questions. I wish to acknowledge the tolerance and patience of my wife, Frances, who also typed the final manuscript.

Laurence W. Fredrick

Contents

To the Student

The history of man and the advances of civilized man, whatever the culture, can almost be traced in the history of astronomy. It is apparent that from the earliest times man viewed and held the heavens in awe. He reflected upon this magnificent celestial scene, and it seemed inconceivable to him that the stars, the moving planets, the sun, and the moon could not affect his life. Because of this he developed the art of studying the stars, which was called astrology by the ancient Greeks.

During later times the stars and the planets were studied somewhat more dispassionately; indeed, the early philosophers clearly distinguished between the study of the stars to predict the course of one's life and the study of the stars to explain the physical world in which we live. This distinction slowly grew until it was complete by the seventeenth century. These observers insisted that there is no influence from heavenly bodies on man's life and *astronomy* developed into the branch of science that studied the sun, the planets, the stars, and their motions in an effort to understand the universe around us.

The domain of study for astronomers begins where the atmosphere of the earth ends. It includes the sun and the planets, the comets and the stars, and later those assemblies of stars that we will learn are galaxies, and all of the galaxies that comprise the entire universe.

In general, astronomy is a passive science—the astronomer must look for the event that he wishes to study or wait for such an event to occur. He cannot set up an experiment in his laboratory and vary the conditions of the experiment seeking answers to his questions. He must wait, sometimes for decades, until the proper events occur for him to study. Astronomy is the study of grand events in the universe and of single events almost too great for one to comprehend. It is the study of normal long-living stars and the cataclysmic events that happen to stars often resulting in strange objects such as white dwarfs, or neutron stars, or even black holes.

In this textbook we begin with the familiar, that is, the earth and its immediate surroundings and its atmosphere and ionosphere that affect our observations. We then proceed systematically away from the earth until we encompass the universe and try to understand why the universe is expanding, what was its origin and evolution. Then we go back and look at the origin and evolution of life itself since, after all, we are occupants, albeit miniscule, of this grand universe.

The text is perforce incomplete, first, although most areas of astronomy are treated, none could be treated extensively in an introductory survey, and, second, the history of the science and the people who made and are making the science could be dealt with only sketchily. Astronomers from the ancients through Kepler, Bode, Slipher, Struve, and Shapley to the present time are interesting and even fascinating people, some of whom deserve separate texts. Thus, we have, in a sense, provided an outline, and it is for specific classroom programs to explore further those subjects that can be treated in greater depth, the historical events that may be emphasized, and those personalities that merit elaboration.

AN INTRODUCTION TO ASTRONOMY

The Earth and the Sky

1

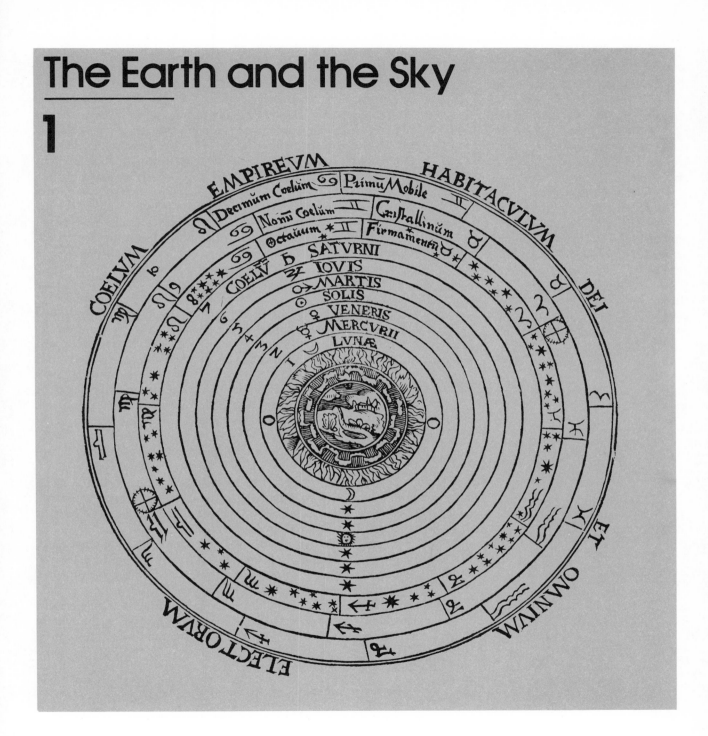

1

A relatively small planet attending the sun, which itself is an average star among the multitude of stars making up the Milky Way galaxy, the earth owes its importance to the fact that we live here. It is from the earth and its immediate environs that we view the celestial scene around us. Therefore, in order to interpret the scene correctly, we must first consider the earth that looms large in the foreground. Our study of astronomy begins with the globe of the earth at the center of the apparent globe of the heavens and encompassed by the atmosphere through which we look out at the celestial bodies.

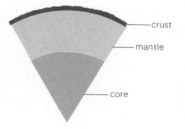

crust

mantle

core

THE GLOBULAR EARTH

The earth revolves around the sun once each year and, at the same time, rotates on its axis once each day. It is a dark globe illuminated by the sunlight having nearly the form of an oblate spheroid—a sphere flattened at the poles and bulged at the equator. The earth's diameter is about 12,700 kilometers,* and it is 43 kilometers greater at the equator than from pole to pole. If its actual form is represented by a globe 0.5 meter in diameter, the equatorial diameter of the globe would be 1.7 millimeters greater than the polar diameter and the highest mountain would rise only 0.3 millimeters above sea level. The earth is actually more round than a good ball bearing or an excellent bowling ball! The irregular surface of the earth is covered largely by water and is enveloped by an atmosphere to a height of several hundred kilometers.

The earth is the only planet known to have large water areas. It is likewise unique among the planets of the sun's family in having an abundance of free oxygen in its atmosphere. With these features and its sufficiently moderate range in surface temperature, the earth is the only planet known to us that seems inviting to life and particularly to human life. Planets attending other stars are difficult to detect with our present means of observation. Such planets might indeed be very numerous; many of them might be abodes of life and perhaps of intelligent beings who can observe and interpret the celestial scene. The opinion that life may be a widespread and important feature of the universe is frequently hypothesized. We will return to this fascinating subject in Chapter 19.

Aside from its water areas and atmosphere, the earth is a globe of rock that consists essentially of two parts: the **mantle**, extending 2900 kilometers below the surface, and the **core**. The crust is the outermost 5 to 40 kilometers of the mantle, being thinner under the oceans than under the continents. The upper levels of the crest are composed of igneous rocks (those formed from the cooling of molten lava), such as granite and basalt, generally overlain with sedimentary rocks (those formed by the layering of sediments under water), such as sandstone and limestone; these rocks are about three times as dense as water. The rest of the mantle is composed of heavier silicates of magnesium and iron. Knowl-

---

* 1 kilometer (km) = 1000 meters = 0.62 miles
1 meter (m) = 100 centimeters = 3.28 feet
1 centimeter (cm) = 10 millimeters (mm) = 0.4 inches

edge of the earth's interior is obtained almost entirely from the way it transmits earthquake waves at different depths to distant seismographs. An effort (called Project Mohole) to drill through the crust promises to yield direct information about the mantle. For the time being, however, this project has been suspended.

The core of the earth begins about 3400 kilometers from the center. Here the material behaves like a liquid in that it transmits compressional (or longitudinal) waves, but not transverse waves. This feature allows a seismologist to distinguish between near and very distant earthquakes. The inner core, within 1600 kilometers from the center, is 18 times as dense as water and is presumably very hot. Its composition may be mainly nickel–iron, like the material in many meteorites; or in another view it may be similar to the composition of the core at higher levels but under pressure here that is high enough to compress the molecules. The marked difference in chemical composition between the earth and the sun, which is composed mainly of hydrogen and helium, will be noted in a later chapter.

1.1 The planet earth

One of the greatest triumphs of man has been the observation of the earth from the moon (Fig. 1.1), where we see the earth as the celestial body that it is. The earth appears as a globe with visible surface features; seas and large lakes contrast markedly with the land. The large bodies reflect the sun specularly; bright areas of snowfields and drifting clouds add variety to the scene. The man-made features, while not immediately obvious, can be recorded over a period of time. Observed from the moon with the unaided eye, the earth appears in the lunar sky as a distinctively marked disk (often obscured by clouds) four times larger than the moon appears to us and goes through the whole cycle of phases from new to full and back to new again. The markings clearly reveal the earth's rotation period.

From the nearest planets the earth would appear to the unaided eye as a bright star. From Mars it would be a fine evening and morning star accompanied by the moon as a fainter star, both showing cycles of similar phases with the aid of a telescope. From the outermost planets the earth would be lost in the glare of the sun. From the nearest star the earth and all the other planets would be invisible even with the largest telescope, and the sun would appear only as one of the stars.

A considerable amount of information about the earth as a planet has been gathered with the help of rocket technology. These studies began in the late 1940s, and information has accu-

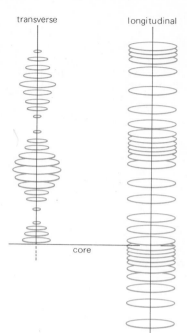

transverse   longitudinal

core

FIGURE 1.1
The earth in crescent phase from a returning Apollo spacecraft. Note that the terminator (line between day and night) is not sharp. (National Aeronautics and Space Administration photograph.)

mulated at an ever increasing pace. The use of staged rockets in 1957 during the International Geophysical Year allowed the launching of earth satellites containing instruments for the study of the earth's environs and its upper atmosphere. More recently manned earth satellites have been launched with greater and greater frequency and promise a whole new era of observation, not just of the earth, but of interplanetary space and of the stars as well.

The purpose of various satellites and probes has been to measure such quantities as the radiation field, density of micrometeorites, atmospheric density, and the anomalies of the earth's gravitational field. Satellites are now being used to relay communications (**passive satellites,** such as Echo I and II, reflect earth-based transmissions, and **active satellites,** such as Telestar and Syncom, receive and retransmit earth transmissions), to observe weather and cloud patterns, to observe earth sesources for economic and environmental purposes, and to experiment in remote-area education programs.

Spacecraft have been launched at speeds, called **escape velocity,** that are sufficient to escape the gravitational pull of the earth. These spacecraft have been used to study and explore the moon, to study the space between the planets, which we will refer to as the **interplanetary medium,** and to study Venus, Mars, and Jupiter.

1.2 The earth's magnetic field

The earth has a magnetic field that resembles the field of a bar magnet with a north and south pole; it is thus referred to as a **dipole field.** The axis of the field passes through the earth's center but is inclined at a considerable angle to the rotational axis. The north geomagnetic pole, toward which the north-seeking pole of a magnetic compass needle is directed, has been located in the Canadian archipelago several hundred kilometers from the earth's pole of rotation. The compass needle oscillates continuously and becomes especially unsteady during a **geomagnetic storm** (see Section 10.10).

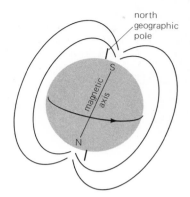

The existence of the dipole field led C. Størmer to predict the existence of trapped (charged) particles, such as electrons and protons, and hence explain the **aurora.** One of the early achievements of the International Geophysical Year (July 1957 through December 1958) was the discovery of two belts with an unexpectedly high concentration of charged particles trapped by the earth's magnetic field. These belts are referred to as the Van Allen belts after J. A. Van Allen and his associates who confirmed their existence. This group of investigators correctly interpreted the Geiger counter readings recorded on several spaceflights by probes in highly elliptical orbits. Actually the radiation zone (which includes the Van Allen belts extends from about 800 kilometers to perhaps 50,000 kilometers above the earth's surface and the Van Allen belts are two regions of high-energy charged particles centered at heights of approximately 3200 and 16,000 kilometers (Fig. 1.2). The source of these charged particles is ascribed to the solar wind (see p. 228).

1.3 Positions on the earth

One way of denoting positions on the earth's surface is with reference to natural or conventional areas. It is often satisfactory to the inquirer if we say, for example, that Cleveland is in Ohio. A second way, especially where positions are required more accurately, is with reference to circles imagined on the conventional terrestrial sphere (Fig. 1.3) that best represents the earth's surface. These familiar circles of our globes and maps are mentioned here so that the resemblance to systems of circles imagined in the sky may be noted later.

The earth's **equator** is the *great circle** halfway between its

---

* This is a circle going around the globe, on the largest circumference of the globe, that is formed by a plane going through the center of the globe. Because a great circle passes through the center of the earth, it divides it into two equal hemispheres.

FIGURE 1.2
Radiation intensity contours resembling a doughnut centered upon the earth's magnetic axis. The two heavy concentration rings are called the Van Allan belts.

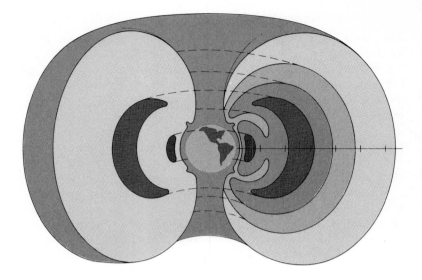

rotational north and south poles. **Parallels of latitude** are *small circles** parallel to the equator. **Meridians** pass from pole to pole and are accordingly at right angles to the equator; they are slightly elliptical but are considered as half-circles for our purposes. The **meridian of Greenwich,** or **prime meridian,** passes through the original site of the Royal Observatory at Greenwich, England. It crosses the equator in the Gulf of Guinea at the point where the longitude and latitude are zero.

If we draw a great circle going through the Greenwich meridian, then the angular distance of any place on the globe is measured by the angles viewed from the center of the globe, between the place and the point where the Greenwich meridian and equator cross, along the great circles. The **longitude** of a place is its angular distance along the equator east or west from the Greenwich meridian, from 0° to 180° either way. If the longitude is 60°W, the place is somewhere on the meridian 60° west of Greenwich meridian. The **latitude** of a place is its angular distance in degrees north or south from the equator, from 0° to 90° either way. If the latitude is 50°N, the place is somewhere on the parallel of latitude 50° north of the equator. (See Fig. 1.3.) When the longitude and latitude are given, the position of a place is uniquely defined. As an example, the longitude of Yerkes Observatory at Williams Bay, Wisconsin, is 88°33′W, and the latitude is 42°34′N.

Here we are using the customary symbols for angular measure:

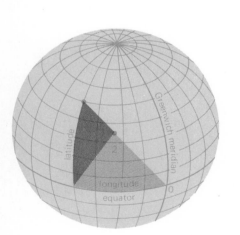

FIGURE 1.3
The coordinate grid on the terrestrial sphere. The basic coordinates are longitude, measured east and west of the Greenwich meridian, and latitude, measured north and south of the equator.

* A small circle is any other circle on the surface of the globe.

degrees (°), minutes ('), and, later, seconds ("). It should be clear when the symbol (°) is used to indicate degrees of temperature because we will always use it in conjunction with a capital letter. Thus °F, °C, °K are degrees of temperature on the Fahrenheit, centigrade, and Kelvin scales, respectively. The abbreviations for feet (') and inches (") will never be used in this text.

THE CONVENTIONAL GLOBE OF THE SKY

A very early picture of the earth and sky represents their appearance so well that it is often employed in diagrams, as shown in Fig. 1.5. Here the earth is imagined as a circular plane on which the sky rests like an inverted bowl. A somewhat later picture, proposed by Greek scholars in the fifth century B.C., is useful in other diagrams (Fig. 2.7). It shows the sky as a spherical shell surrounding a spherical earth at its center. The earth is relatively so small that it may be only a point in the diagrams at the center of the celestial sphere.

1.4 The celestial sphere

Although the stars are scattered through space at various distances from the earth, the difference in their distances is not perceptible to ordinary observation. All the stars seem equally remote. As we view the evening sky, we may imagine that the celestial bodies are set like jewels on the inner surface of a vast spherical shell. This **celestial sphere,** long regarded as a tangible surface, survives only as a convenient means of representing the heavens for many purposes. By this convention the stars can be shown on the surface of a globe or in projection on a plane map. Their positions are then denoted in the same ways that places are located on the globe of the earth.

The center of the celestial sphere may be the center of the earth, the observer's place on the earth's surface, the sun, or anywhere else we choose. The size of the sphere is as great as we care to imagine it. Because the sphere is infinitely large, parallel lines, regardless of their actual distance apart when nearby, will appear to converge at some point a great distance away. This is a familiar axiom from our high-school geometry studies. It is clearly demonstrated by the parallel railroad tracks, which seem to converge in the distance, or the parallel contrails of a highflying aircraft. Similarly, parallel lines in space, regardless of their distance apart, are directed toward the same point on the remote celestial sphere.

The **apparent place** of a star is its position on the celestial

FIGURE 1.4
The Big Dipper's pointer stars are very nearly 5° apart and make an excellent measuring reference as shown.

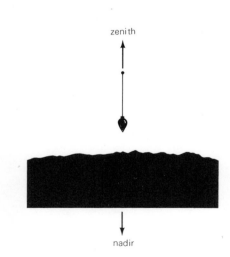

sphere. It denotes the star's direction, and nothing else about its location in space. Where two stars have nearly the same direction, although one may be more remote than the other, they have nearly the same apparent place. In addition, the apparent places of the sun, moon, and planets refer to their positions among the stars projected on the celestial sphere, as though all were at the same distance from us. We say that the sun is entering the constellation Leo and remark on the nearness of the moon to a bright star.

The **apparent distance** between two celestial bodies is accordingly their difference in direction; it is often called the **distance** (i.e., the angular distance) between them when there is no chance for ambiguity. Such distances are expressed in degrees. The distance between the pointers of the Big Dipper is somewhat more than 5°; it is a convenient measuring stick for estimating other distances in the sky (Fig. 1.4).

How may the place of a star be described so that other people will know where to look for it? One way is to specify the constellation in which the star appears. If the star is in the constellation Perseus, anyone who can recognize the different constellations knows approximately where this star is situated. A second way of denoting the apparent place of a star is with reference to circles of the celestial sphere, such as the horizon.

1.5 The horizon

A cord by which a weight is suspended provides a vertical line when the weight comes to rest. This line leads upward to the **zenith,** the point directly overhead in the sky, and it leads downward through the earth to the **nadir,** the point directly underfoot.

The **celestial horizon,** or simply the **horizon,** is the great circle of the celestial sphere that is halfway between the zenith and nadir, and therefore 90° from each. The direction of the horizon is observed by sighting along a level surface, perhaps a table top. Obviously the positions of the zenith, nadir, and horizon among the stars are different at the same time in different parts of the world.

The **visible horizon,** the line where the earth and sky seem to meet, is rarely the same as the horizon of astronomy. On land it is usually irregular and above the celestial horizon. At sea in calm weather it is a circle that lies below the celestial horizon; this **dip** of the sea horizon increases with increasing height of the observer's eye above the level of the sea, caused by the curvature of the earth's surface. Thus,

$$d = 1.8 \sqrt{h}$$

where d is the dip in minutes of arc and h is the altitude in meters. If the observer is at sea and his eye is at sea level he sees the true horizon. If he is above the level of the sea he sees a depressed horizon caused by the dip of the horizon. Further, the observer's visible horizon recedes as he increases his altitude. We will note later that local sunrise, sunset, etc., are effected by the dip of the horizon, that is, the observer's altitude.

1.6 The celestial meridian

Vertical circles are great circles of the celestial sphere that pass through the zenith and nadir, and accordingly cross the horizon vertically. The most useful of these circles is the observer's **celestial meridian**, the vertical circle passing through the north and south poles of the heavens (see Section 2.6). The meridian determines the positions of the four **cardinal points** of the horizon: north, east, south, and west.

North and **south** are the opposite points where the celestial meridian crosses the horizon. The **east** and **west** points are midway between them; as we face north, east is to the right and west is to the left. When the cardinal points are already located, it is proper to define the celestial meridian as the vertical circle passing through the north and south points of the horizon and the zenith.

1.7 Azimuth and altitude

The **azimuth** of a star can be measured in degrees along the horizon from the north point eastward to the foot of the vertical circle through the star. Thus the azimuth of a star is 0° if the star is directly in the north, 90° in the east, 180° in the south, and 270° in the west. This coordinate is often reckoned around from the south point instead. (See Fig. 1.5.)

The **altitude** of a star is its distance in degrees from the horizon, measured along the vertical circle of the star. The altitude is 0° if the star is rising or setting, 45° if it is halfway from the horizon to the zenith, and 90° if it is in the zenith. The **zenith distance**, or the star's distance in degrees from the zenith, is the complement of the altitude.

This is one way of denoting positions on the celestial sphere. If, for example, a star is in azimuth 90° and altitude 45°, the star is directly east and halfway from the horizon to the zenith; if it is in azimuth 180° and altitude 30°, the star is directly south and a third of the way from the horizon to the zenith. Certain instruments operate in this system based on the horizon. The engineer's transit is an azimuth–altitude instrument. The navigator's sextant is employed mainly for measuring the altitudes of celestial bodies.

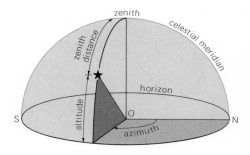

FIGURE 1.5
Azimuth and altitude of a star. Azimuth is measured along the horizon from the north point around toward the east. Altitude is angular distance vertically above (or below) the horizon. Zenith distance is the complement of altitude.

Positions of stars denoted in this way are accurate only for a particular instant and for a particular place on the earth. The azimuths and altitudes of the celestial bodies are always changing with the daily motion of the heavens, and they are different at the same time in different parts of the world. More permanent positions relative to the celestial equator rather than the horizon are defined in Chapter 2. Meanwhile, we turn our attention to the earth's atmosphere and note particularly the effect it has on our view of the sky.

EFFECTS OF THE ATMOSPHERE

1.8 The region of the clouds

The earth's atmosphere is a mixture of gases surrounding the earth. From its pressure of 1.013×10^6 dynes per square centimeter (14.7 pounds per square inch) at sea level, the mass of the entire atmosphere is calculated to be 5.3×10^{21} grams, or somewhat less than a millionth of the mass of the earth itself.* The air becomes rarefied so rapidly with increasing elevation that half of it by weight is within 5.6 kilometers from sea level. The lower atmosphere is divided into two layers, the troposphere and the stratosphere.

The **troposphere** extends from the surface to heights ranging from 16 kilometers at the equator to 8 kilometers at the poles. It consists mainly of nitrogen and oxygen molecules in the proportion of 4 parts to 1 by volume; it also contains water vapor, carbon dioxide, and other gases in relatively small amounts, as well as dust in variable quantity. The water vapor and carbon dioxide are serviceable as a blanket to conserve the surface heat at night. Like the glass roof of the greenhouse, they let the sunlight (short, ultraviolet waves) through to warm the ground and then by strong absorption in the infrared region, the long wavelength side of the spectrum (see Section 5.11), prevent rapid loss of the heat by reradiation, hence the term the **greenhouse effect.**

The troposphere is a region of turbulent air and of clouds. Foglike stratus clouds begin to form at an average elevation of 1 kilometer. Cumulus clouds rise from flat bases about 2 kilometers above the ground. The cirrocumulus clouds of the "mackerel sky" have an average height of 6 kilometers (Fig. 1.6). The filmy cirrus clouds of ice crystals may be as high as 11 kilometers or more.

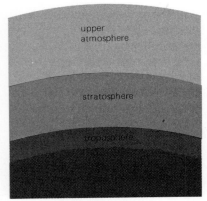

upper atmosphere

stratosphere

troposphere

earth

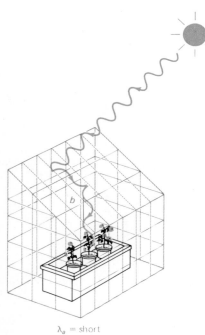

λ_a = short
λ_b = long

---

* 1 U.S. ton $= 9.07 \times 10^5$ grams
1 gram weight $= 981$ dynes

FIGURE 1.6
A mackerel sky showing a well defined pattern.

Even higher clouds, called noctilucent clouds because they are observed after sunset and before sunrise, occasionally occur in the base of the stratosphere at heights of 90 kilometers.

The powers of ten notation, which has been used above, is a great convenience in science, especially astronomy, where numbers are often quite large or small. For example, one one-thousandth (0.001) can be written 10^{-3} and one thousand (1000) can be written 10^3 (see Appendix).

The **stratosphere** extends from the troposphere to a height of 70 kilometers. Its constituents are in about the same proportions as found in the troposphere, except that there is less water vapor and more ozone. Ozone, composed of molecules containing three atoms of oxygen instead of two, is formed mainly by action of the sun's ultraviolet radiations on ordinary oxygen molecules.

Ozone is most abundant at elevations of from 15 to 30 kilometers. It helps to protect us from the extreme ultraviolet rays of the sun, which would be injurious to life of all kinds if they could penetrate to the earth's surface. These rays are being studied in photographs of the solar spectrum taken from above the ozone levels to determine what additional information they contain about the sun itself.

The stratosphere contains one-fifth of the entire air mass. Even at its upper limit it is still dense enough to make some contribution to twilight, as is known from the duration of this light.

250 km F_2

170 km F_1

120 km E

70 km D

Twilight (see Section 2.11) is sunlight diffused by the air onto a region of the earth's surface where the sun has already set or has not yet risen.

1.9 The upper atmosphere

The upper atmosphere reaches from the stratosphere to an altitude of more than 800 kilometers. Here the rarefied gases are most exposed to impacts of high-frequency radiations and high-speed particles from outside. The molecules are largely reduced to separate atoms, and the atoms themselves are shattered into electrically charged components. The **ionosphere**, the region ranging from an altitude of 70 to 320 kilometers, contains at least four fluctuating layers (D, E, F_1, F_2) where the ionized gas is concentrated. By successive reflections between these layers and the ground, radio waves can travel long distances before they are dissipated. When the layers are disrupted during a geomagnetic storm, communication by radio in the higher frequencies is disturbed.

The impact of particles from space on the gases of the upper atmosphere makes these gases luminous in the varied colors of the aurora and in the airglow. In the lower ionosphere the resistance of the denser air to the swift flights of meteors heats these intruders to incandescence, so that they produce bright trails across the sky.

1.10 The daytime sky

The stars are invisible to the unaided eye in the daytime because the atmosphere diffuses the more intense sunlight that reaches us from all parts of the sky. The sunlit air also conceals most of the other celestial bodies by outshining them. Other than the sun, the moon is the only heavenly body ordinarily conspicuous in the daytime. The planet Venus near the times of its greatest brilliancy can be seen without a telescope as a star in the blue sky. The bright planets and stars, however, are easily visible in the daytime sky with a telescope, and with special devices they can be photographed even near the edge of the sun (Fig. 1.7).

Why is the clear sky of the daytime blue, whereas the sunlight itself is yellow? Sunlight is composed of many colors, as we observe when the light passes through a prism, or through raindrops or the spray of a waterfall; it contains all the colors of the rainbow. As sunlight comes through the atmosphere, the violet and blue light is most scattered by the air molecules, and the red light is least affected. Hence on a clear day the sky takes on the blue color of the light that is scattered down to us most profusely.

FIGURE 1.7
*Jupiter near the limb of the eclipsed sun.
This picture was taken from a high-altitude
jet aircraft. (Photograph by M. Waldmeier,
Arosa, Switzerland.)*

When the sun is near the horizon, most of the blue of its direct
light is scattered away before it can reach us through the greater
thickness of air that then intervenes. Thus the sun appears reddened at its rising and setting.

1.11 Apparent flattening of the sun

Another aspect of the sun near the horizon is as familiar as its
reddened color. There the sun sometimes appears so noticeably
flattened at the top and bottom (Fig. 1.8) that its disk appears
elliptical. This appearance is caused by refraction of the sunlight
in the atmosphere. Refraction of light is the change in the direction of a ray of light when it passes from one medium into another, as from rarer into denser air. The effect is described in
Chapter 5 in connection with the operation of the refracting
telescope.

As the light of a celestial body comes through the air, the rays
are bent downward toward the earth by refraction (Fig. 1.9), so
that the object appears higher in the sky than it actually is. The

FIGURE 1.8
Apparent flattening of the sun due to atmospheric refraction. (Yerkes Observatory photograph.)

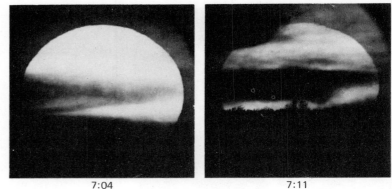

7:04   7:11

FIGURE 1.9
Apparent raising of a star due to atmospheric refraction.

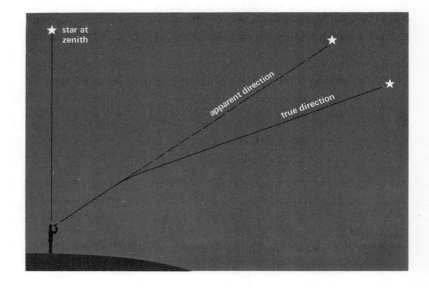

amount of the apparent elevation increases with the zenith distance of the object, but so gradually at first that for more than half the distance to the horizon the difference is not enough to be detected without a telescope. Near the horizon the increase is rapid, until the effect becomes conspicuous to the unaided eye. An object on the horizon is raised above its true place more than half a degree, or more than the apparent diameter of the sun or full moon. Because of the rapid increase in the amount of refraction as the horizon is approached, the circular disk of the sun appears elliptical; the lower edge is raised considerably more than the upper edge.

In addition to its flattened appearance the sun's disk seems to be larger near the horizon than when it is higher in the sky. This is an illusion having nothing to do with refraction. By the same illusion, the moon and the star figures, such as the Big Dipper, seem magnified near the horizon.

1.12 Twinkling of the stars

A familiar feature of the clear night sky is the twinkling, or **scintillation,** of the stars; their light is unsteady, and the fluctuations in apparent brightness are especially noticeable near the horizon. The lower air is often turbulent; warmer currents are rising, cooler currents are descending, and horizontal movements of layers of different densities add to the confusion. Viewed through this turmoil the stars twinkle because of variable refraction and interference of their light, just as the landscape seems to be affected by the "heat waves" over the highway on a summer day.

The bright planets usually do not twinkle. They are disks instead of point sources of light, as a telescope shows. Each point of the disk may twinkle like a star, but the different parts do not do so in unison because their rays take slightly different paths through the disturbed air. Similarly, the moon shines with a steady light.

Viewed with a telescope, the stars appear blurred when the air is especially turbulent. The rays in each beam of starlight that enter the telescope have been diverted from perfect parallelism, so that they are not brought to the same focus by the lens. Features of the moon and planets also appear blurred in these conditions. The **astronomical seeing** is bad, and there is not much to be done about it except to wait for better seeing. With small apertures the image is less blurred but may shift about somewhat.

1.13 Lunar and solar halos

Bright rings that sometimes appear around the moon and sun have no special astronomical significance; they are noticed and commented on by watchers of the skies and are fine examples of refraction effects in the atmosphere. The rings are produced by ice needles and snowflakes in the cirrus (thin, feather-like clouds that are high in the troposphere) and cirrostratus clouds. These six-sided crystals refract the moonlight and sunlight, concentrating it in certain directions. Although many effects are possible, the most common one is a single ring (Fig. 1.10) having a radius of 22° around the moon or sun. The ring often shows rainbow colors, with the red on the inner, sharper edge the most prominent.

FIGURE 1.10
*Solar halos. (Bottom) Common ring halo.
(Top) The lower portion of an unusual
complex halo. (Photographs by Dr. E.
Everhart.)*

"Moon dogs" and "sun dogs" are two enlargements of the ring on opposite sides of the moon or sun. These appear when many snowflakes in the clouds float with their bases in a horizontal position. A second ring having a radius of 46° and parts of other rings are seen less frequently (Fig. 1.10). The impression that the appearance of a ring around the moon or sun gives warning of an approaching storm has some factual basis because the filmy clouds that form the halos are likely to be blown ahead of storm clouds.

1.14 The aurora and the airglow

Two natural illuminations of the night sky in addition to the light of the celestial bodies are the aurora and the airglow. The **aurora** is characterized in middle northern and southern latitudes by a luminous arch across the northern sky, with its apex in the direction of the magnetic pole. Rays like searchlight beams rise above the arch, drifting, dissolving, and reforming, and draperies may appear in other parts of the sky. The rays are usually green, but may be red and yellow. Especially in latitudes farther north and south ribbons spread across the sky (Fig. 1.11). The aurora is referred to as the Aurora Borealis in the northern hemisphere and as the Aurora Australis in the southern hemisphere. Geomagnetic storms result from the influx of charged particles from unusual solar activity being trapped in the earth's magnetic field (see

FIGURE 1.11
A beautiful looped curtain aurora over Alaska. (Geophysical Institute, University of Alaska photograph.)

Section 1.2). When the magnetic field dips down into the ionosphere at the magnetic poles, these particles interact with the atmosphere causing the aurora. The displays are most intense in two zones centered around 23° from the magnetic poles. South of latitude 35° in the northern hemisphere, or south of a line through San Francisco, Memphis, and Atlanta, they appear only during geomagnetic storms of considerable intensity.

The **airglow** is an illumination suffused over the sky. Its light is also caused by energy coming primarily from the sun. It is faintest overhead and brightest not far from the horizon, which shows that the glow is from the atmosphere. It is undetected by the eye but is effectively studied with the photoelectric cell and color filters. Because the airglow is always present, it is the greater menace to celestial photography. The airglow fogs astronomers' photographs, placing a severe limit to the faintest objects that can be reached by increasing the exposure times. Only by carefully selecting a photographic plate and filter combination that eliminate the light from airglow or by using very high orbiting telescopes can this problem be circumvented.

QUESTIONS

1 If the globe mentioned in Section 1.1 were 1 meter in diameter at the equator how much shorter would the polar diameter be? What would the height of the highest mountain be?
2 How do we study the earth's interior?
3 Why was Project Mohole planned so that its drilling would be at sea?
4 Draw up a list of useful studies from earth satellites. Discuss those that effect your everyday life and those that have effected the entire world.
5 Define apparent place and apparent distance.
6 Describe a crude but convenient way to measure the angle of the pole star from the northern horizon. Can you show that this is your latitude? Can you devise a method that will work in the southern hemisphere (include assumptions that you must make)?
7 What are the three basic divisions of the earth's atmosphere?
8 What causes the ionosphere and what is the value of the ionosphere?
9 Why is the daylight sky blue?
10 Why don't the planets twinkle?

FURTHER READINGS

CHRISNALL, G. A. AND G. FIELDER, *Astronomy and Spaceflight,* Harrap, London, 1962, Chaps. 1–4.

NAMOWITZ, S. N. AND D. B. STONE, *Earth Science,* 3rd ed., Van Nostrand, New York, 1965.

STRAHLER, A. N., *The Earth Sciences,* 2nd ed., Harper & Row, New York, 1971.

SUTTON, G., ED., *Global Geophysics,* American Elsevier, New York, 1970.

WINTER, S. S., *The Physical Sciences,* Harper & Row, New York, 1967, Chaps. 4–6.

A drawing of an armillary sphere. Such a sphere is used to show all of the basic circles and the relationships between the horizon, equatorial, and ecliptic coordinate systems.

The Earth's Daily Rotation

2

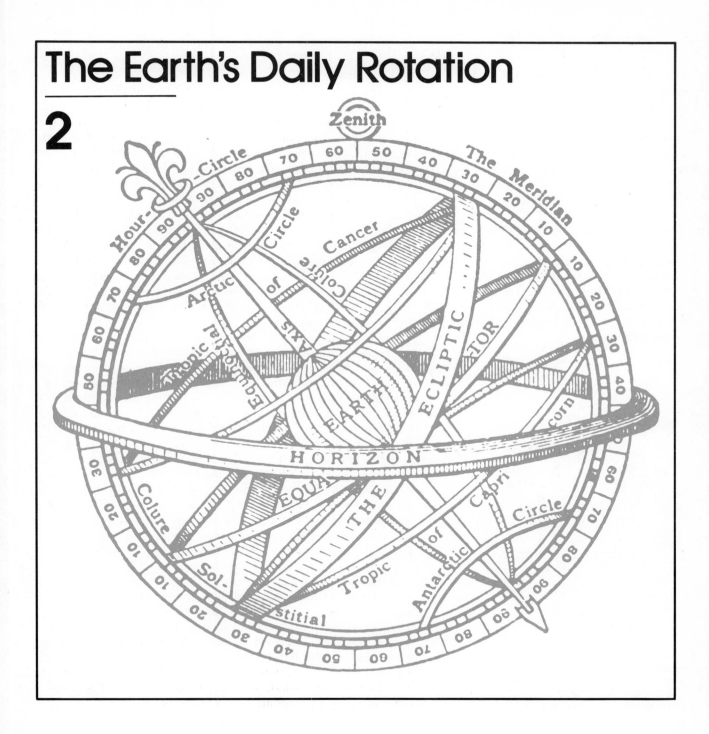

2

The distinction between rotation and revolution is more precise in astronomy than in some other sciences. Rotation is turning on an axis, whereas revolution is motion in an orbit. Thus the earth rotates daily and revolves yearly around the sun. In this chapter we consider the earth's rotation and some of its effects.

The earth rotates from west to east on an axis joining its north and south poles. Among the effects of the rotation are the directions of prevailing winds and cyclones, the behavior of the Foucault pendulum, the bulge of the equator, and the apparent rotation of the heavens.

2.1 The Coriolis effect

Angular speed (or angular velocity) is the rate of turning per unit time. For the earth this is 360° per day, or 15° per hour, or 15 minutes of arc per minute of time, or 15 seconds of arc per second of time. It is the same for all points on and inside the earth, because the earth rotates as a solid body. Linear speed (or linear velocity) is the distance traveled per unit of time. On the turning earth the speed of the earth's rotation decreases with increasing distance from the equator. The speed is nearly 1670 kilometers an hour at the equator; it is reduced to 1280 kilometers an hour in the latitude of New York, 800 kilometers an hour in southern Alaska, and so on, until at the pole there is no velocity at all. Thus we can see that a person at the equator travels faster than a person at the latitude of New York. This is clearly shown in Fig. 2.1 where the distance 1–2 is on the equator and 3–4 is at some latitude l. The speed, s_l, at any latitude l, in degrees, is related to the speed at the equator, s_o, by the simple relation $s_l = s_o \cos l$.

Consider an air current moving north in the northern hemisphere and being carried eastward by the earth's rotation at the same time. Because it is going from a latitude of faster rotation to one of slower rotation, the current forges ahead and is accordingly deflected to the right. If the current is moving south instead, it is going from a place of slower to one of faster rotation and is again deflected to the right. Next consider a current moving either north or south in the southern hemisphere; we see that it is deflected to the left. Thus we have the following rule that applies to these horizontal motions: The earth's rotation deflects moving objects to the right in the northern hemisphere and to the left in the southern hemisphere. This deflection, known as the **Coriolis acceleration** after the scientist who demonstrated it over a century ago, arises because the earth is not an inertial system of reference. That is, it is not a fixed reference system but a rotating one.

The global circulation of the atmosphere is generated by the energy in sunlight, which heats the air most intensely in the tropics. The warmed air rises there and flows toward the poles. Cooled at high elevations the air descends, notably at latitudes 30° where it flows north and south over the surface. Following our rule, these surface winds are deflected to the right in the northern hemisphere and to the left in the southern hemisphere (Fig. 2.2). Thus we have the easterly trade winds of the tropics and the prevailing westerly winds of the temperate zones. The easterly winds of the frigid zones are an associated effect.

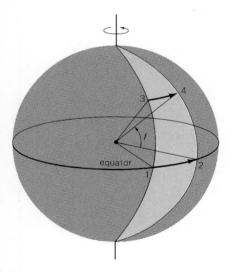

FIGURE 2.1
The surface speed along the equator (1 to 2) is greater than that at some latitude (l) (3 to 4). At the poles it is zero.

Ocean currents follow the prevailing winds in a general way but are complicated by the land barriers. We see the effect of our rule clearly in the Gulf Stream, which flows initially from the southwest.

Eddies in the air circulation, such as cyclones and hurricanes, show the Coriolis effect in the directions of their whirling. Consider a cyclone in the northern hemisphere, where the air is locally rising. The surface currents flowing into this area of low pressure are deflected to the right and reach its center indirectly. Thus the whole area, perhaps 2400 kilometers in diameter, is set whirling in a counterclockwise direction. In the anticyclones, or "highs" of our hemisphere, the air is descending and flowing out over the surface; these accordingly whirl in the clockwise direction. The directions of cyclones in the southern hemisphere are clockwise, and those of anticyclones are counterclockwise. The Coriolis effect can be seen in the global circulation on the planets Mars and Jupiter.

The importance of the Coriolis effect extends beyond its meteorological consequences. Correction for the effect is required in long-range ballistic rocket flights and in determining the altitudes of celestial objects as observed with the bubble octant by the air navigator.

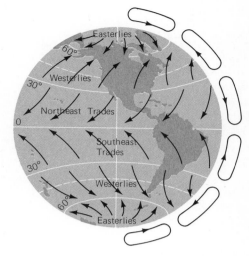

FIGURE 2.2
Prevailing surface winds. The moving air is deflected to the right in the northern hemisphere and to the left in the southern hemisphere.

2.2 The Foucault pendulum

A convincing proof of the earth's rotation was first demonstrated to the public by the physicist J. Foucault in 1851. Under the dome of the Panthéon in Paris, Foucault freely suspended a heavy iron ball by a wire 61 meters long and started it swinging. The audience saw the plane of the oscillation slowly turn in the clockwise direction. They were observing the changing direction of the meridian caused by the earth's rotation relative to the direction of the swing of the pendulum. This celebrated demonstration is often repeated in planetariums and museums. (See Fig. 2.3.)

The rate of change in the direction of the swing of this pendulum depends on the latitude. At the equator, where the direction of the meridian in space is not altered by the earth's rotation, there is no change at all in the direction of the pendulum. In the latitude of Chicago the rate of change is 10° an hour. At the pole it is 15° an hour, so that the plane of the oscillation turns completely around in a day. The period of one complete rotation of the pendulum in hours may be calculated by the formula

$$P = \frac{24}{\sin l}$$

where l is the latitude.

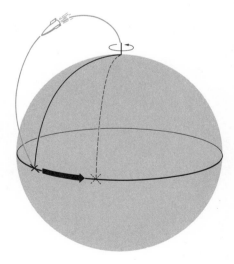

FIGURE 2.3
The rotation of the earth is revealed by the Foucault pendulum. (Photograph by Elliott Kaufman, courtesy of The Franklin Institute.)

Another effect of the earth's rotation is seen in the behavior of the gyrocompass, where the rotor automatically brings its axis into the plane of the geographical meridian, and thus shows the direction of true north. This valuable aid to the navigator would not operate if the earth did not rotate. Still another rotational effect is the bulging of the earth's equator.

2.3 Centrifugal effect of the earth's rotation

The equator is more than 21 kilometers farther from the earth's center than are the poles; in this sense it is downhill toward the poles. Why then doesn't all the water of the oceans assemble in these lowest regions around the poles? Why does the Mississippi River flow "uphill" toward the equator? The reason is found in the earth's rotation.

All parts of the rotating earth have a tendency to move away from the axis. It is the same centrifugal effect that urges a stone to fly away when it is whirled around at the end of a cord. Part of the effect of the earth's rotation on an object at its surface is to slide the object toward the equator (Fig. 2.4). The earth has adjusted its form accordingly so that the upslope toward the equator offsets the tendency to slide toward it.

The second component of the centrifugal effect lifts the object, so that its weight on a spring balance is diminished. An object weighing 86 kilograms* at the pole, where there is no centrifugal effect, becomes almost 0.5 kilogram lighter at the equator, where this effect is greatest, although part of the decrease in weight here results from the greater distance of the equator from the earth's center. The force of gravity is weaker at the equator because the distance from the center of the earth is greater. If the period of the rotation should decrease to 1 hour and 24 minutes by our present clocks, an object at the equator would weigh nothing at all. At this excessive rate of rotation the earth would be unstable and liable to disruption.

2.4 Wanderings of the poles

When the positions of places on the earth are determined repeatedly with high accuracy, it is found that their latitudes do not remain the same. This **variation of latitude** has been observed for many years at a number of international latitude stations distributed over the world, and these results tell us how the equator and therefore the poles have been wandering.

The wanderings of the poles are caused by shifts of the earth relative to its axis, so that the poles change slightly on the surface (see Fig. 2.5). One variation is seasonal, having a period of a year; the other is an oscillation in a period of 14 months. The resulting motion of the poles is irregular and limited. Neither pole is withdrawing much more than 12 meters from its average place; all of the wanderings are now confined to an area smaller than that of a baseball diamond. Some scientists suggest the possibility of wider migrations in the past, which might have caused the marked changes in climate in geological times.

2.5 Changing period of the earth's rotation

The earth's rotation has long set the standard for our timekeeping. The consequent daily rotation of the heavens has provided the master clock by which all other clocks have been corrected. The earth-clock was formerly considered entirely reliable; the period of the earth's rotation was supposed to be uniform, until it finally

* 1 kilogram (kg) = 2.2 pounds

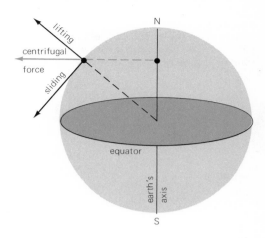

FIGURE 2.4
The earth's rotation tends to diminish the weight of an object and slide it toward the equator.

FIGURE 2.5
The trace of the earth's pole over the last six years. The year mark is for the beginning of that year.

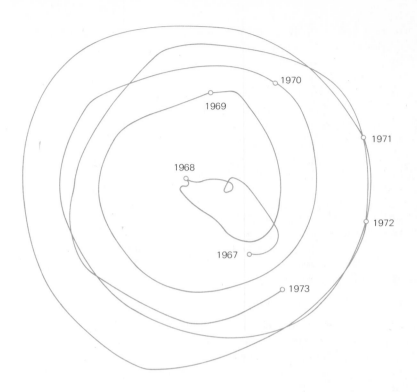

proved otherwise when more accurate timekeeping methods were developed.

Suppose that someone begins with the idea that his watch is always right. As the days go by he is surprised to find that everything is getting ahead of time by his watch. He misses trains that seem to depart too early; the sun rises before it should; the town clock runs faster and faster. Presently he decides that his watch must be running slow. It is so with the earth's rotation. Periodic occurrences, such as the revolution of the moon around the earth, are forging ahead of regular schedules timed by the earth-clock. We conclude that the period of the earth's rotation is increasing, and also that the increase is not perfectly regular (Fig. 2.6).

By comparing the recorded times of early eclipses with the calculated times (the times they would have occurred if the earth had been rotating uniformly), astronomers have concluded that the length of the day is increasing at the rate of 0.0016 second in a century. This would mean that the earth-clock has run slow by more than 3.25 hours during the past 20 centuries. The tides in the oceans are regarded as the brakes responsible for reducing the speed of the rotation. They are caused chiefly by the moon's attraction and follow the moon around the earth once in a month, while the earth itself rotates under the tide figure once a day. Much uncertainty, however, remains in the size of the increase of the period, which should be improved with the present more precise means of making observations.

There are also sudden and not as yet accurately predictable variations in the length of the day. The earth's rotation has run off schedule, both fast and slow, as much as half a minute after allowance for the tidal retardation is made. Small periodic variations, mainly annual and semiannual, are detected by the best clocks; these are ascribed to winds and tides, and they are reliably repeated.

APPARENT ROTATION OF THE HEAVENS

2.6 The celestial coordinates

The stars rise and set, daily circling westward and keeping precisely in step as they go around. The patterns of stars, such as the Big Dipper, look the same night after night and year after year. It is as though the stars were set on the inner surface of a rotating hollow globe. This celestial sphere seems to turn daily from east to west around an axis which is the axis of the earth's rotation projected to the sky.

The **celestial poles** are the two opposite points on the celestial sphere toward which the earth's axis is directed, and around which the stars circle. The north celestial pole is directly in the north, from a third to halfway up in the sky for observers in different

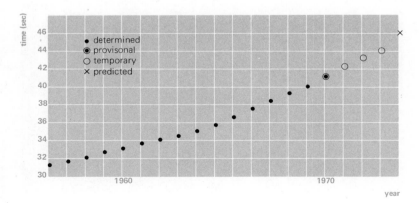

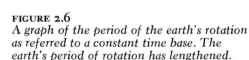

FIGURE 2.6
A graph of the period of the earth's rotation as referred to a constant time base. The earth's period of rotation has lengthened.

parts of the United States. The south celestial pole is similarly depressed below the south horizons of these places.

The pointers of the Big Dipper (Fig. 2.7) direct the eye to Polaris, the **pole star,** at the end of the Little Dipper's handle. This moderately bright star is within 1°, or about two moon-breadths, of the pole itself. It is also the **north star** showing the approximate direction of north. The south celestial pole is not similarly marked by any bright star in its vicinity.

Just as the earth's equator is halfway between the terrestrial poles, so the **celestial equator** is halfway between the north and south celestial poles. (See Fig. 2.8.) This circle crosses the horizon at its east and west points at an angle that is the complement of the latitude. Thus in latitude 40°N, or the latitude of Philadelphia, the celestial equator is inclined 50° to the horizon and has an altitude of 50° at its highest point in the south, the complement of 40° being 90° − 40° or 50°.

Hour circles in the sky are like longitudes on the earth. They are half circles that connect the celestial poles and are therefore perpendicular to the equator. Unlike the circles of the horizon system, which are stationary relative to the observer, these circles are to be considered as sharing in the rotation of the celestial sphere. If 24 hour circles are imagined equally spaced, they will

FIGURE 2.7
The Big Dipper's pointers show the way to the celestial pole.

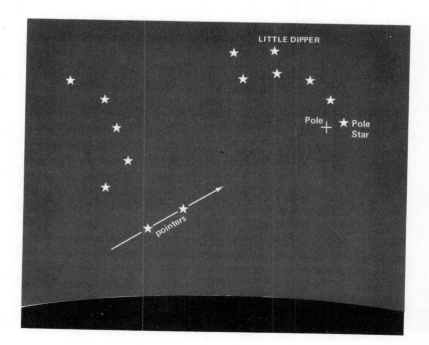

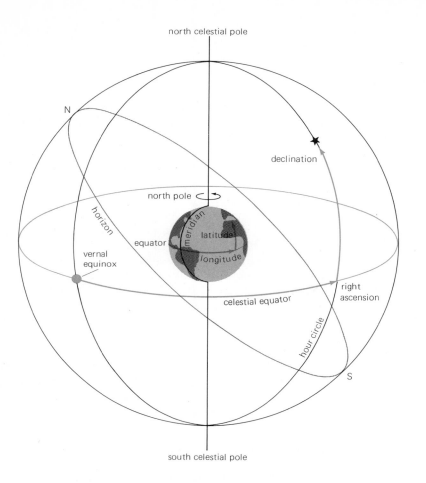

north celestial pole

N

declination

north pole

horizon

meridian

latitude

equator

longitude

vernal
equinox

right
ascension

celestial equator

hour circle

S

south celestial pole

coincide successively with the observer's celestial meridian at intervals of an hour. With reference to the celestial equator and its associated circles, the position of a celestial body is given by its right ascension and declination, which resemble terrestrial longitude and latitude.

The **right ascension** of a star is its angular distance measured eastward along the celestial equator from the **vernal equinox** to the hour circle through the star. (The vernal equinox is the point where the sun's center crosses the celestial equator at the beginning of spring.) Right ascension is expressed in time more often than in angular units. Because a complete rotation of the heavens, through 360°, is made in 24 hours, 15° is equivalent to 1 hour, and 1° to 4 minutes of time. Thus a star's right ascension may be given as 60° or 4 hours.

The **declination** of a star is the star's distance in degrees north or south from the celestial equator, measured along an hour circle through the star. The declination is marked either N or with a plus sign if the star is north of the equator, and S or with a minus sign if it is south.

As an example, the right ascension of the star Arcturus is $14^h 14^m$ and its declination is $19°22'N$. The star is accordingly $213°30'$ east of the vernal equinox and $19°22'$ north of the celestial equator. Notice that right ascension is measured only eastward, whereas terrestrial longitude is measured both east and west. The approximate right ascensions and declinations of the brighter stars can be read from the star maps in Chapter 11.

The hour angles are often employed in the equator system instead of right ascension. The local **hour angle** of a star is reckoned westward along the equator from the observer's celestial meridian through 360° or 24 hours. Unlike right ascension, which remains nearly unchanged during the day, the hour angle of a star increases at the rate of 15° per hour and, at the same instant, has different values for observers in different longitudes.

The hour angle is frequently used in directing a telescope to a celestial object by means of a graduated circle; the hour angle (H.A.) of the object is found by subtracting its right ascension (R.A.) from the sidereal time (S.T.). (See Section 4.2.) Hour angles are useful in celestial navigation. The Greenwich hour angles of celestial bodies are tabulated for this purpose at convenient intervals of the day throughout the year in nautical and air almanacs.

2.7 Latitude equals altitude of celestial pole

The latitude of a place on the earth is its distance in degrees from the equator. For our present purpose it is more conveniently defined as the number of degrees a vertical line at a place is inclined to the plane of the equator, or to the plane of the celestial equator. This vertical line is directed toward the observer's zenith.

In Fig. 2.9 we see that the observer's latitude is the same as the declination (degrees north or south of celestial equator) of his zenith. This angle equals the altitude of the north celestial pole, because both are complements of the same angle between the directions of the zenith and pole. Thus, *the latitude of a place equals the altitude of the celestial pole at that place.* In examining the figure we recall that parallel lines are directed toward the same point in the heavens.

The rule determines the **astronomical latitude.** This observed latitude depends on the vertical line and is affected by anything

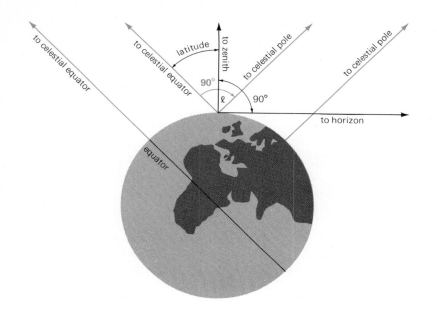

FIGURE 2.9
The latitude of a place on the earth equals the altitude of the celestial pole at that place. Astronomical latitude is defined as the angle between the vertical line and the equator plane. This angle equals the altitude of the celestial pole because both are complements of the angle between the pole and the zenith.

that alters the vertical. A mountain near the place of observation would cause the plumb line to incline a little in its direction. Such "station errors" may make a difference of nearly a minute of arc in the latitude, although they are usually much smaller. **Geographical latitude** is the observed latitude corrected for station error; it is the latitude that would be observed if the earth were perfectly smooth and uniform.

The latitude of a place on the earth is determined by the rule mentioned above. If there were a bright star precisely at the celestial pole, the latitude could be found simply by measuring the altitude of that star. The pole star itself may be employed if correction is made for its distance from the pole. More often the observations are made to determine the declination of the zenith, which we have seen is the same as the latitude. A simple calculation gives the required value when the altitude of a celestial body is observed at its upper transit of the celestial meridian (see Section 4.1).

Suppose that a navigator sights the sun at its crossing of the meridian south of the zenith and determines its true altitude as 51°10′, so that its zenith distance is 38°50′. The sun's declination obtained from an almanac for the time of the sight is 22°0′N. The latitude equals the sun's zenith distance plus its declination or 60°50′N (Fig. 2.10). If the sun had crossed the meridian the same distance north of the zenith, the zenith distance would receive

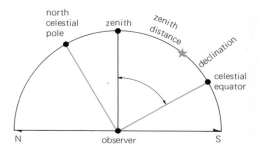

FIGURE 2.10
The latitude of a place equals the zenith distance of a celestial body at its upper transit of the meridian of the place plus its declination at that time.

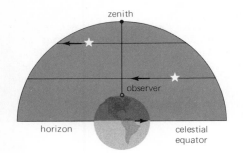

FIGURE 2.11
At the pole the stars circle parallel to the horizon.

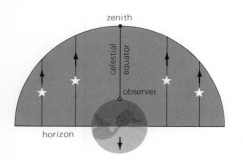

FIGURE 2.12
At the equator the daily courses of the stars are perpendicular to the horizon.

the minus sign, and the latitude would have been 16°50′S. In operations of the Coast and Geodetic Survey and in some observatories the latitude is obtained very accurately by observing stars with special instruments.

Let us now turn the latitude rule around: When the latitude of a place is given, we know the altitude of the north celestial pole at that place and how the daily courses of the stars are related to the horizon.

2.8 Stars observed from earth

At the north pole, latitude 90°, the north celestial pole is in the zenith, and the stars go around it daily in circles parallel to the horizon. Stars north of the celestial equator never set, while those south of the equator never come into view (Fig. 2.11). In this statement we avoid confusion by referring to the true positions of the stars and not as they are elevated by atmospheric refraction (Section 1.11). The sun, moon, and planets rise and set as seen from the north pole whenever they cross the celestial equator. The sun rises about 21 March and sets about 23 September. The moon rises and sets about once a month.

At the south pole everything is reversed. There the south celestial pole stands in the zenith, and the stars of the south celestial hemisphere never set. The long period of sunshine begins with the rising of the sun about 23 September and ends about 21 March.

At the equator, where the latitude is zero, the altitude of the celestial pole is zero by our rule. The north celestial pole is situated at the north point of the horizon, and even the pole star now rises and sets as it circles near this pole (Fig. 2.12). The south celestial pole is at the south point. The celestial equator passes directly overhead from east to west.

At the equator, all parts of the heavens are brought into view by the apparent daily rotation. All stars rise and set. Their courses cross the horizon at right angles and are bisected by it, so that every star is above the horizon 12 hours daily if we neglect refraction. This applies to the sun as well; the duration of sunshine is 12 hours at the equator throughout the year.

The heavens turn obliquely for us who live between the pole and equator. Suppose that we are observing from latitude 40°N. Here the north celestial pole is 40° above the north point of the horizon, and the south pole is the same distance below the south point. The celestial equator arches across from the east point to the west, inclined so that its highest point is 40° south of the zenith. The daily courses of the stars are likewise inclined toward the south. Half the celestial equator is above the horizon (Fig. 2.13). North of the equator the daily courses of the stars come up more and more above the horizon until they are entirely above it. Southward from the celestial equator they are depressed more and more until they disappear completely.

In this oblique arrangement the celestial sphere is divided into three parts: (1) a circular area around the elevated celestial pole contains the stars that never set; (2) a similar area around the depressed pole contains the stars that never rise; (3) the remaining band of the heavens symmetrical with the celestial equator contains the stars that rise and set. Hence, in latitude 40°N the area of the heavens that is always above the horizon is within a radius of 40° around the north celestial pole, and the area that never comes into view is within a radius of 40° around the south celestial pole. The remainder of the heavens, a band extending 50° on either side of the celestial equator, rises and sets.

If we travel north, the celestial poles move farther from the horizon. The polar areas grow larger until they come together when the pole is reached; there as we have seen no stars rise and set. If we travel south, the celestial poles approach the horizon. The polar areas grow smaller until they disappear when the equator is reached, where all stars rise and set.

2.9 Circumpolar stars

Circumpolar stars (Fig. 2.14) go around the celestial poles without crossing the horizon. These stars either never set or never rise, because they are closer to one of the celestial poles than the distance of that pole from the horizon, a distance that is equal to the observer's latitude. We accordingly make this rule about them for observers in the northern hemisphere:

A star having a distance from the north celestial pole (90° minus the star's declination) less than the latitude of a place never sets at that place. A star having a distance from the south celestial pole less than the latitude never rises.

Suppose that the observer is in latitude 40°N, and consider as an example the bowl of the Big Dipper, declination about 58°N. It never sets here because its north polar distance of 32° is less than the latitude. As a second example, consider the star Canopus, declination 53°S. This brilliant star is not visible in latitude 40°N because its south polar distance of 37° is less than the latitude.

2.10 The midnight sun

The sun on 22 June is as far north of the equator as it ever gets. The declination of its center is 23.5°N, so that its north polar

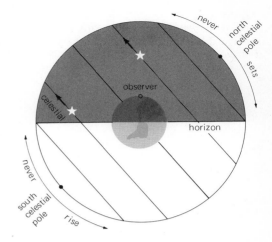

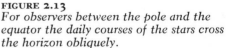

FIGURE 2.13
For observers between the pole and the equator the daily courses of the stars cross the horizon obliquely.

FIGURE 2.14
Circumpolar star trails taken with a meteor camera. The north star is the short bright trail near the center. A bright meteor is seen as dashes across the picture.

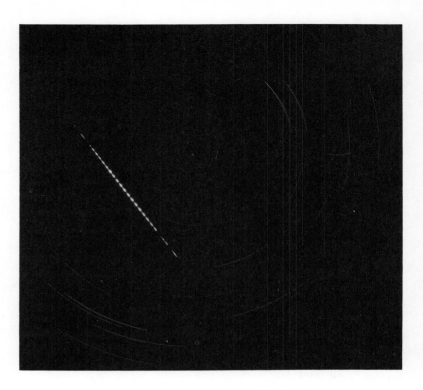

distance is 66.5°. How far north must we go on 22 June to see the sun circle around the pole without setting? According to our rule we must go beyond latitude 66.5°. When we take into account the refraction effect and the size of the sun's disk, however, we can expect to see the sun at midnight on this date as far south as 65.75°N.

The **midnight sun** is seen wherever the sun becomes circumpolar (Fig. 2.15). In far northern latitudes during spring the sun enters the area of the heavens that is always above the horizon; it becomes circumpolar in far southern latitudes at the opposite season.

2.11 The duration of twilight

The gradual transition between daylight and the darkness of night, which we call **twilight,** occurs during the time that the sun, after its setting and before its rising for us, can shine on the atmosphere above us. What is said here about the evening twilight applies in reverse to the morning twilight.

Civil twilight ends when the sun's center has sunk 6° below the horizon. Then it is no longer possible without artificial illumination to continue outdoor operations that require good light. **Nautical twilight** ends when the sun's center is 12° below the horizon. Then the sea horizon is likely to be too dim for the navigator's sextant sights. **Astronomical twilight** ends when the sun's center is 18° below the horizon; by that time the fainter stars have come out overhead. The times of sunset and sunrise and the duration of twilight can be found in some of the almanacs for any date and latitude. The minimum times for the three critical twilights on clear days can be estimated from the fact that the sun, by definition, moves 15° per hour giving 24, 48, and 72 minutes, respectively.

The duration of twilight varies with the time of year and the latitude. Twilight is shortest at the equator where the sun descends vertically and so reaches the limiting distance below the horizon in the shortest time. Astronomical twilight lasts somewhat more than an hour at the equator, and an hour and a half or more in the latitude of New York. On 22 June it does not end at all north of 48°N, about the latitude of Victoria, British Columbia. Civil twilight persists from sunset to sunrise on this date from latitude 60°, or the latitude of Oslo, Norway, to nearly 66°, where the midnight sun is seen. On this same date Tierra del Fuego, Argentina, catches only a brief glimpse of the sun. However, on 21 December the sun is almost circumpolar for the residents of Tierra del Fuego while in Oslo the sun is barely seen.

FIGURE 2.15
The sun is seen here as a circumpolar star. The multiple exposures were made over midnight at which time the sun is referred to as the midnight sun.

QUESTIONS

1 Some locations on earth are often in a high pressure system, so often that the condition receives the name "The-So-and-So" High. Where do we expect these to occur?

2 Does the period of 84 minutes mentioned in paragraph 3 of Section 2.3 bring to mind any recent events?

3 What causes the wanderings of the earth's poles and the changing period of the earth's rotation?

4 Can you give any examples where the nonuniform rotation of the earth makes it unsuitable for timekeeping? The irregularity is only on the order of a fraction of a millisecond (10^{-3} second).

5 How do we locate the pole on the sky in the northern hemisphere? Can you suggest a method for the southern hemisphere (Map 6, Chapter 11)?

6 If the right ascension of the local meridian is 12^h30^m and the right ascension of a certain star is 14^h00^m, what is its hour angle?

7 The hour angle of a star is 12 hours, and it has a declination of 29.5°S, your latitude is 40°S, can you observe the star? Why?

8 What is the difference between astronomical latitude and geographical latitude?

9 At what latitude can you catch at least a glimpse of Canopus (53°S) once a year?

10 The declination of the moon is occasionally as great as 28°N and 28°S, at what latitudes must you be to see the moon circumpolar?

FURTHER READINGS

BLAIR, T. A., *Weather Elements*, Prentice-Hall, Englewood Cliffs, N.J., 1942.

DAVIDSON, MARTIN, *Elements of Mathematical Astronomy*, 3rd ed., Macmillan, New York, 1962.

SMART, WILLIAM M., *Spherical Astronomy*, Cambridge University Press, New York, 1962; paperback edition, Dover, New York.

A 1676 print of the earth's orbit around the sun and the moon's orbit around the earth. The tilt of the earth is faithfully represented to explain the seasons.

3

While the stars are circling westward around us as though they were set on the rotating celestial sphere, the sun seems to lag behind gradually. The sun shifts slowly toward the east against the turning background of the heavens. If we could readily view the stars in the daytime sky, we would then see that the sun moves eastward about twice its breadth in a day, and that it circles completely around the heavens in this direction in the course of a year. Although it is not easily possible to observe the sun's progress directly, this movement was recognized and charted by early watchers of the skies, for it is clearly revealed by the steady procession of the constellations toward the west from night to night at the same hour through the year.

3.1 Annual motion of the constellations

At the same hour each night as the seasons go around, the constellations move slowly across the evening sky and down past the sun's position below the horizon. If we look at a particular group of stars at the same time night after night, we eventually observe this westward movement. We notice, for example, Orion's part in the unending parade (Fig. 3.1); it is the brightest and among the most familiar of the constellations.

Orion, a tilted oblong figure with three stars in line near its center, comes up over the east horizon early in the evening in the late autumn. At the approach of spring this constellation appears upended in the south. As spring advances Orion comes out in the twilight farther and farther west, with its oblong figure now inclined the other way, until it follows the sun too closely to be visible. Then at dawn in midsummer Orion appears in the east again, on the other side of the sun.

This westward procession of Orion and other constellations past the sun's position shows that the sun is moving toward the east among the stars. Its motion is a consequence but not a proof of the earth's annual revolution around the sun. The same effect would occur if the sun revolved around the earth, as most people before the time of Copernicus thought.

Everyone knows today that the earth revolves around the sun. We learn this fact at an early age and accept it as an item of common knowledge. We are told, too, that the sun is something like 149.6 million kilometers away, so that the earth must be speeding along at the average rate of 29.8 kilometers a second to travel completely around it in a year. There is nothing in our everyday experiences, however, to convince us that the earth is moving in this way. Displacements of the stars that prove the earth's revolution beyond reasonable doubt are too minute to be detected without a telescope. One proof, however, is the aberration of starlight.

3.2 Aberration of starlight

Raindrops fall vertically when there is no wind, yet they seem to slant if one walks rapidly through the rain, and to slant even more if one runs instead. As the observer's speed increases, the place from which the raindrops seem to fall shifts farther in the direction he is going. This is the aberration of raindrops. See Fig. 3.2.

There is a similar effect on the rays of light from a star. The

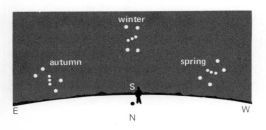

FIGURE 3.1
Orion in the evening at different seasons.

aberration of starlight (α) is the apparent displacement of the star in the direction in which the observer is moving:

$$\frac{29.8}{3 \times 10^5} = \frac{\alpha}{206,265}$$

where 29.8 is the velocity of the earth in its orbit, 3×10^5 is an approximation of the velocity of light, and 206,265 is the number of seconds of arc in a radian. It is a much smaller displacement, to be sure; starlight comes down so much faster than the rain that its direction is not appreciably altered by ordinary speeds. Even the earth's swift flight around the sun of 29.8 kilometers per second displaces the stars only 20.5″ at the most, an amount too slight to be detected by the unaided eye. This displacement of the stars from their true positions is readily observed with the telescope, and here we have convincing evidence of the earth's revolution.

If the earth is taking us around the sun, the direction of the revolution must change continuously. The direction of the star's aberration displacement must also change continuously, for it is always in the direction of the observer's motion. If the earth revolves once in a year, the stars must seem to move in small annual orbits around their true places. This is what they seem to be doing, as observed through a telescope.

In 1727, nearly two centuries after the death of Copernicus (who correctly predicted the earth's movement around the sun, but had no proof for his theory), J. Bradley first observed the aberration of starlight at Oxford and explained its important meaning. Henceforth, there could be no reasonable doubt that the earth revolves around the sun. In a later chapter (see Section 12.1) we

FIGURE 3.2

Aberration of raindrops and starlight. Just as raindrops come down slanting to one who is running, so the stars are apparently displaced always ahead of us as we go around the sun. Each star seems to describe a small orbit around its true position.

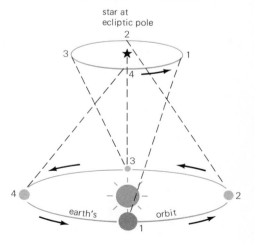

larger
eccentricity

smaller
eccentricity

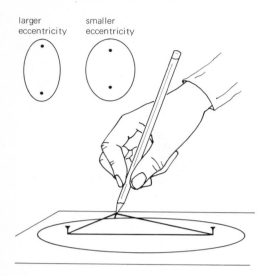

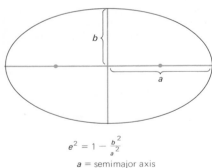

FIGURE 3.3
An ellipse can be drawn by looping a string around two nails as shown.

$$e^2 = 1 - \frac{b^2}{a^2}$$

a = semimajor axis
b = semiminor axis

consider the parallax effect caused by the earth's revolution, which is much smaller and affects each star differently.

3.3 The earth's orbit

If the earth's orbit around the sun were a circle with the sun at its center, the earth's distance from the sun would remain the same throughout the year, so that the sun would always appear to be the same size. Although the difference is not noticeable to the unaided eye, the sun's apparent diameter does vary during the year. It is greatest early in January and smallest early in July; the difference between the two is one-thirtieth of the average diameter. Thus the earth is nearest the sun about the first of January and farthest from the sun about the first of July, which might seem surprising to anyone who had forgotten that the seasons are not caused by the earth's varying distance from the sun.

The earth's distance from the sun averages about 1.496×10^8 kilometers; it varies from 1.469×10^8 to 1.52×10^8 kilometers.

The earth's orbit is an ellipse of small eccentricity with the sun at one focus. The terms used in this statement are defined as follows:

The **ellipse** is a plane curve where the sum of the distances from any point on its circumference to two points within, called the **foci,** is constant and equal to the longest diameter, or **major axis,** of the ellipse. The definition suggests a way to draw an ellipse (Fig. 3.3), and the drawing of several ellipses in this way makes clear the significance of the eccentricity.

The **eccentricity** (e) of the ellipse denotes its degree of flattening. It is represented by the fraction of the major axis that lies between the two foci. If the eccentricity is zero, the foci are together at the center and the curve is a circle. The ellipse flattens more and more as the eccentricity increases, until at an eccentricity of 1 the curve becomes a parabola. The earth's orbit is nearly a circle; its eccentricity is only 0.017, or 1/60. If Fig. 3.4 were drawn to scale, the sun's center would be 0.5 millimeter from the center of the ellipse, and the ellipse itself could be scarcely distinguished from a circle.

Perihelion and **aphelion** are the points on the earth's orbit which are, respectively, nearest and farthest from the sun; they are at opposite ends of the major axis. The earth arrives at perihelion about 1 January and at aphelion about 1 July, as we have noticed before. These times vary a little in the calendar and also advance an average of 25 minutes a year, because the major axis of the earth's orbit moves slowly around the sun in the direction of the earth's revolution. The earth's **mean distance,** 1.496×10^8

kilometers, from the sun is half the length of the major axis, or the average of the perihelion and aphelion distances. The earth is at this distance from the sun in early April and again in early October; this distance is referred to as the **astronomical unit** (AU).

3.4 The ecliptic

The earth revolves eastward around the sun once in a year and at the same time rotates on its axis in the same direction once in a day. The earth's axis is inclined 23.5° from the perpendicular to the plane of its orbit, or we may say that the earth's equator is inclined 23.5° to the plane of the orbit. This determines the relation (Fig. 3.5) between the celestial equator and the eastward path the sun seems to describe around the heavens as we revolve around the sun.

The **ecliptic** is the sun's apparent annual path around the celestial sphere; it is a great circle inclined 23.5° to the celestial equator. The ecliptic is in the plane of the earth's orbit, and the celestial equator is in the plane of the earth's equator. The **north** and **south ecliptic poles** are 90° from the ecliptic, and are, respectively, 23.5° from the north and south celestial poles. The north ecliptic pole (R.A. 18ʰ, Decl. 66.5°N) is in the constellation Draco (Map 1, Chapter 11).

The **equinoxes** are two opposite points on the celestial sphere where the ecliptic crosses the celestial equator. They are so named because days and nights are said to be equal in length when the sun arrives at an equinox, although atmospheric refraction makes the duration of sunlight slightly the longer on such occasions. The **solstices** are two opposite points midway between the equinoxes, where the ecliptic is farthest north or south from the celestial equator. Here the "sun stands," so far as its north and south motion is concerned, as it turns back toward the equator.

The equinoxes and solstices are points on the celestial sphere; their positions in the constellations are shown in the star maps in Chapter 11. The **vernal equinox** (R.A. 0ʰ, Decl. 0°) is the point where the sun crosses the celestial equator on its way north, about 21 March. The **summer solstice** (R.A. 6ʰ, Decl. 23.5°N) is the northernmost point of the ecliptic; the sun arrives here about 22 June. The **autumnal equinox** (R.A. 12ʰ, Decl. 0°) is the point where the sun crosses the celestial equator on its way south, about 23 September. The **winter solstice** (R.A. 18ʰ, Decl. 23.5°S) is the southernmost point of the ecliptic; the sun arrives here about 22 December. These dates vary a little from year to year owing to the plan of leap years. In each case the positions of the sun refer to the sun's center.

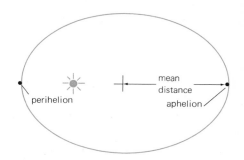

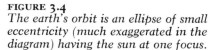

FIGURE 3.4
The earth's orbit is an ellipse of small eccentricity (much exaggerated in the diagram) having the sun at one focus.

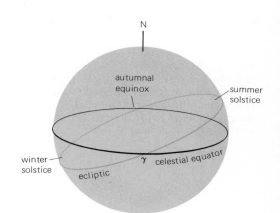

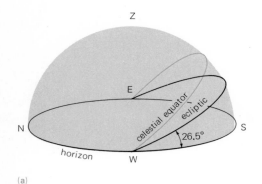

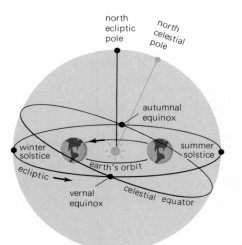

The celestial equator is inclined to the ecliptic. It is inclined by the same amount that the earth's equator is inclined to the plane of its orbit around the sun.

FIGURE 3.6
Relation between ecliptic and horizon. (a) The ecliptic is least inclined to the horizon in our latitudes at sunset at the beginning of autumn. (b) It is most inclined at sunset at the beginning of spring.

The celestial equator keeps the same position in the sky throughout the year. It is inclined to the horizon at an angle that is the complement of the latitude. Thus in latitude 40°N the equator crosses the horizon at the east and west points at an angle of 50°, and is 50° above the horizon in the south. The ecliptic, however, takes different positions in the evening sky during the year.

Because the ecliptic is inclined 23.5° to the celestial equator, its inclination to the horizon can differ as much as 23.5° either way from that of the equator. At sunset at the beginning of autumn (Fig. 3.6) in middle northern latitudes the ecliptic is least inclined to the horizon; the moon and bright planets that may be visible at the time are seen rather low in the south. At sunset at the beginning of spring the ecliptic is most inclined to the horizon; the moon and planets are then crossing more nearly overhead.

A number of features familiar to watchers of the skies are affected by the varying angle between the ecliptic and horizon. Among these are the harvest moon, the direction of the horns of the crescent moon, the favorable times for seeing the planet Mercury as evening or morning star, and the favorable seasons for viewing the zodiacal light.

3.5 Celestial longitude and latitude

The moon and bright planets never depart very far from the sun's path around the heavens. For this reason, astronomers of early times who were especially interested in the motions of the sun, moon, and planets, referred the places of these and other celestial bodies to the ecliptic, and they named the coordinates celestial longitude and latitude. **Celestial longitude** is measured in degrees eastward from the vernal equinox along the ecliptic. **Celestial latitude** is measured north or south from the ecliptic along a circle at right angles to it.

The ecliptic coordinates have only limited use today. For most purposes the position of a celestial body is now referred to circles based on the celestial equator by means of right ascension and declination, which closely resemble terrestrial longitude and latitude. These newer coordinates themselves would doubtless have been called celestial longitude and latitude if the terms had not already been assigned.

THE EARTH'S PRECESSION

If a top is spinning with its axis inclined to the vertical, the axis moves around the vertical line in the direction of the spin. The rotation of the top resists the effort of gravity to tip it over, and the conical motion of the axis results, until the spin is so reduced by friction that the top falls over.

The earth is rotating similarly on an axis that is inclined to the ecliptic plane, which is nearly the plane of the moon's motion around us (Fig. 3.7). The attractions of the moon and sun on the earth's bulging equator tend to bring the earth into the plane with themselves, and thus to straighten up the earth's axis relative to its orbit. Their efforts are resisted by the rotation, so that the earth's axis moves slowly around the line joining the ecliptic poles in the direction opposite to that of the rotation, once around in approximately 26,000 years at the present rate. This is the **earth's precession.**

As the earth's axis goes around in the precessional motion, the celestial poles, toward which the axis is directed, move among the constellations. The poles describe circles 23.5° in radius around the ecliptic poles (Fig. 3.8), bringing successively to bright stars along their paths the distinction of being the pole star for a time. Thus alpha Draconis was the pole star in the north 5000 years ago, the predecessor of our present pole star.

As shown in Fig. 3.8, the north celestial pole is now approaching Polaris and will pass nearest it about the year 2100 at half its present distance. Thereafter, Polaris will describe larger and larger daily circles around the pole, and its important place will at length be taken by stars of Cepheus in succession. In the year 7000, alpha Cephei will mark the north pole closely. In 14,000, Vega in Lyra will be a brilliant, although not very close, marker.

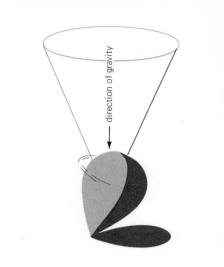

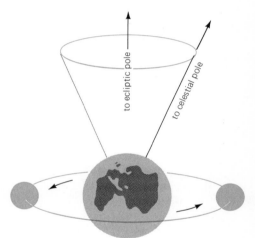

FIGURE 3.7
The earth resembles a spinning top. The pull of the moon on the bulging equator of the rotation earth is the chief cause of the precessional motion.

FIGURE 3.8
Precessional path of the north celestial pole.
The celestial pole describes a circle of 23.5°
radius around the ecliptic pole.

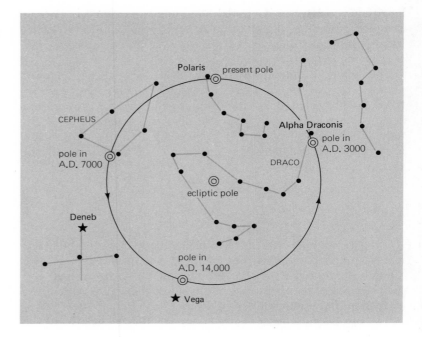

In our skies the constellations are slowly shifted by precession relative to the circumpolar areas, while the poles remain in the same places at the distance from the horizon equal to the observer's latitude. About 6000 years ago the Southern Cross rose and set everywhere in the United States. Now it is not visible except in the southernmost parts of the country.

The earth's precession, as we have seen, causes the line joining the celestial poles to describe the surface of a right circular cone around the line joining the ecliptic poles. This motion is counterclockwise as we face north. Meanwhile the celestial equator slides westward along the ecliptic, keeping about the same angle between them. The equinoxes, where the two inclined circles intersect, slide westward along the ecliptic. This is the **precession of the equinoxes.** Their motion, including smaller effects of planetary attractions of the earth, is at the rate of 50″ a year.

The precession of the vernal equinox (Fig. 3.9) causes continuous variations in the right ascensions and declinations of the stars. It accounts for the lack of agreement between the constellations and signs of the zodiac, and it makes the year of the seasons shorter than the period of the earth's revolution around the sun (see Section 3.7).

3.6 The zodiac: its signs and constellations

The *zodiac* (from Greek, meaning "band of animals") is the band of the heavens 18° wide, through which the ecliptic runs centrally. It contains the sun, moon, and bright planets at all times, with the occasional exception of the planet Venus. This is the reason for the special importance of the zodiac in the astronomy of early times and in the ancient pseudoscience of astrology, in which the places of the planetary bodies had great significance.

Twelve constellations of the zodiac lie along this band of the heavens. Because these constellations are of unequal size, the **signs of the zodiac** were introduced in early times in the interest of uniformity; these are 12 equal divisions of the zodiac, each 30° long, marked off eastward from the vernal equinox. Each block, or sign, of the zodiac has the name of the constellation it contained 20 centuries ago.

The names of the 12 constellations and signs of the zodiac are: Aries, Taurus, Gemini, Cancer, Leo, Virgo, Libra, Scorpius, Sagittarius, Capricornus, Aquarius, and Pisces.

In the meantime, the vernal equinox has moved westward among the stars of the zodiac, and the whole train of signs has followed along, because the signs are counted from the equinox. Each sign has shifted out of the constellation for which the sign was named and into the adjoining figure to the west. Thus, when the sun arrives on 21 March at the vernal equinox, or "first of Aries," the sun is entering the zodiacal sign Aries. The sun is then in the constellation Pisces, however, and will not enter the constellation Aries itself for another month.

3.7 The year of the seasons

The year of the seasons is about 20 minutes shorter than the true period of the earth's revolution. The difference is caused by the precession of the equinox.

The **sidereal year** is the interval of time in which the sun

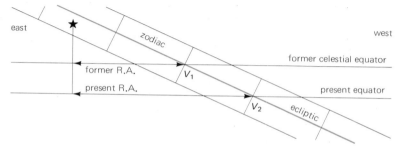

FIGURE 3.9
Precession of the equinoxes. The westward motion of the vernal equinox from V_1 to V_2 causes the signs of the zodiac to slide westward away from the corresponding constellation of antiquity. The right ascensions (RA) and declinations (δ) of the stars are altered by precession.

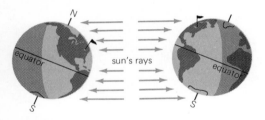

FIGURE 3.10
In the northern hemisphere the weather is warmer in summer because the daily duration of sunshine is greater and the sun's rays are more nearly direct.

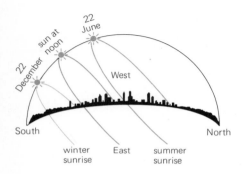

FIGURE 3.11
The daily circles of the sun for a northern hemisphere location in the summer and winter.

appears to perform a complete revolution around the heavens with respect to the stars. This is the true period of the earth's revolution. The length of the sidereal year is $365^d6^h9^m10^s$ of mean solar time.

The **tropical year** is the interval between two successive arrivals of the sun's center at the vernal equinox. Because the equinox is shifting westward to meet the sun, this year of the seasons, from the beginning of spring to the beginning of spring again, is shortened. The length of the tropical year is $365^d5^h48^m46^s$. This is about the average length of the present calendar year.

THE SEASONS

Why is the weather warmer in summer than in winter? It is not because we are nearer the sun in summer, for we have seen that the earth is more than 5 million kilometers farther from the sun in summer in the northern hemisphere than in winter. The cause of the seasons is found in the inclination of the earth's equator to the plane of its orbit (Fig. 3.10).

In summer the northern end of the earth's axis is inclined toward the sun. Not only is the daily duration of sunlight longer at any place in our northern latitudes than it is in winter, but the sun also climbs higher, so that its rays are more nearly vertical and more concentrated on our part of the world. In winter the northern end of the axis is inclined away from the sun. The daily duration of sunlight is then shorter for us and the sun is lower at noon. The sun's rays come down more slanting in the winter, so that they are more spread out over the ground and are also obstructed more by the greater thickness of the intervening air. Thus the weather is warmer in summer than in winter because the sun shines for a longer time each day and reaches a greater height in the sky (Fig. 3.11).

3.8 The lag of the seasons

Summer in our northern latitude begins about 22 June, when the sun arrives at the summer solstice and turns back toward the south. The duration of sunlight is longest on that day, and the sun climbs highest in the sky. Yet the hottest part of the summer is likely to be delayed several weeks until the sun is well on its way south. Why does the peak of the summer come after the time of the solstice?

As the sun goes south, its rays bring less and less heat to our part of the world from day to day. For some time, however, the diminishing receipts still exceed the amounts of heat we are losing by the earth's radiation into space. Summer does not reach its peak until the rate of heating is reduced to the rate of cooling. Similarly our winter weather is likely to be more severe several weeks after the time of the winter solstice. After the sun turns north, about 22 December, the temperature continues to fall generally until the heat we receive from the sun becomes as great as the daily loss.

3.9 The seasons in the two hemispheres

Although the seasons are not caused by the earth's varying distance from the sun, the earth is farther from the sun in our summer and nearer the sun in our winter than at similar seasons in the southern hemisphere (Fig. 3.12). Summers in the northern hemisphere might well be a little cooler than southern summers, and northern winters might be milder than southern winters. Thus the northern hemisphere might seem to have the more agreeable climates. Yet the variation in the earth's distance from the sun is only 3 percent of the distance itself, and there is more water in the southern hemisphere to modify extremes of temperature.

It would be not valid to make a general comparison of the weather in corresponding latitudes north and south of the equator. Differences in elevation and in the effects of air and ocean currents would have to be taken into account. Consider, for example, the climates near the two poles. There is mainly water around the north pole. On the other hand, a great expanse of land surrounds the south pole, and high land too, so that winters are more severe.

3.10 The climate zones

The positions and widths of the climatic zones are determined by the inclination of the earth's equator to the plane of its orbit. Because this inclination is 23.5°, the **torrid zone** extends 23.5° north and south from the equator; everywhere in this region the sun is directly overhead on some days of the year. The **Tropic of Cancer,** 23.5° north of the equator, forms the northern boundary of the torrid zone. Here the sun is overhead at noon on 22 June, when it arrives at the summer solstice and enters the sign Cancer of the zodiac. The **Tropic of Capricorn,** 23.5° south of the equator, is the southern boundary of the torrid zone. Here the sun is overhead at noon on 22 December, when it enters the sign Capricornus.

Similarly, the north and south **frigid zones** extend 23.5° from the poles and are bounded by the Arctic and Antarctic Circles, respectively. Although they are modified a little by the effect of refraction near their borders, the frigid zones are the regions where the sun may become visible at midnight during part of the year, and may not appear even at noon at the opposite time of the year.

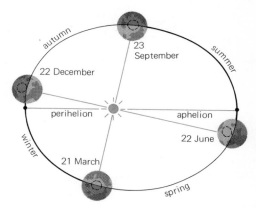

FIGURE 3.12
The northern hemisphere is inclined most toward the sun at the summer solstice (22 June) and inclined most away at the winter solstice although the earth is nearest the sun in early January.

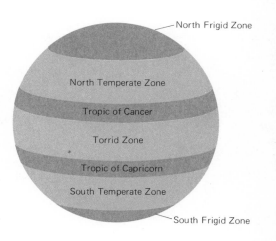

Extreme conditions occur at the poles themselves, where the sun shines continuously for six months and remains out of sight for the following six months.

The north and south **temperate zones** lie between the torrid and frigid zones. Here the sun never reaches the point overhead, nor does it ever fail to appear above the horizon at noon.

QUESTIONS

1 Astronomers use the terms rotation and revolution in a very strict way. Define these terms.
2 Explain the aberration of starlight. Why do we neglect the rotation of the earth in our simple discussion?
3 In Section 3.2 we say that Bradley's explanation of the aberration of starlight leaves no reasonable doubt that the earth revolves about the sun. Is this true? Why?
4 What are the definitions of perihelion and aphelion?
5 The word *solstice* is not fully explained in the sense of its origin. Where does the word originate?
6 Why and how does the ecliptic change its position in the sky at your latitude?
7 Having an observatory on the equator of the earth has some real advantages, a number of which are implied in this chapter. How many can you give?
8 What is the cause of precession in the case of a top? Why is this different for the earth?
9 How has the failure to accept precession affected astrology?
10 Why do the physical effects of the seasons lag behind the geometrically defined seasons?

FURTHER READINGS

DAVIDSON, MARTIN, *Elements of Mathematical Astronomy*, 3rd ed., Macmillan, New York, 1962.

SMART, WILLIAM M., *Spherical Astronomy*, Cambridge University Press, New York, 1962; paperback edition, Dover, New York.

STRAHLER, A. N., *The Earth Sciences*, 2nd ed., Harper & Row, New York, 1971.

TRICKER, R. A. R., *The Paths of the Planets*, American Elsevier, New York, 1967, Chaps. V and XII.

An old geographical representation of the earth. North is at the top and east to the right. The central line marks the equator. Lines parallel to the equator are the latitude circles of various places. Line one is Meroes in the upper Nile valley. Line four is Rhodes and five is Rome. The latitudes of the latter two places are wrong as are those of places located farther north.

Timekeeping

4

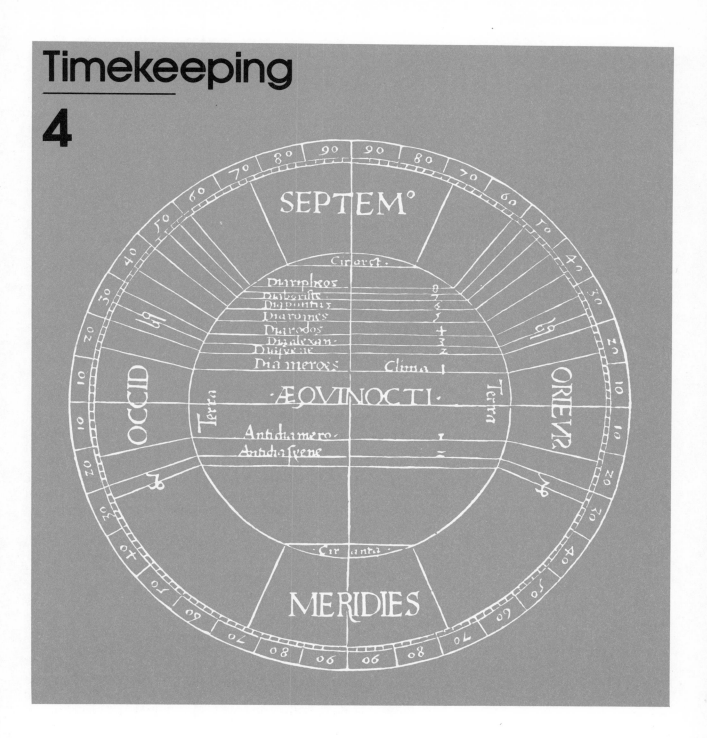

4

Our watches are likely to be more reliable timepieces if they are compared frequently with clocks or time signals that are set according to a clock in an observatory. The observatory clock may be regulated by comparison with the master clock in the sky, which itself is operated by the daily rotation of the earth. The time of day in ordinary use will be more clearly explained after we have considered the kinds of time that astronomers read from the celestial clock in order to derive the correct standard time. For the timing of extremely precise intervals we must rely upon atomic clocks.

4.1 The clock in the sky

Any one of the stars or any other object on the celestial sphere might be chosen as the time reckoner, which can be placed at the end of an hour hand of the clock in the sky. The hour hand would be that part of an hour circle connecting the time reckoner and the celestial pole. This hour hand would go around once each day, telling the time of day to those who can read it. The following definitions are true for any point that is selected as the time reckoner.

A **day** is the interval between two successive transits of the time reckoner over the same portion of the celestial meridian. A star **transits** when it crosses the meridian. Because a star does so twice a day, the distinction is made between **upper transit,** over the half of the circle through the celestial poles that includes the observer's zenith, and **lower transit,** over the other half that includes the nadir. It is noon when the time reckoner is at upper transit. Local time of day is the hour angle (see Section 2.6) of the time reckoner if the day begins at noon, or it is 12 hours plus the hour angle if the day begins at midnight.

4.2 Sidereal time

Instead of selecting one of the stars as the time reckoner for sidereal time, or "star time," astronomers chose the vernal equinox. The **sidereal day** is accordingly the interval between two successive upper transits of the vernal equinox. This interval is 0.008 second shorter than the period of the earth's rotation, because of the precession of the equinox. **Sidereal time** is the hour angle of the vernal equinox, which might have been more correctly called equinoctial time. It is reckoned from local sidereal noon through 24 hours to the next noon.

Sidereal time is kept by special clocks in the observatories, which may be set directly from the clock in the sky. The principle, illustrated in Fig. 4.1, is as follows: The sidereal time at any instant is the same as the right ascension of a star that is at upper transit at that instant. Evidently what is needed is something to show the place of the celestial meridian precisely, so that the instant of the star's crossing can be correctly observed. The astronomical transit instrument will serve to explain how the observation may be made.

The **astronomical transit instrument** is a small telescope that can be rotated on a single axis set horizontally in the east and west direction. As the telescope is turned, it always points toward the celestial meridian, which is represented by a vertical thread at the focus of the eyepiece. When the observer directs the telescope to the place where a star is about to transit, he can watch the star move across the field of view until it reaches the meridian line.

Suppose that the sidereal clock reads $3^h42^m22^s.7$ at the instant the star is at upper transit and that the star's right ascension as given in a catalog is $3^h42^m26^s.2$, which by the rule is the correct sidereal time at that instant. The sidereal clock, therefore, is 3.5 seconds slow. By comparison of clocks and a simple calculation, the error of a standard time clock can be found as well.

Recording devices are employed to time the transits of stars because they are more accurate than direct visual observations. The **photographic zenith tube** has replaced the simple transit instrument at the U.S. Naval Observatory and elsewhere (Fig. 4.2). It is a fixed vertical telescope that photographs stars as they cross the meridian nearly overhead. The converging beam of star-light formed by the lens of the photographic zenith tube is reflected by a mercury surface before coming to focus on a small photographic plate below the lens. With this device an error of only 0.003 second is expected in a time determination from a set of 18 stars.

4.3 The solar day is longer than the sidereal day

Suppose that the sun's center has just reached the vernal equinox and they both are at upper transit on the celestial meridian of the observer at O (Fig. 4.3). It is sidereal noon and also solar noon, that is, noon by the sun. The sidereal day will end when the earth has made a complete rotation relative to the vernal equinox, bringing the equinox again to upper transit. Meanwhile, the earth has advanced in its orbit around the sun an average of $360°/365.25$, or a little less than $1°$, so that after the end of the sidereal day the earth must rotate through this additional angle to bring the sun on the meridian again. Because the earth rotates at the rate of $1°$ in 4 minutes, the solar day is longer than the sidereal day by a little less than 4 minutes. The difference is more nearly 3^m56^s.

Because the apparent rotation of the celestial sphere is completed in slightly more than a sidereal day, a star rises at nearly the same sidereal time throughout the year. On solar time, however, it rises about 4 minutes earlier from night to night, or 2 hours earlier from month to month, so that at the same time on successive nights the star appears a little farther west than on the previous night. Thus at the same hour by our watches the stars march slowly westward across the sky as the year advances (see Section 3.1). Each season brings its own display of constellations.

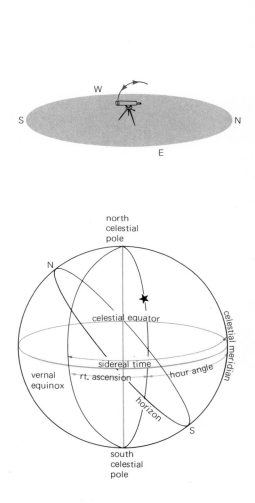

FIGURE 4.1
Relation between the sidereal time and the right ascension and hour angle of a star. Sidereal time equals the right ascension of a star plus its hour angle. When the star is at upper transit (hour angle zero), the sidereal time equals the star's right ascension.

FIGURE 4.2
The 8-inch photographic zenith tube at the U.S. Naval Observatory (Official U.S. Navy photograph.)

4.4 Apparent and mean solar time

Because our activities are regulated by the sun and not by the stars, we prefer to keep solar time rather than sidereal time. Sidereal noon, for example, comes at night during half of the year. **Apparent solar time** is time by the **apparent sun,** that is, the sun we see; it is reckoned from local apparent midnight through 24 hours to the next midnight. The apparent sun, however, is not a uniform timekeeper. The apparent solar day, measured by the sundial, varies in length for two principal reasons:

1 The earth's revolution around the sun is not at a uniform rate. The earth revolves in its elliptical orbit in accordance with

the **law of equal areas** (see Section 7.3): the line joining the earth and sun in winter in the northern hemisphere must go around farther than the longer line in summer to sweep over the same area (Fig. 4.4). Accordingly, the earth revolves farther in a day in winter and must rotate farther after the ending of the sidereal day to bring the sun again to the observer's meridian. For this reason the apparent solar day is longer in winter than in summer.

2 The ecliptic is inclined to the celestial equator. When the sun is near a solstice, where the ecliptic is parallel to the celestial equator, its daily eastward displacement by the earth's revolution has the full effect in delaying the sun's return to the observer's meridian. When it is near an equinox, however, part of the sun's daily displacement is north or south, and this part does not delay the return. For this reason the apparent solar day is longer in

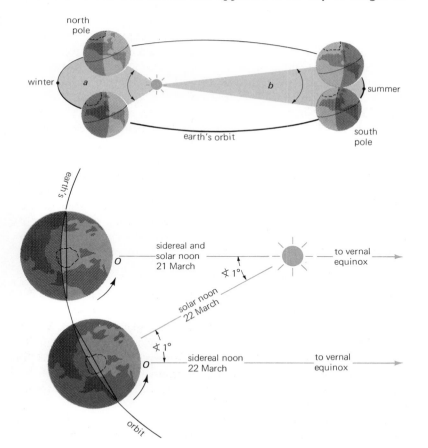

FIGURE 4.3
The solar day is longer than the sidereal day.

FIGURE 4.4
The earth's variable revolution. The earth revolves faster in the northern winter, because it is then nearer the sun.

TABLE 4.1
Equation of time[a]

1 Jan.	−	3^{m}22^s
1 Feb.	−	13 35
1 Mar.	−	12 29
1 Apr.	−	4 02
1 May	+	2 54
1 Jun.	+	2 21
1 Jul.	−	3 39
1 Aug.	−	6 16
1 Sept.	−	0 06
1 Oct.	+	10 12
1 Nov.	+	16 22
1 Dec.	+	11 05

[a] Apparent solar time is faster (+) or slower (−) than mean solar time.

summer and winter than in spring and fall. Both effects conspire to make the day by the apparent sun longer in winter in the northern hemisphere.

Mean solar time is time determined by the mean sun; it is reckoned from local mean midnight through 24 hours to the next midnight. The **mean sun** may be regarded as a point moving eastward along the celestial equator at a rate equal to the average rate of the apparent sun's motion along the ecliptic. This conventional sun would be a smooth running timekeeper if the earth's rotation were precisely uniform. Its day, which is the average of all apparent solar days through the year, would then have constant length.

The **equation of time** is the difference at any instant between apparent and mean solar time; its value can be found from tables in astronomical almanacs. Table 4.1 shows how much the apparent time is fast or slow with respect to mean solar time on the first of each month. The values are for midnight at the Greenwich meridian averaged over 18 years. The table holds for any other meridian, but in any given year the actual value on the day given may differ by as much as 10 seconds either way except on 1 November, when the equation of time is changing very little. The maximum discord in time occurs on 3 or 4 November and may reach 16^{m}25^s. In 1973, for example, the discord was greatest (16^{m}23^s) on 3 November.

The rapid change in the equation of time near the beginning of the year has an effect that is noticed by everyone. At this time of year the earth is nearest the sun and is accordingly revolving fastest. The apparent sun is then moving eastward fastest along the ecliptic, delaying its rising and setting as timed by the mean sun. For this reason the sun does not begin to rise earlier in the morning by our watches, which keep mean time, until about 2 weeks after the date of the winter solstice, although it begins to set later in the evening about two weeks before that date.

4.5 Universal time and ephemeris time

The international unit of time is the **atomic second** defined by transitions between the two hyperfine-structure, ground-state levels of the cesium-138 atom in the absence of a magnetic field. What this means is that the cesium atom has two energy levels in its ground, or lowest, state. When the atom goes from one level to the other it will absorb or emit a photon of a very precise wavelength or frequency, and the atomic second is defined in terms of this frequency (see Section 4.9).

Universal time is the local mean solar time at the Greenwich meridian. It is the basis of all ordinary timekeeping and is likely to continue to be, because observatories that transmit or monitor mean solar time signals keep their clocks corrected by frequent sights on meridian transits of the stars. These observations are said to be universal time zero (UT 0). For certain purposes this time must be corrected for the variation of the observer's meridian resulting from the motion of the poles (see Section 2.4). This time is called UT 1. A third universal time is UT 2, which is UT 1 corrected for variations in the rate of rotation of the earth (see Section 2.5). The time actually broadcast (see Section 4.8) and hence used in civil affairs is UT C, which is a smoothed approximation of UT 2. One can clearly see that the matter of timekeeping is not to be taken lightly. In forecasting the universal times of astronomical events, however, the irregularities in the earth's rotation make precise predictions difficult. For this purpose we define a more uniform time.

Beginning with the year 1960, the *American Ephemeris* and *Nautical Almanac* and the *British Astronomical Ephemeris*, which now conform in other respects as well, have tabulated the fundamental ephemerides* of the sun, moon, and planets for intervals of ephemeris time. **Ephemeris time** goes on uniformly without regard for irregularities in the earth's rotation. Its constant arbitrary unit equals the length of the tropical year at the beginning of the year 1900 divided by 31,556,925.9747, which was the number of mean solar seconds in the year at that epoch.

Ephemeris time has run increasingly fast with respect to universal time during the present century. The difference between the two was 44 seconds in 1973 and it is expected to be 45 seconds at the beginning of 1975. These differences can be determined approximately by extrapolation from past events listed in the almanac, however, they await precise determination nearer the time of the new event. For this purpose the observed universal time of a particular position of the moon among the stars (see Section 6.3) is compared frequently with the predicted ephemeris time of that position.

Unlike sidereal time and apparent solar time, which are always local times, mean solar time is most often employed in the conventional forms of zone time and standard time now to be considered.

4.6 Time zones

The difference between the local times of two places at the same instant is equal to the difference between their longitudes ex-

* Ephemeris is derived from the Greek word meaning diary and is a table of predicted positions of a planet, the moon, or the sun.

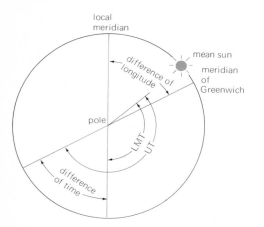

FIGURE 4.5
Difference of local time between two places equals their difference in longitude.

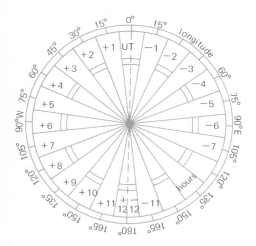

FIGURE 4.6
Time zone diagram. The earth is viewed from above the south pole. The numbers outside the circle are the longitudes of the standard meridians. The numbers inside are corrections in hours to universal time.

pressed in time (Fig. 4.5). This is true because the earth makes a complete rotation with respect to a celestial time reckoner in 24 hours of its kind of time, and there are 360° or 24 hours of longitude around the world. The rule applies to any kind of local time, whether it is sidereal, apparent solar, or mean solar, where the same kind of time is denoted at the two places. When the local time at one place is given and the corresponding local time at another place is required, add the difference between their longitudes if the second place is east of the first, subtract if the second place is west.

In the time diagram of Fig. 4.5 we are looking at the earth from above its south pole and are projecting the sun onto the equator in the longitude where the sun is overhead; the east to west direction is counterclockwise. The observer on the earth is in longitude 60°W, or 4^hW. The local mean time (LMT) there is 9^h and the Greenwich mean time (GMT) or universal time is 13^h, so that the difference of 4^h in the local times of the two places is the difference of their longitudes. In this case the time of day is counted through 24 hours continuously: 9^h is 9 A.M. and 13^h is 1 P.M.

Conversely, the difference between the longitudes of two places is equal to the difference between the local times of the places at the same instant. This is the basic rule for determining longitudes.

The local mean solar times at a particular instant are the same only for places on the same meridian. The time becomes progressively later toward the east and earlier toward the west; the rate of change at latitude 40° is 1 minute of time for a distance of a little more than 21 kilometers. The confusion that would ensue if every place kept its own local time is avoided by use of time zones. In its simplest form the plan is as follows:

Standard meridians are marked off around the world at 15° intervals, or 1-hour intervals, east and west from the Greenwich meridian (Fig. 4.6). Accordingly, the local times on these meridians differ successively by whole hours. The time to be kept at any other place between the meridians is the local mean time of the standard meridian nearest that place. Thus the world is divided by boundary meridians into 24 equal time zones, each 15° wide and having one of the standard meridians running centrally through it. The time is the same throughout each zone; it is 1 hour earlier than the time in the adjacent zone to the east and 1 hour later than the time in the zone to the west. The rule to be followed when crossing a boundary between two zones on an eastward voyage is to set your watch ahead 1 hour, the minutes and seconds remaining the same as before, and on a westward voyage to set your watch back 1 hour.

Zone time is used in the operation of ships at sea. This uniform plan is occasionally modified near land so that the ship's clock may agree with the standard time kept ashore. Because of the speed of modern aircraft it has become the practice to use Greenwich mean time as the standard operations time. This is so even within the continental United States, although the prevailing local time is used for public schedules.

Standard time divisions on land generally follow the pattern of the time zones at sea. Their boundaries are often irregular, they are affected by local preference, and they are subject to change. In some areas the legal time differs from the times in adjacent belts by a fraction of an hour. There is also the arbitrary and non-uniform practice of setting the clocks ahead of the accepted standard time for part of the year, called daylight saving time and often referred to as "fast" time. In the United States daylight saving time has been made uniform, beginning at 2 A.M. on the fourth Sunday in April and ending at 2 A.M. on the fourth Sunday in October and has been adopted on a state-by-state basis.

Four standard times (Fig. 4.7) are employed in the greater parts of the continental United States and Canada, namely, Eastern, Central, Mountain, and Pacific standard times. They are, respectively, the local mean times of the meridians 75°, 90°, 105°, and 120° west of the Greenwich meridian and are therefore 5, 6, 7, and 8 hours slower than universal time. The standard time in Newfoundland and Labrador is 3½ hours instead of 4 hours slow, and in Alaska is generally 10 hours slower than universal time. Certain changes for Canada are shown in the Dominion Observatories Operations Map of 1958. Apart from Quebec, Ontario, and the Northwest Territories, each province of Canada has adopted a single standard time.

Clearly, if we work our way around the earth the date must change abruptly by one day. By international agreement a line approximating the meridian 12 hours east (and west) of the Greenwich meridian is referred to as the **international date line** and is the meridian where the date changes by one day. To visualize this, suppose that an airplane leaves San Francisco at noon on Monday and proceeds due west around the world with the speed of the earth's rotation relative to the sun in that latitude. The sun accordingly remains almost stationary in the sky during the voyage, and the watches aboard are set back to noon whenever a boundary between the time divisions is crossed. When the plane comes around again to San Francisco, the crew might be

FIGURE 4·7
Standard time zones for the western hemisphere. The time for each zone at an arbitrary time is shown at the top, and the numbers below show the corrections from zone time to universal time.

surprised to find that it is now Tuesday noon if they had forgotten the rule for the change of date.

The rule is to change the date generally at the 180° meridian. When this meridian is crossed on a westward voyage, as in going from San Francisco to Manila, the date is advanced; if the line is crossed on Monday noon, it then becomes Tuesday noon. On an eastward crossing the date is set back. Where the 180° meridian traverses land areas, the international date line may depart from the meridian to avoid populated areas and so that a politically associated region is not divided. Thus it bends to the east around Siberia and to the west around the Aleutian Islands.

4.7 Conversion of time

When the time at one place is given, one may wish to know the corresponding time at another place. The conversion is often to or from universal time, which is often used in almanacs. In passing from zone time to universal time, the correction is found by dividing the longitude of the place in degrees by 15 and taking the nearest whole number of hours. In passing from the standard time of a place to universal time, the correction may not be the same as from zone time in the same longitude, because of irregularities in the standard time zones and the possible use of daylight saving time at the place. The following examples illustrate some conversions.

1 The zone time is 21^h22^m, 25 May, at a place in longitude 103°W. Find the corresponding universal time and date.

ANSWER: The nearest standard meridian would be at 105°W so the local zone time is 7 hours behind Greenwich. Then 21^h22^m plus $7^h = 28^h22^m$. The day changes at 24^h00^m so the day is 26 May and the UT is 4^h22^m.

2 The diary of the *American Ephemeris and Nautical Almanac* gives the time of the beginning of autumn in 1960 as 23^d1^h September, universal time. Find the corresponding Central standard time and date.

ANSWER: Central standard time corresponds to the standard meridian which is 6 hours west of Greenwich. The time at Greenwich is the same as 25^h00^m on 22 September. Thus the Central standard time is 19^h00^m on 22 September.

3 The daylight saving time in San Francisco is 7:30 P.M. Find the corresponding time at Galveston, Texas, which remains on Central standard time.

ANSWER: If the daylight saving time in San Francisco is 7:30 P.M., then Pacific standard time is 6:30 P.M. The Mountain time zone intervenes between the Pacific and Central time zones, so the difference is two hours. Therefore it is 8:30 P.M. in Galveston, Texas.

4.8 Radio time service

Time signals from the National Bureau of Standards (NBS) transmitters at Fort Collins, Colorado and the Dominion Observatory in Ottawa supply convenient Standard time service to all of North America. Throughout North America the time from NBS can be obtained by telephone as well by dialing 303-447-1192. The time given is UT C (see Section 4.5). These signals are monitored by the U.S. Naval Observatory and compared with observations of the stars. The time broadcast is allowed to accumulate an error of 0.7 second and then the signals are corrected by a whole second; these changes are announced in advance and detailed corrections for the entire year past are published.

The broadcasts from the U.S. are by radio WWV on 2.5, 5, 10, 15, 20, and 25 megahertz* while from Canada they are by radio CHU on 3.33, 7.335, and 14.670 megahertz. These broadcasts are on the "short-wave" bands. In addition time can be obtained from the extensive Loran network if one has the proper receivers and ancillary equipment. In the near future it is hoped that the U.S. television channels will carry accurate time in a coded form. In order to obtain the time from these transmissions one will need a special converter, but this broadcast will in no way interfer with the video portion of the broadcast. In fact, this type of service is already available in France today.

U.S. television stations do broadcast a well controlled 10-megahertz signal for those requiring intervals of time accurate to 10^{-7} second. In the future we can expect that all of the time services of the world will be interlocked, or at the very least intercompared, by a communications satellite system.

4.9 Atomic clocks

We have seen (see Section 2.5) that the earth's rotation is not exactly uniform and that the day is increasing in length. While the small changes cause no inconvenience to our everyday affairs, they do make it difficult to time intervals with precision. As the need for greater and greater accuracy in time measurements grew, it became necessary to rely on the inherent accuracy of **atomic clocks.** Clocks based upon the vibration of the ammonia molecule have proven highly satisfactory for the measurement of precise intervals of time. In this clock, microwaves from a well controlled and correctable oscillator are transmitted into ammonia gas and are absorbed if they vibrate at the exact frequency of the gas. If the frequency of the microwaves drift, the amount absorbed decreases and this is sensed by a circuit that then corrects the oscillator. Such clocks are accurate to within a few seconds in 50 years.

Even more accurate atomic clocks use cesium atoms (Fig. 4.8) sorted out by a series of magnets and irradiated by microwaves of the proper frequency. The spin axes of cesium atoms are distributed at random. A first set of magnets serves to align the atoms, one-half spinning one direction and the other half spinning the other direction. The microwave radiation then serves to flip the atoms over. If the microwaves are at the exact frequency all flip over and the numbers with each spin are the same, but if the microwave frequency drifts even the least amount more atoms with

* 1 megahertz = 10^6 cycles per second.

FIGURE 4.8
*A cesium beam atomic clock. This clock is
a recent major advance in timekeeping and
it is accurate to better than a millionth of
a second over a period of three weeks.
(Official U.S. Naval Observatory photograph.)*

one spin flip over than do those of the other. A second series of
magnets serve to separate cesium atoms of one spin to one detector
and those of the opposite spin to another detector. Thus if the
microwave frequency has drifted, one detector or the other gets
an increased number of cesium atoms and the microwave genera-
tor must be corrected accordingly. Such clocks are accurate to a
few seconds over several centuries.

Actually the average of many cesium clocks is uniform to 1 part
in 10^{14} seconds although the individual clocks show random
variations as large as 2 parts in 10^{12} seconds.

Atomic time referred to as A 1 is defined in terms of the vibrations of the cesium atom. The second of atomic time is agreed
to be 9,192,631,770 vibrations of the cesium atom. Atomic time
agreed exactly with UT 2 at midnight on 1 January 1958 by international agreement. At that time UT 2 differed from ephemeris
time by 32.15 seconds, so ephemeris time differs from atomic time
by this amount and the difference remains constant.

CALENDARS

Calendars have been in use since the beginnings of civilization.
They have tried to combine natural measures of time, the solar

day, the lunar month, and the year of the seasons, in the most convenient ways and have encountered difficulties because these measures do not fit evenly one into another. Calendars have been of three types: the lunar, the lunisolar, and the solar calendar.

The **lunar calendar** is the simplest of the three types, and it was the earliest to be used by almost all civilizations. Each month began originally with the "new moon," the first appearance for the month of the crescent moon after sunset. Long controlled only by observation of the crescent, this calendar was eventually operated by fixed rules. In the fixed lunar calendar the 12 months of the common lunar year are alternately 30 and 29 days long, making 354 days in all. The Mohammedan calendar is a survivor of this type.

The **lunisolar calendar** tries to keep in step with both the moon's phases and the seasons, and is the most complex of the three types. It began by occasionally adding a thirteenth month to the short lunar year to round out the year of the seasons. The extra month was later inserted at fixed intervals. The Hebrew calendar is the principal survivor of the lunisolar type.

The **solar calendar** makes the year conform as nearly as possible to the year of the seasons, and neglects the moon's phases; its 12 months are generally longer than the lunar month. Only a few early nations, notably the Chinese, Egyptians, and eventually the Romans, adopted this type of calendar. The early Mayan calendar was based partially on the period of Venus but evolved into a highly developed solar calendar. This calendar was passed along to the Aztecs (Fig. 4.9) with its novel plan of leap years to keep it in step with the seasons.

The early **Roman calendar** formally begins with the founding of Rome in 753 B.C. It was originally a lunar calendar of a sort, beginning in the spring and having 10 months. The names of the months, if we use mainly our own style instead of the Latin, were: March, April, May, June, Quintilis, Sextilis, September, October, November, and December. The years for many centuries thereafter were counted from 753 and were designated A.U.C. ("in the year of the founding of the City"). Two months, January and February, were added later and were eventually placed at the beginning, so that the number months have appeared in the calendar out of their proper order ever since that time.

In its 12-month form the Roman calendar was of the lunisolar type. Unlike most early people, the Roman day began at midnight instead of at sunset. An occasional extra month was added to keep the calendar in step with the seasons. The calendar was managed so unwisely, however, that it fell into confusion; its dates

FIGURE 4.9
An Aztec calendar for the year 1790.
(Photograph courtesy of OAS.)

drifted back into different seasons from the ones they were supposed to represent.

When Julius Caesar became the ruler of Rome, he was disturbed by the bad condition of the calendar and took steps to correct it. He particularly wished to discard the lunisolar form with its troublesome extra months. Caesar was impressed with the simplicity of the solar calendar the Egyptians were using, and he knew of their discovery that the length of the tropical year is very nearly 365¼ days. He accordingly formulated his reform with the advice of the astronomer Sosigenes of Alexandria. In preparation for the new calendar the "year of confusion," 46 B.C., was made 445 days long in order to correct the accumulated error of the old one. The date of the vernal equinox was thereby brought to 25 March. The Julian calendar began on 1 January 45 B.C.

The **Julian calendar** was of the solar type, and so neglected the moon's phases. Its chief feature was the adoption of 365¼ days as the average length of the calendar year. This was accomplished conveniently by the plan of leap years. Three common years of

365 days are followed by a fourth year containing 366 days; this **leap year** in our era is divisible by 4.

In lengthening the calendar year from the 355 days of the old lunisolar plan to the common year of 365 days, Caesar distributed the additional 10 days among the months. With further changes made in the reign of Augustus, the months assumed their present lengths. After Caesar's death in 44 B.C., the month Quintilis was renamed July in honor of the founder of the new calendar. The month Sextilis was later renamed August in honor of Augustus.

Because its average year of 365^d6^h was 11^m14^s longer than the tropical year, the Julian calendar fell behind with respect to the seasons about 3 days in 400 years. When the council of churchmen convened at Nicaea in A.D. 325, the vernal equinox had fallen back to about 21 March. It was at that convention that previous confusion about the date of Easter was ended.

As the date of the vernal equinox fell back in the calendar, 21 March and Easter, which is reckoned from it, came later and later in the season. Toward the end of the sixteenth century the equinox had retreated to 11 March. Another reform of the calendar was proposed by Pope Gregory XIII.

Two rather obvious corrections were made in the Gregorian reform. First, 10 days were suppressed from the calendar of that year; the day following 4 October 1582, became 15 October for those who wished to adopt the new plan. The date of the vernal equinox was restored in this way to 21 March. The second correction made the average length of the calendar year more nearly equal to the tropical year, so that the calendar would not again get so quickly out of step with the seasons. Evidently the thing to do was to omit the 3 days in 400 years by which the Julian calendar year was too long. This was done conveniently by making common years of the century years having numbers not evenly divisible by 400. Thus the years 1700, 1800, and 1900 became common years of 365 days instead of leap years of 366 days, whereas the year 2000 remains a leap year as in the former calendar. The average year of the new calendar is still too long by 26 seconds, which is hardly enough to be troublesome for a long time to come.

The **Gregorian calendar** was gradually adopted, until it is now in use, at least for civil purposes, in practically all nations. England and its colonies including America made the change in 1752. By that time there were 11 days to be suppressed; that century year was a leap year in the old calendar and a common year in the new one. To correct this, 2 September 1752 was followed by 14 September. This abrupt adjustment led to many civil disturbances because the working man felt that he had somehow been cheated out of twelve days' wages. The fact that he hadn't worked those missing days seemed not to be understood. The countries of eastern Europe were the last to make the change, when the difference had become 13 days.

4.10 Easter

Easter was originally celebrated by some early churches on whatever day the Passover began, and by others on the Sunday included in the Passover week. The Council of Nicaea decided in favor of the Sunday observance and left it to the church at Alexandria to formulate the rule, which is as follows.

Easter is the first Sunday after the fourteenth day of the moon (nearly the full moon) that occurs on or immediately after 21 March. Thus if the fourteenth day of the moon occurs on Sunday, Easter is observed one week later. Unlike Christmas, Easter is a movable feast because it depends on the moon's phases; its date can range from 22 March to 25 April.

Dates of Easter

14	April	1974
30	March	1975
18	April	1976
10	April	1977
26	March	1978
15	April	1979
6	April	1980
19	April	1981
11	April	1982
3	April	1983
22	April	1984
7	April	1985

QUESTIONS

1 What is the time reckoner for astronomers?
2 What is meant by the term "apparent sun"? Why does the time it keeps vary throughout the year?
3 The star eta Cassiopea (R.A. 0^h47^m) is at lower transit. What is the sidereal time?
4 The equation of time changes most rapidly from about 3 November to about 8 February. Can you explain why?
5 What is universal time? What is ephemeris time?
6 Your clock keeping universal time reads 16:20 when the sun is on your meridian. In what time zone are you located?
7 In the text in Section 4.9 we assume there is a need for very accurate time in order to be able to measure time intervals precisely. Can you think of any examples where this might be necessary?
8 Why is a lunar calendar unacceptable to present day societies?
9 What was the principal improvement of the Gregorian calendar?
10 When will we need to adjust the Gregorian calendar? What did you use as your criterion for requiring the adjustment? How would you accomplish the adjustment?

FURTHER READINGS

WHITROW, G. J., *What Is Time?*, Thames and Hudson, London, Chaps. 1–4.

WILSON, P. W., *The Romance of the Calendar*, Norton, New York, 1937

An old astrolabe. The astrolabe was invented by Hipparchus to facilitate the determination of sidereal time, the times that the sun and stars will reach various positions, the latitude of the observer, etc. The instrument depicted is very elegant compared to Hipparchus' original.

Telescopes and Instrumentation

5

5

The chief feature of the optical telescope is its objective, which receives the light of a celestial object and focuses the light to form an image of the object. The image may be formed either by refraction of the light by a lens or by reflection from a curved mirror. Optical telescopes are accordingly of two general types: refracting telescopes and reflecting telescopes. Other telescopes are a combination of both types. In radio telescopes the radio radiations from a celestial source are received directly by an antenna or are concentrated on the antenna by a reflecting paraboloid or other surface of large aperture.

REFRACTING TELESCOPES

5.1 Light

Light reaches us in waves that spread from a source, such as the sun. These waves can be compared to ripples spreading over the surface of a pond when a stone is dropped into the water. **Light** is the sensation produced when waves of appropriate lengths enter the eye. Light can also be described in terms of its wave motion.

The **wavelength** (λ) is the distance from crest to crest of successive waves.

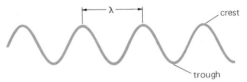

The light that produces a sensation to the human eye, known as visible light, falls in a limited range. These wavelengths vary from 3800 (violet light) to 7500 Ångstrom units* (red light). The total **radiation** from a source may have a far greater range of wavelengths than the eye can detect, from gamma rays that have a billion waves per centimeter to radio waves that are many kilometers long (see Appendix). The **frequency**, v, of the radiation is the number of waves emitted by the source in a second; it equals the velocity of light, c, divided by the wavelength, λ.

$$v = \frac{c}{\lambda}$$

The **velocity of light** is about 299,792 kilometers a second; it is the speed of all electromagnetic radiation (light, radio waves, X-rays, etc.) in a vacuum. For ease of calculation we may use 3×10^5 kilometers per second for the velocity of light. The speed is reduced in a medium, such as air or glass, depending on the density of the medium and the wavelength. This is the reason for the refraction of light.

Although the wave nature of light suffices to explain the functions of telescopes and most of their instrumentation there is an alternative explanation in the form of the particle nature of light first proposed and expounded by Newton. In this explanation light travels in little packets of energy called **photons** in accordance with quantum theory developed only a half century ago. When we

* 1 micron (μ) = 0.001 mm
  1 Ångstrom unit (Å) = 10^{-4}

consider the emission and absorption of light, we will base our assumptions on the particle nature of light. For now, the wave theory allows us to visualize more easily phenomena such as refraction and interference.

Refraction of light is the change in the direction of a ray of light when it passes obliquely from one medium into another of different density, as from air into glass. A "ray" of light denotes the direction in which a narrow section of the wave system is moving.

When a ray of light obliquely enters a denser medium, where its speed is reduced (Fig. 5.1), one side is retarded before the other. The wave front is accordingly swung around, and the ray becomes more nearly perpendicular to the boundary between the two media. When the oblique ray enters a rarer, thinner medium instead, it is refracted away from the perpendicular. Evidently the direction of the ray is not altered if it is originally perpendicular to the boundary.

Figure 5.2 shows how a double convex lens forms by refraction a real inverted image (that is, an image that can be projected on a screen) of an object that is farther from the lens than is the focal point, F. This is the point where rays parallel to the axis of the lens are focused. Rays passing through the center of the lens are unchanged in direction. When the image formed by this lens is brought within the focal distance of a similar lens, the second lens serves as an eyepiece with which to view and magnify the image, leaving it inverted. A single lens, however, does not give a sharp image, mainly because it refracts shorter waves more than longer ones, violet light more than red, so that it does not bring them together at the same focus. This is one major shortcoming of a refracting telescope and is called **chromatic aberration**. The confusion of colors is partly correctable by using a compound lens.

The objective of a refracting telescope is generally a combination of two lenses. Such an objective having two or more elements of various types of glass in order to minimize chromatic aberration is referred to as an **achromat**. Its **aperture**, or clear diameter, is given when denoting the size of the telescope. The **focal length**, or distance from the objective to its focus, is often about 15 times the aperture.

$$\text{focal ratio} = \frac{\text{focal length}}{\text{aperture}}$$

Thus a 60-centimeter telescope has an aperture of 60 centimeters and may have a focal length of 9 meters. For visual use the inverted image formed by the objective is viewed with an **eyepiece**,

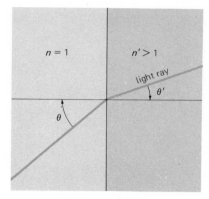

$$n \sin \theta = n' \sin \theta'$$

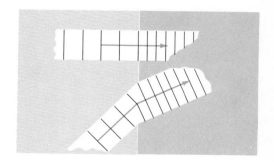

FIGURE 5.1
Refraction of light. A ray of light passing obliquely from one medium into another is changed in direction.

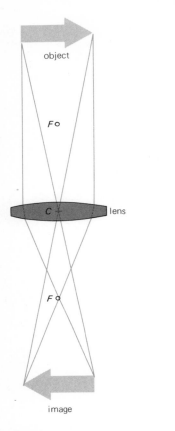

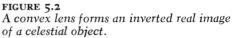

FIGURE 5.2
A convex lens forms an inverted real image of a celestial object.

a magnifier constructed of small lenses set in a sliding tube. The objective of a visual refractor focuses together the yellow and adjacent colors of the spectrum, to which the eye is especially sensitive, but not the blue and violet light that most affects the ordinary photographic plate. Without a correcting device it does not serve well for general photography.

Where a telescope intended for visual purposes is used as a camera, a plate holder replaces the eyepiece. A yellow filter and a yellow-sensitive plate may be combined to utilize the light that is sharply focused, or a correcting lens may be introduced for a particular kind of plate. We note from simple geometrical optics that the object distance D_o is infinity, and therefore the image distance D_i is equal to the focal length of the telescope *f.l.* Hence, the plate is located at the focus.

$$\frac{1}{D_o} + \frac{1}{D_i} = \frac{1}{f.l.}$$

Refracting telescopes intended only for photography often have more than two lenses in the objective; they generally have shorter focal lengths and give clear pictures of greater areas of the sky. If the ratio of the focal length to the aperture is less than 8 it is referred to as an **astrograph** (see Section 5.5).

5.2 Large refractors

The 40-inch telescope of Yerkes Observatory at Williams Bay, Wisconsin, is the largest refracting telescope; its focal length is 21 meters. (Here and in Section 5.4 we follow the common practice of using English units for indicating the aperture size of telescopes.) It has about the greatest permissible aperture for an instrument of this kind, where the objective can be supported only at the edge. A larger lens might sag seriously under its own weight.

The **equatorial mounting** of the Sproul telescope (Fig. 5.3) is an example of the type generally used for the larger refracting telescopes. The **polar axis** is parallel to the earth's axis; around it the telescope is turned parallel to the celestial equator. The **declination axis** is supported by the polar axis; around it the telescope is turned along an hour circle, from one declination to another.

The polar axis carries a graduated circle showing the hour angle of the star toward which the telescope is pointing. There is also a dial on the pier, or on a console, which indicates the star's right ascension. A circle on the declination axis shows the declination of the star. An electronic system usually displays this on the console also. By the use of these circles the telescopes can be pointed

FIGURE 5.3
The equatorial mounting of the Sproul Observatory's 60-centimeter refractor. The polar axis appears horizontal because of the angle of the picture. The large ring is the declination circle. (Sproul Observatory photograph.)

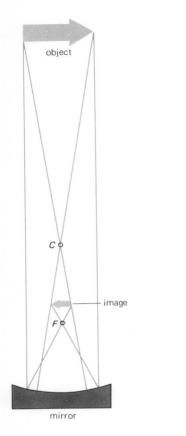

object

C

image

F

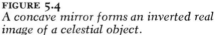

mirror

FIGURE 5.4
A concave mirror forms an inverted real image of a celestial object.

toward a celestial object of known right ascension and declination; it is then kept pointing at the object by a driving mechanism in the pier. The dome is turned by motor, so that the telescope may look out in any direction through the opened slit.

The 36-inch telescope of Lick Observatory on Mount Hamilton in California is second in size among refracting telescopes. About 40 refracting telescopes have apertures of 50 centimeters or more (see Appendix).

REFLECTING TELESCOPES

5.3 Reflection of light

The concave spherical mirror represented in Fig. 5.4 has its center of curvature at C and its focal point at F, to which rays parallel to the axis of the mirror are reflected. Rays passing through the center strike the mirror normally and return without change in direction. The mirror forms an inverted real image of a distant object.

All mirrors are achromatic because reflection does not disperse the light according to wavelength as does refraction. The spherical mirror, however, does not form a clear image; its focal points are not the same for its inner and outer parts. This defect is called **spherical aberration**. This is remedied by using a paraboloidal mirror instead of a spherical one. A good image is then formed, which may be viewed with an eyepiece or photographed. Thus a concave mirror can serve as the objective of a telescope.

The reflecting telescope has as its objective a concave mirror at the lower end of the tube; this is a circular disk of glass having its upper surface ground to suitable curvature and coated with a film of metal, such as aluminum. The glass serves only to give the desired form to the metal surface, and it need not have the high optical quality required for a lens. The entire back of the disk may be supported. The focal length of the mirror is often about 5 times its aperture; sometimes, however, it is less. Because the reflecting telescope is shorter than the ordinary refractor of the same diameter, it is less costly to construct and to house. For these reasons most modern telescopes are of the reflecting type.

The large mirror reflects the light of the celestial object to the **prime focus** in the middle of the tube near the upper end, where the image is accessible to the observer only in the very largest telescopes. In the **Newtonian form** a small plane mirror at a 45° angle near the top of the tube reflects the converging beam from the large mirror to focus at the side of the tube. In the **Casse-**

grainian form a small convex mirror replaces the plane mirror; it reflects the beam back through an opening in the large mirror to focus below it. Where the large mirror has no opening, which is the case with the 100-inch Mount Wilson telescope, the returning beam is reflected to the **Nasmyth focus** at the side by a plane mirror in front of the large one. This and some other large telescopes provide another place of observation, the **coudé focus**, by reflection of the beam through the polar axis to a laboratory below. See Fig. 5.5.

5.4 Large reflectors

The 200-inch Hale telescope of the Hale Observatories (Fig. 5.6), began in 1932 and completed in 1948, is the world's largest optical telescope. At the prime focus, 16.8 meters above the aluminum-coated mirror, the observer is carried in a cage that is 1.5 meters in diameter, obstructing less than 10 percent of the incoming light. At this focus the field of view is about 30 minutes of arc and the eye can detect a star 400,000 times fainter than the unaided eye. When the 1-meter, convex, secondary mirror is set in the converging beam it reflects the light to the Cassegrainian focus yielding an effective focal length of 81.4 meters. The observer is also carried in a cage at this focus as well. With the aid of a tertiary mirror the light can be directed to the coudé focus in a laboratory below the observing floor yielding an effective focal length of 155.4 meters. This great telescope has been in the forefront of astronomical research for a quarter of a century.

Next in size among the active reflectors is the 120-inch telescope of the Lick Observatory. Its primary focal length is 15.2 meters, and it too has a coudé focus. This telescope is followed in size by the 107-inch telescope of the University of Texas, the 2.57-meter telescope of the Crimean Astrophysical Observatory, and the 100-inch telescope of the Hale Observatories.

These great optical telescopes are soon to be joined by an array of large radio telescopes (see Appendix). Under construction and scheduled for completion before 1978 are two 4-meter reflectors, one for the Kitt Peak National Observatory in Arizona and one for the Cerro Tololo Inter-American Observatory in Chile, a 3.8-meter telescope for the Max-Planck Institute in Germany at a site to be designated, a 3.5-meter telescope for the European Southern Observatories in Chile, a 3.5-meter telescope for the joint Anglo-Australian Observatory in Siding Springs, a 3.6-meter French telescope at a site yet to be selected, a 3.5-meter Italian telescope at a site to be located in Hawaii, and a 2.5-meter telescope for the Hale Observatories in Chile. All of these telescopes

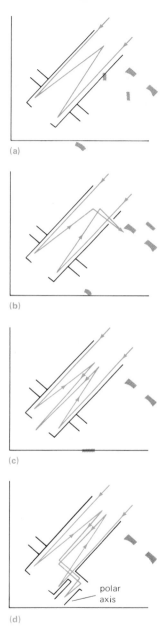

(a)

(b)

(c)

polar
axis

(d)

FIGURE 5.5
Four possible forms for a reflecting telescope. (a) the prime focus, (b) the Newtonian focus, (c) the Cassegrainian focus, (d) the coudé focus.

FIGURE 5.6
The 200-inch Hale telescope on Mt. Palomar. Top row, left to right, shows the telescope looking east, in the zenith, and west. Bottom row, left to right, shows the telescope looking north, slightly south of the zenith, and south (Hale Observatories photographs).

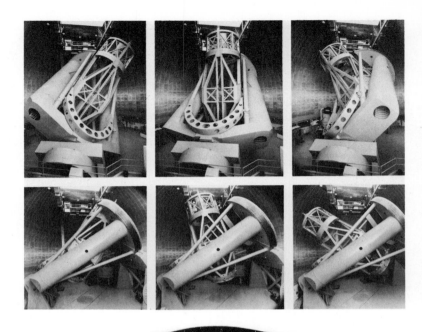

FIGURE 5.7
The region around sigma Persei taken with a wide field Schmidt telescope. The field of view is about six degrees.

are designed to press forward research in extragalactic areas. A great reflector of almost 6 meters in diameter has been under construction near Zelenchukskaya since 1948.

These telescopes are ideally suited for photographing small areas of the sky or studying individual objects with specialized instruments, such as photometers and spectrographs, soon to be described. Other telescopes commonly referred to as astrographs are more effective in recording large areas of the sky on photographic plates.

5.5 Astrographs

The large-field telescope plays a major role in present-day astronomy. The field size of these **astrographs** is often 10° by 10°, hence, hundreds and thousands of stars are recorded at one time (Fig. 5.8). Plates taken with various filters allow rapid rough star color estimates, and plates repeated after an appropriate interval allow studies of the motions and variability of stars. These plates also serve to locate objects of interest for the large telescopes.

The earliest telescopes of this type employed objectives made up of three or more lenses. The purpose of the compound lens was to focus the large field on a reasonable size plate with as few aberrations as possible. Such telescopes often had focal ratios of or somewhat less than 8. Recently, it has been possible to combine lenses with mirrors to achieve the same purpose; very large apertures and short focal lengths makes them very "fast" telescopes. The smaller the focal ratio the faster a telescope is said to be, just as with camera lenses. Focal ratios of 1.5 are not uncommon. The **Schmidt telescope** and the **Maksutov-Bouwers telescope** are examples.

As its main element the Schmidt telescope has a spherical mirror, which is easy to make in an optical shop but is not by itself suitable for a telescope. Parallel rays reflected by the central part of such a mirror are focused farther away from the focal point than those rays reflected from its outer zones. The appropriate correction is effected by a special type of thin lens, the **correcting plate**, at the center of curvature of the mirror. The lens slightly causes the outer parts of the entering beam to diverge with respect to the middle, so that the entire beam is focused on a slightly curved surface. The photographic plate, suitably curved by springs in the plate holder, faces the mirror between it and the correcting plate. The size of this type of telescope is denoted by the diameter of the correcting plate.

One of the largest telescopes of this kind is the 48-inch Schmidt

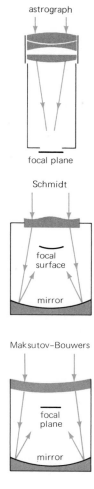

FIGURE 5.8
A schematic drawing of the optics used in wide-field telescopes. The traditional astrograph uses 3 or more lenses to image the field on a flat plate at the focus. The images suffer chromatic aberration. The Schmidt uses a spherical mirror and corrector plate to image an undistorted field on a curved focal surface. The plates, usually glass, must be bent and the corrector plate introduces very little chromatic aberration. The Maksutov-Bouwers system uses a spherical mirror and a spherical lens (meniscus lens) to image an undistorted field on a flat plane. This system has a little more chromatic aberration than the Schmidt, but it is negligible.

of the Palomar Observatory (Fig. 5.9). Its 1.83-meter mirror has a radius of curvature of 6.1 meters, which is about the length of the tube. The focal length is 3.05 meters giving a focal ratio of about 1.7. An important achievement of this telescope is the National Geographic Society–Palomar Observatory Sky Survey, a photographic atlas of the heavens north of declination 27°S. The atlas consists of 935 pairs of negative prints from blue- and red-sensitive plates; each print is 35.6 centimeters square and covers an area 6°.6 on a side. The survey reaches stars of the twentieth visual magnitude and exterior galaxies of the 19.5 magnitude. The largest telescope of this design is the 1.35-meter Schmidt telescope of the Karl Schwarzschild Observatory in Tautenburg, Germany.

FIGURE 5.9
The Hale Observatories 48-inch Schmidt telescope located on Mt. Palomar in California. (Hale Observatories photograph.)

The Maksutov-Bouwers telescope employs a spherical mirror and a relatively thick lens having spherical surfaces referred to as a meniscus lens. These surfaces are easy to generate, the only problem being that the glass used to make the lens must be of excellent optical quality. The large field of view is imaged upon a relatively small, flat plate. The largest telescope of this type has a 1-meter aperture and is operated at a field station in Chile by astronomers of the Pulkova Observatory. This optical system can be given very long effective focal lengths and still be compact, so that in the Cassegrainian form it has found extensive use by amateur astronomers and as a telephoto camera lens.

There are a host of specialized telescopes in use, for example in solar observing, and we will describe a few of these at the proper place. The Further Reading list includes several interesting and readable works dealing with modern telescopes, including radio telescopes, and their designs.

5.6 Advantages of a large telescope

The **light-gathering power** of a telescope increases in direct proportion to the area of the objective, or the square of its diameter (Fig. 5.10). This is the most important function of a large telescope. A particular star is accordingly 400 times as bright with the 200-inch telescope as with a 25.4-centimeter (10-inch) telescope. Thus stars can be observed with the former telescope that are too faint to be detected with the latter.

The **magnifying power** with a given eyepiece increases with the focal length of the telescope. When the telescope is employed visually, the magnifying power (M) equals the focal length (F) of the objective divided by the focal length (f) of the eyepiece that is used, that is, $M = F/f$. When the telescope is employed as a camera, the diameter of a celestial object having a noticeable size in the photograph equals the focal length of the objective times the angular diameter of the object in degrees, divided by $57°.3$. Thus the moon, having an angular diameter of about $0.5°$, appears 14.6 centimeters in diameter in a photograph at the prime focus of the 200-inch telescope, and about 3.4 centimeters with a 25-centimeter refracting telescope of the usual type. In either case, the size can, of course, be increased by enlargement of the photograph. We should note, however, that stars are point sources and their size on a photograph is the result of their apparent brightness and has nothing to do with their true dimensions.

The **resolving power** of a telescope is the angular distance between two stars that can be just separated visually with the telescope in the best conditions. This least distance, d in seconds

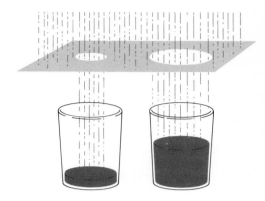

FIGURE 5.10
The number of photons collected by a telescope is proportional to the square of the aperture. Note that in a steady rain hole (right) which is twice as large as the left, the hole collects four times as much rain water. The function of a large telescope is to collect more photons.

of arc, is related to the wavelength, λ, and the aperture, *a*, in the same units by the formula:

$$d'' = 1.03 \times 206{,}265 \times \frac{\lambda}{a}$$

For visual telescopes the formula becomes:

$$d'' = \frac{11.7}{a}$$

where the aperture is expressed in centimeters. The value is $0''.023$ for the 200-inch telescope, and $0''.47$ for a 25-centimeter telescope. We notice presently how the much greater wavelengths employed in radio telescopes make these telescopes less effective in separating fine detail when they are used singly. We should point out that the great resolving power of the large telescopes is never used. Even in the Andes Mountains where the atmosphere allows stellar images of only $0''.5$ or even less, a 25-centimeter-aperture telescope is sufficient. In space the quality of the optical surfaces will be the limiting factor rather than the atmosphere.

RADIO ASTRONOMY

Radio astronomy is the study of the heavens by reception of cosmic radiations at radio wavelengths. The radiations that can come through the atmosphere to the ground range in wavelengths from less than 1 centimeter to about 30 meters. Waves much shorter than 1 centimeter are absorbed by molecules of our atmosphere, and those longer than 30 meters cannot ordinarily penetrate the electrified ionosphere. Selected parts of this wavelength range are known as **windows** because these specific wavelengths can penetrate the atmosphere more easily than others. The best known is the 21-centimeter wavelength, which is useful in determining the velocity of the source of cosmic radiation (see Section 5.12).

There are two forms of cosmic radiation at radio wavelengths. The first type is **thermal radiation,** which would be expected only from the temperature of the source. It may come, for example, from an emission nebula, where the gas has been highly heated and ionized by a blue star in the vicinity. This kind of radiation is stronger in the centimeter than in the meter lengths. The second type is **nonthermal radiation,** which may be the sort of radiation emitted by fast-moving ionized particles revolving in the synchrotron of the radiation laboratory. It is much stronger in the meter wavelengths than in the shorter ones, and it is polarized (see Chapter 17).

Although radio reception from the heavens was achieved as early as 1931 by Karl Jansky, extensive activity in radio telescope building and operation did not begin until 1946, after the end of World War II.

5.7 The radio telescope

The radio telescope is analogous to the optical telescope in principle and purpose. In the most common type of telescope there is a large paraboloidal reflector that concentrates the radiation at the feed or focus of the concave surface, just as light rays are gathered by lenses or mirrors in an optical telescope. The focus is referred to as the feed because the first such telescopes were wartime radar reflectors that were "fed" the pulses from the focus. Located at the focus is an antenna or a waveguide that receives the signal and sends it to an amplifier and from there to a recorder.

Another type of so-called telescope is the multielement array where series of antennas are interconnected in various ways to achieve a given purpose. Such arrays may be a combination of simple dipole antennas, helixes, or even paraboloids.

The radio telescope is effective during the day as well as at night; the long-wave radiations it receives from the heavens are also not seriously obstructed by the clouds of our atmosphere nor by the interstellar dust that conceals much of the universe from the optical view. In addition to its reception from outside sources, the radio telescope is employed for recording radio pulses emitted from stations on the earth and reflected back by the nearer celestial bodies; thus far the moon and sun and the planets Mercury, Venus, and Mars have been observed by **radar** techniques.

5.8 Large radio telescopes

The 91.4-meter transit telescope is a fine example of a very large paraboloid. It is called a transit telescope because, like its optical counterpart, it can look only along its meridian (Fig. 5.11). This and many other telescopes including a 42.7-meter steerable telescope (Fig. 5.12) are operated as a national facility by the National Radio Astronomy Observatory and are located at Green Bank, West Virginia.

The University of Manchester in England has a 76.2-meter telescope at the Jodrell Bank Experimental Station. The Radiophysics Laboratory in Australia operates a 64-meter telescope at Parkes, 322 kilometers west of Sydney. Stanford University has a 45.7-meter telescope. A very large fixed reflector near Arecibo, Puerto Rico, was designed by Cornell University scientists. This reflector is a 304.8-meter spherical dish placed in a natural depres-

FIGURE 5.11
The large 91.4-meter transit telescope located in Greenbank, West Virginia. This giant telescope can be moved only up and down along the meridian. Its surface is composed of an open mesh wire. (National Radio Astronomy Observatory photograph.)

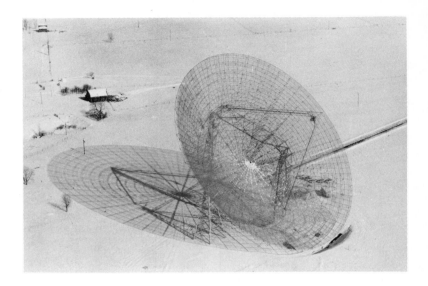

FIGURE 5.12
The 42.7-meter fully steerable radio telescope of the National Radio Astronomy Observatory. The sloping shaft is the polar axis and points to the north pole. The surface is a carefully formed series of metal sheets allowing the telescope to be used at short radio wavelengths. (National Radio Astronomy Observatory photograph.)

sion in the mountains. The shift in its pointing direction to as much as 20° from the zenith and the correction for the spherical form of the reflector are accomplished at the antenna feed supported above the dish. A fixed parabolic cylindrical reflector 182.9 meters long and 122 meters wide set in a stream bed serves as the reflector of a radio telescope at the University of Illinois.

5.9 Resolving power

The resolving power of a single radio telescope is a measure of the fineness of detail it can record. Calculated by the same formula (see Section 5.6) as for the optical telescope, the resolving power is the smallest angular distance required to separate two radio point sources with a particular telescope. The value of this critical angle is directly proportional to the wavelength of the radiation and inversely proportional to the diameter of the antenna. Because it operates with the longer wavelengths, a paraboloidal radio telescope is less effective in separating fine detail than is an optical telescope of the same aperture. Thus the critical separation for a 21-meter (827-inch) radio dish at the wavelength of 21 centimeters is 35' as compared with a least separation of only 0".023 at visual wavelengths with the objective of the 200-inch optical telescope.

A partial remedy for the deficiency in resolving power of radio telescopes is the use of larger paraboloids or interference methods promoted by multielement antennas, or both.

Two large paraboloids properly connected together (Fig. 5.13) but separated by a known distance can be made to yield the resolution of a single telescope having a diameter equal to the separation. Thus, by our formula, two such telescopes separated by 2 kilometers and observing at a wavelength of 21 centimeters will have a resolution of 22". This is called long baseline interferometry. With the introduction of very accurate atomic clocks (see Section 4.9) there is no need for the telescopes to be physically connected; their individual signals can be recorded along with the time on magnetic tapes and then the magnetic tapes can be brought together. This allows the telescopes to be separated by intercontinental distances and very high resolution is achieved. For example, suppose the two telescopes are 2000 kilometers apart and again working at a wavelength of 21 centimeters; the resolving power is now 0".002! This is called very long baseline interferometry.

An example of the high resolution achieved by the multielement fixed array is the Mills Cross named after its designer, B. Y. Mills (Fig. 5.14). An east–west line of fixed-dipole antennas will give a long east–west pattern that is very narrow in the north–

FIGURE 5.13
The two radio telescopes of the California Institute of Technology's Owen's Valley Observatory, when properly connected form the two basic elements of a radio interferometer. Two such telescopes have the resolution of a single telescope as large as the spacing between them. (California Institute of Technology photograph.)

FIGURE 5.14
The great Mills Cross telescope in Australia. Each bar of the telescope is 1.6 kilometers long. The cylindrical reflector of the east–west bar can be rotated to mechanically shift the east–west beam north and south. (Photograph courtesy of the Australian News and Information Bureau.)

south direction. A similar line of fixed-dipole antennas in the north–south direction gives a long north–south pattern but a very narrow pattern in the east–west direction. The signals from this cross pattern can be added in phase and out of phase. Subtracting the two leaves only the small pencil beam (Fig. 5.15). By changing the phase of the whole east–west line the pattern can be made to move north–south and similarly for the north–south line, hence the pencil beam can be made to look at various parts of the sky.

A third form of a high-resolution telescope is the aperture synthesis telescope (Fig. 5.16). The interferometer mentioned earlier can be made to synthesize a large telescope if one of the paraboloids is movable. If the baseline is not east–west the rotation of the earth allows the "filling in" of the aperture to a certain extent. If both telescopes are movable, one east–west and the other north–south this filling in takes place more rapidly. It is even more efficient if there are numerous paraboloids on a given baseline, one or two of which are movable. This is called a super-synthesis telescope and uses the rotation of the earth for filling in as well. An instrument of this type is located at Westerbork in Holland (Fig. 5.17).

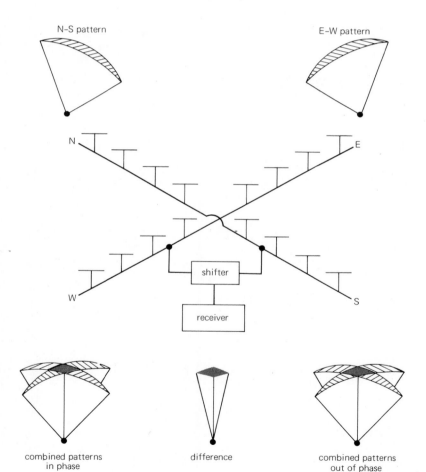

FIGURE 5.15
A pictorial representation of how a Mills Cross telescope achieves a narrow beam and, hence, high resolution. The two orange slice patterns are combined both in phase and out of phase. The difference between the in phase and out of phase signal is the narrow central portion common to both.

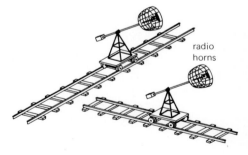

FIGURE 5.16
A schematic representation of an aperture synthesis telescope. The effective aperture is a circle covering the entire tracks. By moving the telescopes the intensity over the entire pattern can be sampled and a radio picture constructed.

FIGURE 5.17
The great supersynthesis telescope at Westerbork, Holland. The telescope samples the entire pattern by being rotated by the earth instead of moving the telescopes around.

THE SPECTROGRAPH

5.10 Dispersion of light

When a beam of light passes obliquely from one medium to another, as from air into glass, its direction is altered (see Section 5.1). Because the amount of the change in direction increases progressively with decrease in wavelength of the light, from red to violet, the beam is **dispersed** by refraction into a **spectrum,** or an array of the components of light. (Recall that wavelength is the distance between the crests of successive waves; see Section 5.1.) Thus the rainbow is produced when sunlight is dispersed by raindrops. The spectrum obtained by passing the visible light through a glass **prism** (Fig. 5.18) is made up of the colors red, orange, yellow, green, blue, indigo, and violet. Radiation beyond the visible region extends into the ultraviolet in one direction and the infrared in the other. These wavelengths may be recorded by other means including photography.

A **grating** is often used as the dispersing element instead of a prism. The ridges of the grating act to disperse the light into many spectra called the orders of the grating. The chief advantage of a grating is that its spectra are linear with wavelength, which makes the subsequent analysis more simple. A grating, however, spreads the light among its many orders so it is very inefficient. In the early part of this century opticians learned to shape the ridges so that most of the light could be concentrated into a single order. Such a grating is called a **blazed grating** and looks very much like a saw tooth if sliced across the ridges. In most cases there are at least 600 blazed faces to the millimeter and some of these gratings are as large as 25 centimeters on a side. Because of the very great efficiency of a blazed grating it has essentially replaced the prism as the dispersing element in the spectrographs used by astronomers.

An instrument that makes use of the dispersion of light to record the spectrum is called a **spectrograph.** The photographic record or picture is a spectrogram. If the spectrograph is fitted with an eyepiece for direct viewing of the spectrum it is called a spectroscope. A familiar type of spectroscope consists of a blazed grating (or sometimes a prism) toward which a **collimator** and a telescopic camera are directed. The light enters the collimator (an inverted telescope) through a narrow slit formed by the sharpened and polished edges of two metal plates. The slit is at the focal point of the collimator. After passing through the collimator, the rays are accordingly parallel in all wavelengths as they fall upon the grating. The grating disperses this beam into a spec-

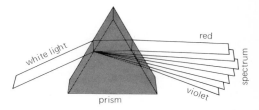

FIGURE 5.18
Light is refracted by a prism, blue light more than red. This effect is the basic principle upon which the spectrograph was originally based.

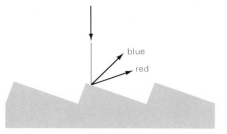

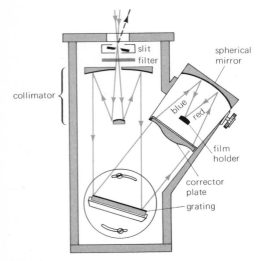

FIGURE 5.19
Schematic drawing of a grating spectrograph using a Schmidt camera (left). A grating spectrograph mounted on a 1-meter telescope (right).

trum that is brought to focus by a telescopic camera (Schmidt telescope) (Fig. 5.19 left). For the purposes of astronomy the spectroscope is attached to the telescope (Fig. 5.19 right), which serves to concentrate the light of a celestial body and to focus it on the slit. Here, and in the laboratory, the eyepiece of a spectroscope is generally replaced by a plate holder and hence it is referred to as a spectrograph.

A refinement of the conventional spectrograph as presented here is the echelle spectrograph, which is essentially the spectrograph described with the addition of a predispersing element, such as a thin prism or a coarse grating. Other "spectrographs" make use of the principles of interferometry and are especially adaptable to photoelectric methods of recording, but will not be discussed except as needed later.

5.11 Emission and absorption spectra

The **bright line spectrum,** or **emission line spectrum,** is an array of bright lines. The source of the light is a glowing gas at low pressure that radiates in a limited number of wavelengths. Each gaseous chemical element emits its characteristic selection of wavelengths and can therefore be identified by the pattern of lines of its spectrum. The glowing gas of a neon tube, for example, produces a bright line spectrum.

The **continuous spectrum** is a continuous emission in all wavelengths. The source of the light is a luminous solid or liquid, or it may be a gas under conditions where it does not emit selectively. The glowing filament of a lamp and the sun's surface produce continuous spectra. There may also be emission and absorption continuums of limited extent in the spectra of certain gases.

The **dark line spectrum,** or **absorption line spectrum,** is an array of dark lines on a bright continuous background. Gas at low pressure intervenes between the observer and the source of light, which by itself may produce a bright line spectrum. The gas abstracts from the light the pattern of wavelengths it emits in the same conditions, and thus reveals its own chemical composition. The spectrum of the sun's atmosphere is essentially a dark line spectrum, because the sunlight has filtered through the atmospheres of the sun and earth before reaching us.

These three types of spectra are often called Kirchhoff's laws (Fig. 5.20).

Where the gas consists of molecules, such as carbon dioxide or methane, the spectrum shows **bands** of bright or dark lines characteristic of these molecules.

It is important to note that the λ 21-centimeter line (1420 MHz) used by the radio astronomer is essentially a bright (emission) line of hydrogen arising from the spin-flip mechanism (see Section 12.10) within the atom noticed by van de Hulst. This occurs because the neutral hydrogen clouds are warmer than interstellar space. Occasionally the λ 21-centimeter line is observed in absorption. Recently, the multiplet from the vibrations of the hydroxyl molecule (OH) have been observed in absorption and in one case in emission. The OH lines at λ 18 centimeters are much weaker than the usual interstellar H line. These are a few

FIGURE 5.20
The formation of the basic types of spectra. A glowing filament yields a continuous spectrum as shown on the left. If the filament is viewed through a thin cool gas we see an absorption line (dark line) spectrum as shown on the right. If we view only the gas we see a bright line (emission line) spectrum as shown at the bottom. The bright line spectrum is the complement of the dark line spectrum.

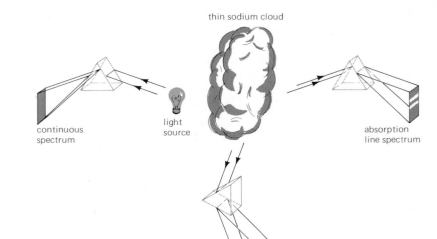

selected "radio" lines. We will note several important additional examples later.

5.12 The Doppler effect

When a source of sound waves, such as the whistle of a locomotive, is approaching the observer, the waves come to him crowded together, so that the pitch of the sound is raised. When the source is receding, the pitch is lowered. In 1842, the physicist C. J. Doppler pointed out that a similar effect is required by the wave theory of light; thus a star should appear bluer when it is approaching and redder when it is receding from us. In 1848 the physicist H. Fizeau explained that the perceptible effect would be a displacement of the lines in the star's spectrum. The **Doppler effect** in the spectrum can be stated as follows.

When a source of light is relatively approaching or receding from the observer, the lines in its spectrum are displaced to shorter or longer wavelengths, respectively, by an amount that is directly proportional to the speed of approach or recession. Thus,

$$\frac{\triangle\lambda}{\lambda} = \frac{v}{c}$$

$$\lambda - \lambda' = \triangle\lambda$$

where $\triangle\lambda$ is the shift from wavelength λ, v is the velocity of the source, and c is the velocity of light.

The displacement of the spectrum lines to the violet $(-)$ or red $(+)$ from their normal positions is often our only means of observing the motion of a celestial body relative to the earth. Applications of this useful effect are noted in later chapters.

PHOTOELECTRIC INSTRUMENTS

Because of the inability of the human eye to accurately compare brightnesses and positions of stars, astronomers turned to photographic techniques for these purposes. Photographs have the advantage of recording numerous stars and objects on a single plate. For great accuracy in measuring brightness, however, the photographic plate has some drawbacks; in particular it does not store light linearly and it has a low efficiency. For this purpose astronomers have turned to photoelectric devices, and such devices may eventually replace the photographic plate entirely.

5.13 Photoelectric photometer

The photoelectric effect explained theoretically by Einstein in 1905, is the release of electrons from a surface when it is illuminated by light. This effect varies with intensity and surfaces can be made to have efficiencies at least 20 times greater than photographic emulsions. Efficiencies 30 times greater are not unusual. When placed in tubes, such surfaces lend themselves to **photometry** (the measurement of brightness) at the telescope. The small current from the phototube is amplified and recorded by some convenient means (Fig. 5.21). Figure 5.22 shows a **photometer** on the telescope.

The trained human eye can generally give magnitudes or brightness (see Section 12.18) accurate to one tenth of a magnitude.

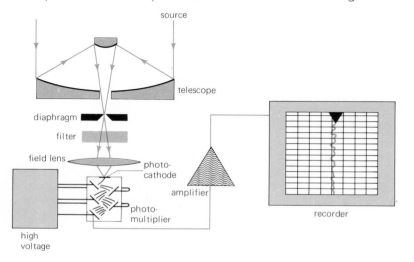

FIGURE 5.21
Schematic drawing of a photoelectric photometer. It is typical of many astronomical instruments. The background signal is limited by narrowing down the field by a diaphragm. The wavelength is limited by a filter. The photomultiplier tube multiplies the signal many fold after which it is amplified and recorded.

FIGURE 5.22
A two channel photoelectric photometer mounted on a small telescope. (Leander McCormick Observatory photograph.)

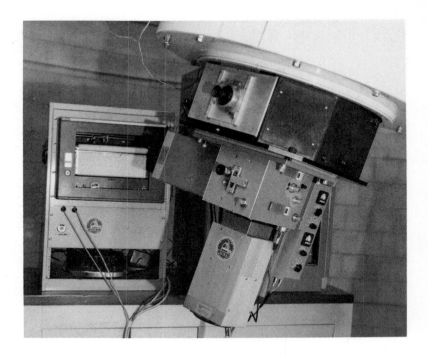

Using a photographic plate magnitudes accurate to a few hundredths of a magnitude can be obtained. With the phototube we can measure magnitudes to a few thousandths, and this accuracy explains the value of the photoelectric technique.

5.14 Image tubes

Tubes using the photoelectric effect and forming images are now being used on the telescope. The television pickup tube (vidicon or orthicon) is such a tube. The **image tubes** that astronomers use are generally simpler than television tubes, however. In these simple tubes the photoelectrons are focused on a phosphor by means of a magnet and the glowing phosphor is then photographed (Fig. 5.23).

Image tubes are used in conjunction with the photographic plate and hence the system can record many stars, the entire field of view, at one time. The secret here is to take advantage of the efficient photocathode and then impart energy to the resulting electrons. Two tubes in one unit (a cascade tube) result in a considerable gain in time at the final photograph. A saving of a factor of 10 or more is regularly achieved by such **image intensifiers** (Fig. 5.24).

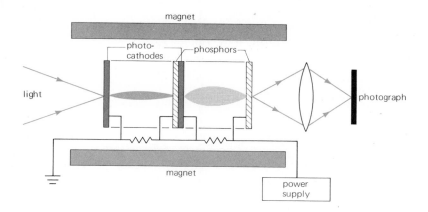

FIGURE 5.23
A schematic drawing of a simple cascade image tube system. The light falls on the tube on the left. It is converted and amplified and forms an image to be photographed on the back of the tube.

Another advantage of these tubes is that their cathodes can be made to respond to infrared light. Such a tube still produces a blue image (because the phosphor is blue) to be photographed, and is therefore known as an **image converter**. Image converters have gains of more than 1000 over comparable photographic techniques where the wavelength of the radiation is 0.0001 centimeter (infrared). This is chiefly because of the extreme slowness of infrared-sensitive films.

The use of image tubes of a great variety of types is increasing. Indeed, most spectrograms of very faint objects, such as those discussed in Chapter 18, are now taken with spectrographs equipped with image tubes. It is not inconceivable that within 10 years all spectra of faint objects will be recorded using some form of image tube.

A large variety of photoelectric devices are finding increasing use in astronomy; the analog in the radio region is the **maser**, which is again a center of interest because it promises greater efficiency than present amplifiers. The reason is not hard to find. Astronomy, except for planetary and solar physics, is in the main a passive science because experiments are beyond the control of the scientist. These experiments, at least the interesting ones, always seem to occur at very great distances, hence, the amount of light arriving at the earth per square centimeter is very slight. The astronomer must be able to collect and record as much light as possible to interpret these events.

The astronomer must have great patience or at least a long list of interesting phenomena that he would like to observe. Then, when one comes along he must be willing to work under the most extreme circumstances with every possible type of instrument, including telescopes in space.

FIGURE 5.24
A two-stage image intensifier. This tube is fabricated by RCA and is approximately 4 inches in diameter including the rubber casing. (Photograph by Carnegie Institution of Washington, DTM.)

QUESTIONS

1 What is the cause of the refraction of light?
2 The word achromat is derived from Greek. What does it mean when applied to telescopes? Is its use when referring to refractors proper?
3 Why are most large telescopes and many radio telescopes mounted in the equatorial system?
4 Why are all of the great telescopes being built today reflectors?
5 List the five useful focii of a reflecting telescope.
6 Where does the word coudé come from and what does it mean? (You will have to do some outside checking.)
7 What is meant by a "fast" telescope?
8 One telescope has a focal length of 200 centimeters and an eyepiece of 2-centimeter focal length. A second telescope has a focal length of 100 centimeters and an eyepiece of 0.75 centimeter. Which has the greatest magnification?
9 Why is the focus of a large paraboloid in radio telescopes referred to as the feed?
10 How do radio telescopes make up for low resolution?

FURTHER READINGS

BROWN, R. HANBURY, AND A. C. B. LOVELL, *The Exploration of Space by Radio*, Wiley, New York, 1958.

CHRISTIANSON, W. N. AND HOGBORN, J. A., *Radio Telescopes*, Cambridge University Press, London, 1969.

HEYWOOD, J., *Radio Astronomy and How to Build Your Own Telescope*, ARC Books, New York, 1963.

INGALLS, ALBERT G., ED., *Amateur Telescope Making, Book 1; Amateur Telescope Making, Advanced; Amateur Telescope Making, Book 3*, Scientific American, New York.

KING, HENRY C., *The History of the Telescope*, Sky Publishing Corp., Harvard Observatory, Cambridge, Mass., 1955.

MICZAIKA, G. R., AND WILLIAM M. SINTON, *Tools of the Astronomer*, Harvard University Press, Cambridge, Mass., 1961.

STEINBERG, J. L., AND J. LEQUEUX, *Radio Astronomy*, R. N. Bracewell, transl., McGraw-Hill, New York, 1963.

WOOD, F. BRADSHAW, ED., *Photoelectric Astronomy for Amateurs*, Macmillan, New York, 1963, Chaps. 1–4.

A 1482 cut by J. S. Bosco depicting the conditions for lunar and solar eclipses. The basic concept presented is correct, but the artist has taken great liberty to get the point across.

The Moon in Its Phases

6

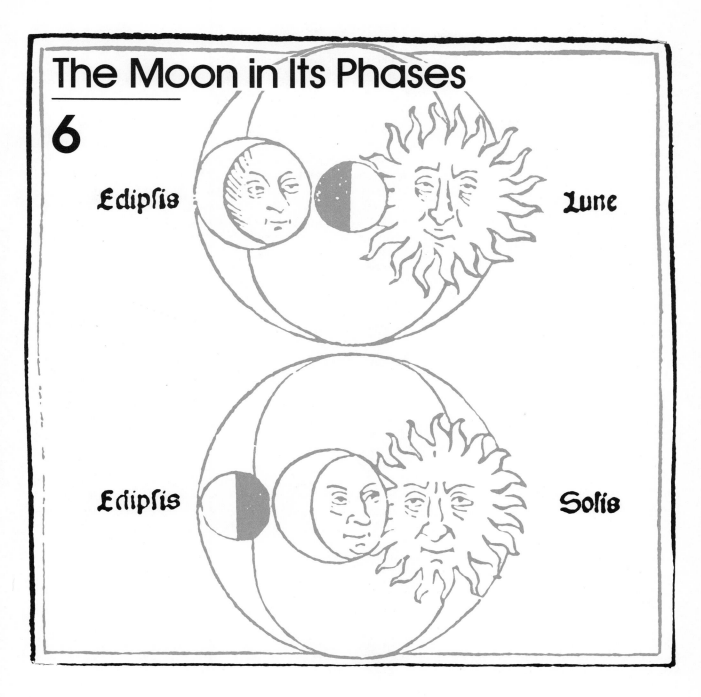

Eclipsis   Lune

Eclipsis   Solis

6

*The moon is the earth's satellite.
Except for the sun, it is the most conspicuous object in the sky because it
is the nearest celestial body to the
earth. The moon and the earth revolve
together around the sun and at the
same time mutually revolve around a
point between their centers. Some
satellites orbiting other planets are
larger than the moon, but none of
these reflects as much sunlight to its
planet as does the moon to the earth.
In this chapter we consider the moon's
motion relative to the sun and to the
earth, its surface features, which are in
marked contrast with those of the
earth, and its eclipses when it enters
the earth's shadow.*

6.1 The moon's distance and orbit

The distance of a celestial body from the earth is frequently measured by observing its parallax. **Parallax** is the difference in the
direction of an object when it is viewed from two different places.
Notice how a nearby object seems to jump back and forth against
a distant background when it is viewed alternately with each eye.
The nearer the object, the greater is its change in direction; the
amount of the parallax would be greater, of course, if our eyes
were farther apart. When the parallax has been determined and
the distance between the two points of observation is known, the
distance of the object is readily calculated.

The difference in the moon's directions from two widely separated places on the earth is so great that the parallax can be
accurately measured. Observed at the same instant from New York
and San Francisco, the moon's positions among the stars differ
by nearly the full breadth of the moon, or about half a degree.
When we speak of the moon's parallax ordinarily, it is as though
one observer were at the center of the earth and the other at the
equator with the moon on his horizon. It is this **equatorial horizontal parallax** that is given for the sake of uniformity.

The moon's parallax at its average distance from the earth is
nearly 1°. The corresponding distance of the moon from the
center of the earth is 384,400 kilometers, which is about 60 times
the radius of the earth, or less than 10 times its circumference.
Close agreement with this value has been obtained by measuring
the two-way travel times of many radio pulses reflected from the
moon.

The distances that separate the celestial bodies are so great in
comparison with the sizes of the bodies that it is often impractical
to represent the two on the same scale in the diagrams. In Fig. 6.1,
for example, the moon should be placed 60 centimeters from the
2.5-centimeter earth.

Although the moon is not the largest of the **satellites**, it has the

FIGURE 6.1
*The moon's distance can be obtained by
measuring the apparent angular shift against
a distant background. This shift amounts to
almost 0.5 degree, or the diameter of the
moon itself, when viewed from New York
and San Francisco.*

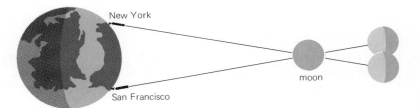

distinction of being the one most nearly comparable with its **primary** in size and mass. The moon's diameter of 3476 kilometers is more than a quarter of the earth's diameter, and its mass is 1/81 as much as the earth's mass.

Imagine the centers of the earth and moon joined by a stout rod. The point of support at which the two bodies would balance is the **center of mass** of the earth–moon system. It is this point around which the earth and the moon mutually revolve monthly, and it is the elliptical path of this point around the sun that we have hitherto called the earth's orbit (Fig. 6.2). The center of mass of the earth–moon system is only 4667 kilometers from the earth's center and is therefore within the earth. Thus the moon revolves around the earth, although not around its center. In the descriptions that follow it is convenient to consider the moon's revolution relative to the earth's center.

The moon's orbit relative to the earth is an ellipse of small eccentricity (e), having the earth at one focus (Fig. 6.3).

$$e = 0.0549$$

At **perigee**, where the moon is nearest to us, the distance between the centers of the moon and earth may be as small as 356,800 kilometers. At **apogee**, where the moon is farthest from us, the distance may be as great as 406,400 kilometers. The resulting variation of more than 10 percent in the moon's apparent diameter is still not enough to be conspicuous to the unaided eye.

The moon's orbit around us is changing continually, mainly because of unequal effects of the sun's attraction for the moon and earth. For example, the major axis of the orbit rotates eastward in a period of about 9 years. This is one of the many variations that make the determination of the moon's motion an intricate problem, but a problem now being so well solved that the moon's course in the heavens is predictable with great accuracy. It is worth commenting that Hipparchus gave the moon's diameter as 31' and its mean distance (using present-day units) as 376,000 kilometers! He was in error by less than 2 percent.

6.2 The phases of the moon

The moon's phases are the different shapes it shows (Fig. 6.4). Records of the phases of the moon have been found dating from 13,000 B.C. The moon is a dark globe like the earth; half is in the sunlight, while the other half turned away from the sun is in the darkness of night. The phases are the varying amounts of the moon's sunlit hemisphere that are turned toward us successively in the course of the month.

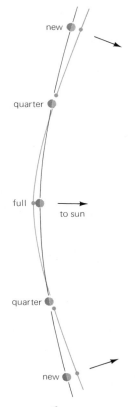

FIGURE 6.2
The orbits of the earth and moon around the sun. It is not readily apparent from the drawing but the orbit of the moon around the sun is actually everywhere concave to the sun.

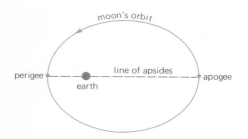

FIGURE 6.3
The moon's orbit relative to the earth is an ellipse.

It is the **new moon** that passes the sun; the dark hemisphere is toward us. The moon is invisible at this phase unless it happens to pass directly across the sun's disk, causing an eclipse of the sun. On the second evening after the new phase the thin **crescent** moon is likely to be seen in the west after sundown. The crescent grows thicker night after night, until the sunrise line runs straight across the disk at the **first quarter**. Then comes the gibbous phase as the bulging sunrise line gives the moon a lopsided appearance. Finally, a round **full moon** is seen rising at about nightfall. The phases are repeated thereafter in reverse order as the sunset line advances over the disk; they are gibbous, **last quarter**, crescent, and new again.

The horns, or **cusps**, of the crescent moon point away from the sun's place, and nearly in the direction of the moon's path among the stars. They are more nearly vertical in the evenings of spring, and more nearly horizontal in the autumn. Their direction in the sky follows the changing direction of the ecliptic with respect to the horizon during the year.

Often when the moon is in the crescent phase we see the rest of the moon dimly illuminated. This appearance has been called "the old moon in the new moon's arms," for the bright crescent seems to be wrapped around the faintly lighted part (Fig. 6.5).

The thin crescent is in the sunlight. The rest of the moon's disk is made visible by sunlight reflected from the earth. Just as the moon tempers the darkness of night for us, so the earth shines on the moon.

As we know from our television observations if anyone lived on the earthward side of the moon long enough, he would see the earth up among the stars in his sky (Fig. 1.1) going through all

FIGURE 6.4
The phases of the moon. The outer figures show the phases as seen from the earth.

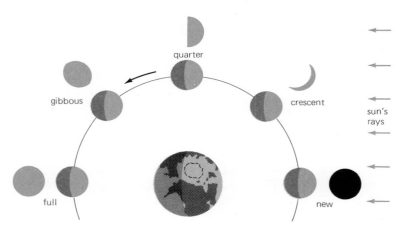

the phases that the moon shows to us. They are, of course, complementary; full earth occurs at new moon. The full earth would look 4 times as great in diameter as the full moon appears to us, and something like 60 times as bright. The earth is not only a larger mirror to reflect the sunshine but, owing to the atmosphere, it is a better reflector as well; it returns a third of the light it receives from the sun, although this varies somewhat depending upon the presence or absence of clouds.

Earthlight is plainest when the moon is a thin crescent, for the earth is then near its full phase in the lunar sky. It is a bluer light than that of the sunlit moon, because it is sunlight reflected by our atmosphere. The air around us scatters the shorter wavelengths more effectively, as the blue sky shows. Figure 6.5 shows the earthlight on the moon at crescent phase in the morning sky. The planet to the upper right is Saturn.

The month of the phases from new moon to new again, averages slightly more than 29.5 days, and varies in length more than half a day. It is the **synodic month,** and the lunar month of the calendars. Because it is shorter than our calendar months, with a single exception, the dates of the different phases are generally earlier in successive months.

The length of the **sidereal month** averages 27.3 days. This is the true period of the moon's revolution around the earth. At the end of this interval the moon has returned to nearly the same place among the stars. In the meantime the sun has been moving eastward as well, so that more than two days elapse after the end of the sidereal month before the synodic month is completed. In Fig. 6.6 between positions 1 and 2 the month has completed one sidereal month. It does not complete the synodic month until it reaches position 3.

6.3 The moon's path among the stars

Two apparent motions of the moon are observed by everyone. First, the moon rises and sets daily; it circles westward around us along with the rest of the celestial scenery because the earth is rotating from west to east. Second, the moon moves eastward against the turning background of the stars, because it is revolving in this direction around the earth. In the course of a day the moon revolves 360°/27.3, or about 13°, moving slightly more than its own diameter in an hour.

The moon's apparent path among the stars is nearly a great circle of the celestial sphere, which is inclined about 5° to the ecliptic. It therefore crosses the ecliptic at two opposite **nodes.** The ascending node is the point where the moon's center crosses

FIGURE 6.5
Earthlight on the moon at crescent phase in the morning sky. The earthlit moon can be seen in Figure 6.7 also. (Yerkes Observatory photograph.)

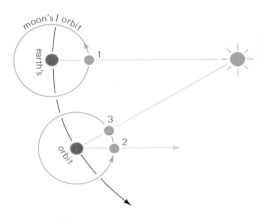

FIGURE 6.6
The month of the phases is longer than the sidereal month. The moon completes a full orbit (360°) moving from 1 to 2. However, since the earth has moved with respect to the sun and stars, the moon must move to position 3 before new moon again occurs.

FIGURE 6.7
*Jupiter and its satellites emerge from occulta-
tion by the moon. Such pictures are striking
evidence of the lack of an atmosphere on
the moon. (Griffith Observatory photograph.)*

the ecliptic going north; the descending node is the point where
it crosses going south. The nodes regress, or slide westward along
the ecliptic. Regression of the nodes goes on at a much faster
rate than does precession of the equinoxes (see page 44). In only
18.6 years the nodes shift completely around the ecliptic. The
moon's path through the constellations of the zodiac is considerably different from month to month. All of this, including the
fact that the moon's orbit is not circular, was known at the time
of Hipparchus.

The moon's revolution around the earth provides an important
means of timekeeping that is independent of the earth's variable
rotation. There is a practical observational problem, however: One
must be able to determine precisely the universal time when the
center of the moon's large disk arrives at a particular place among
the stars. The corresponding uniform ephemeris time (see Section
4.5) when the moon has this position can be determined by the
theory of the moon's motion. Frequent comparisons of this kind
permit accurate corrections from the predicted ephemeris times of
celestial events to the universal times for ordinary use.

Occasions when the moon passes over (**occults**) a star have
been employed for a similar purpose. The star disappears almost
instantly at the moon's eastern edge and reappears as abruptly at
the western edge. Predictions of occultations for stations in the
United States and Canada are published in *Sky and Telescope*.
Approximate times of occultations of some planets and bright stars
are listed in the *American Ephemeris and Nautical Almanac*.
Occultations are interesting to watch with the telescope or with
the unaided eye when the objects are bright enough. The photograph of Jupiter and its satellites emerging from behind the moon
(Fig. 6.7) illustrates a spectacular example.

Occultations of stars by the edge of the moon and their re-
appearance have two scientific uses: (1) they provide a direct
measure of the contour of the edge of the moon, and (2) they
provide a means for measuring the angular diameter, or size, of
the star being occulted. The latter measurements are extremely
valuable and make use of the diffraction effect assuming that the
moon has essentially no atmosphere. These measurements require
ultrafast response electronics and recording techniques since the
entire event takes place in less than 0.5 second.

A photographic device, known as the dual-rate moon position
camera, is employed at the U.S. Naval Observatory and elsewhere for more convenient reading of the moon-clock. Attached
to a telescope of moderate size, it holds the moon in a fixed position relative to the stars during the exposure. In the photographs

the positions of 30 or 40 points on the bright edge of the moon are measured in relation to about 10 neighboring stars. This method not only gives a better control of our timekeeping but is also improving our knowledge of the moon's motions and its variations.

6.4 Daily, monthly, and annual motions

We have seen that the solar day is about 4 minutes longer than the sidereal day because the sun moves eastward relative to the stars. The moon moves eastward still faster than the sun, and therefore the "lunar day" is longer than the solar day. The interval from upper transit of the moon to its next upper transit averages 24^h50^m of solar time, varying as much as 15 minutes either way. This interval is of special importance to those who live beside the ocean, for it is twice the interval between high tides.

The moon's crossing of the meridian as well as its rising and setting are delayed an average of 50 minutes from day to day. The time variation of the moonrise is even more marked, and we notice it particularly in the rising of the moon near its full phase.

The **harvest moon** is the full moon that occurs nearest the time of the autumnal equinox. The moon is then near its full phase and, as observed in our northern latitudes, it rises from night to night with the least delay. In the latitude of New York the least delay is shorter by an hour than the greatest delay. Thus the harvest moon lingers longer in our early evening skies than does the nearly full moon of other seasons, giving more light after sundown for harvesting. It can be shown that the least delay in the moonrise on successive nights occurs when the moon's path among the stars is least inclined to the east horizon, as it is at sunset at the time of the autumnal equinox. The full moon following the harvest moon is known as the **hunter's moon** for similar reasons.

The moon moves north and south of the celestial equator during the month, just as the sun does during the year and for the same reason. The moon's path among the stars is nearly the same as the ecliptic and is, therefore, similarly inclined to the celestial equator.

Consider the full moon. Opposite the sun at this phase, the full moon is farthest north when the sun is farthest south of the equator. The full moon near the time of the winter solstice rises in the northeast, climbs high in the sky, and sets in the northwest, as observed in the middle northern latitudes; it is above the horizon for a longer time than are the full moons of other seasons. In summer it is the other way around. The full moon of June rises in

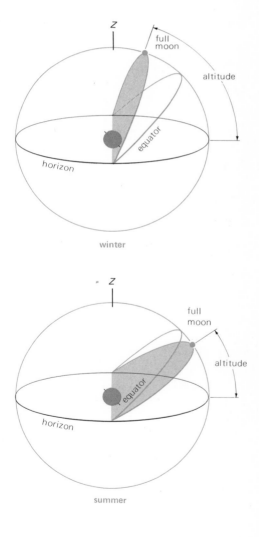

the southeast, transits low in the south, and soon sets in the southwest, like the winter sun.

In some years the moon ranges farther north and south than in other years. The greatest range in its movement in declination occurs at intervals of 18.6 years. This was the case in 1969, when the moon was going fully 5° farther than the sun both north and south from the celestial equator. Many people remarked on it at the time, probably because of the great general interest in the moon due to the impending manned lunar landing. The moon reached its highest point on 25 March 1969 when its declination was +28°43′.

As often as it is revealed in the changing phases, the face of the "man in the moon" is always toward us. This means that the moon rotates on its axis once in a sidereal month, while it is revolving once around the earth. Although the statement is true in the long run, anyone who watches the moon carefully during the month can see that the same hemisphere is not turned precisely toward us at all times. Spots near the moon's edge are sometimes in view and at other times turned out of sight, as Fig. 6.8 clearly shows. The moon seems to rock as it goes around us. These apparent oscillations, or **librations**, arise chiefly from two causes.

1 The moon's equator is inclined about 6.5° to the plane of its orbit. Thus its north pole is brought toward us at one time, and its south pole is toward us two weeks later, just as the earth's poles are presented alternately to the sun during the year.

2 The moon's revolution is not uniform. In its elliptical orbit around us, the law of equal areas (see Section 4.5) applies: The nearer the moon to the earth, the faster is its revolution. Meanwhile the rotation of the moon is practically uniform. Thus the two motions do not keep perfectly in step, although they come out together at the end of the month. The moon rocks in the east and west direction, allowing us to see farther around it in longitude at each edge than we could otherwise.

These librations have been studied for about a century, were well known to the ancient Greeks, and are presently being investigated by means of lasers and reflecting surfaces, called **retroreflectors**, placed on the moon by several of the astronaut teams. This technique measures the distances to the several widely spaced reflectors to within a few centimeters. Fully 59 percent of the moon's surface has faced the earth when the lunar month is completed. The remaining 41 percent is never seen from the earth

except on photographs returned by the astronauts or by various photographic lunar spacecraft.

THE FEATURES OF THE MOON

6.5 The lunar seas

Two features of the moon are plainly visible to the unaided eye. First, the changing phases are among the most conspicuous sights in the heavens and were the first to be correctly explained. Second,

FIGURE 6.8
The libration of the moon is clearly detected in these two photographs. The photographs were taken about the same phase but several years apart. The maria are much closer to the limb in the left-hand picture. (Lick Observatory photographs.)

FIGURE 6.9
A photograph of full moon showing the maria (left). The various maria including the manned landing locations are shown on the right. The numbers refer to the Apollo program numbers. (Photograph by Yerkes Observatory.)

the large dark areas we see on the disk of the moon were as well known in early times, although they were not as well interpreted as they are today. This latter statement is a relative one, because in 200 B.C. scholars held that the moon was roughly round and was composed of mountains, valleys, and rocks like the earth. This is extremely remarkable, and confirmation waited until 1609 (see Section 6.6) and 1969.

The dark areas form the eyes, nose, and mouth of the familiar "man in the moon" and the profile of the "girl in the moon," which is easily seen when Fig. 6.9 is inverted to represent the view of the moon without the telescope. While such a picture looks strange to astronomers, it follows astronautical convention. The "girl reading," the "hare," and the "frog" are all formed by the same group of seas.

Formerly supposed to be water areas, the lunar seas were given fanciful watery Latin names that have now survived their original meanings. One is Mare Serenitatis (Sea of Serenity); others are Mare Imbrium (Sea of Storms) and Sinue Iridum (Bay of Rainbows). The seas cover half the moon's surface visible from the earth and are especially prominent in its northern hemisphere. Roughly circular, they are mostly connected, with the conspicuous exception of Mare Crisium (Fig. 6.9). Although we still call them "seas," the dark spots are relatively smooth plains that are not all at the same level. A frequent explanation is that they are hardened pools of lava, resembling the great basaltic plateaus of the earth covered with a sandlike grit.

Most of the major maria have centers of mass or mascons (a contraction of "mass concentrations") that are located about 100 kilometers below the surface. It is conjectured that their origin stems from great meteoritic impacts late in the moon's history. However, the far side of the moon is essentially devoid of maria and it is hard to believe that these great impacts should selectively strike the side of the moon facing the earth.

Students of the moon concluded before the intensive lunar programs that the maria were relatively smooth undulating planes strewn with rocky rubble. The upper layers were not too strongly packed and beneath this was a layer of broken rock. Deeper, there should be a rather solid base rock of igneous origin. Such deductions came from studies of the reflectivity of the moon in various wavelengths compared with various rocks on the earth, the cooling curve of the moon when it was eclipsed by the earth, its reflectivity of radar pulses of various wavelengths, etc. In general the photographic spacecraft confirmed this and the astronauts have returned rocks that confirm much of our previous ideas in detail. The rocks brought back from the maria are mostly of igneous origin and range in age from 2.5×10^9 years to 4.6×10^9 years.

6.6 The moon through the telescope

There is still another feature of the moon's surface to be detected occasionally with the naked eye; the sunrise and sunset lines are not perfectly smooth. Bright projections into the dark hemisphere show where the sunlight illuminates lofty peaks before the sun has risen or after it has set on the plains around them. The moon is mountainous, a feature that was verified as soon as the telescope came into use.

In 1609 Galileo heard of the discovery by a Dutch spectacle maker that two lenses held at suitable distances before the eye gave a clearer view of the landscape. Galileo fitted two small lenses into a tube and went out to view the heavens; naturally he directed this telescope toward the moon first. Galileo observed the lunar mountains and estimated that they are comparable in height with terrestrial mountains. He could do this by noticing when the

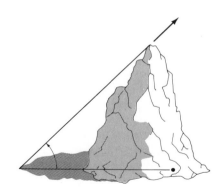

peaks first catch the rays of the sun and determining the distance from the sunrise line into the darkness. Like little stars, the peaks appear at first and then grow larger until they join the sunlit hemisphere. The heights of the lunar mountains may also be calculated from the lengths of their shadows and the altitude of the sun in their sky at that time.

The mountains on the moon are most distinct when they are near the sunrise or the sunset line. There the shadows are long, as they are on earth in the early morning or late afternoon, so that the sunlit peaks stand out in bold relief among the shadows. The view of the mountains is especially good within about two days of the quarter phases.

Compare the two photographs (Fig. 6.10) of the same region at different phases of the moon. In the view at the left the sun is shining more directly on that region; the shadows are shorter, and the surface seems rather flat. In the photograph at the right the region is near the sunset line; the shadows are longer, and the mountains are more conspicuous. The same photographs reveal one of the most interesting features about the moon, that is its reflectivity or **albedo** as a function of phase. The general albedo of the moon is low, being only about 7 percent, and thus one interprets the surface to be composed of a dark rock such as basalt. To reproduce the reflectivity curve as a function of phase (called the moon's **photometric function**), however, we must resort to many special conditions in the material. The surface material must be rather rough (see Section 6.5), in the sense that gravel or coarse sand is, in order to explain the optical reflectivity and it must be very porous to explain the radar data. This has been substantiated in general by manned missions to the moon.

There are only a few irregularities of the moon's surface that remind us at all of our own mountain ranges. The best known of these are the three ridges that form the curving western border of Mare Imbrium, the right eye of the "man in the moon" as we view the moon without the telescope, and that separate it from Mare Serenitatis. These are the Apennines, Caucasus, and Alps— their names are among the few that have survived from Hevelius' map of the moon of 1647, in which the lunar formations have the names of terrestrial ones. With a few exceptions the prominent mountains (and craters) on the moon bear the names of scholars of former times, according to the system introduced by Riccioli in 1651. Examples are the Leibnitz and Doerfel mountains near the moon's south pole; some of their peaks rise as much as 7900 meters above the plains.

The mountains around Mare Imbrium slope abruptly toward

FIGURE 6.10
*The moon under two different angles of
illumination illustrating the moon's photometric function. The region of Mare
Nubium at the moment of full moon is on
top. The same area two days before
full moon is on the bottom. (Leander Mc-
Cormick Observatory photographs.)*

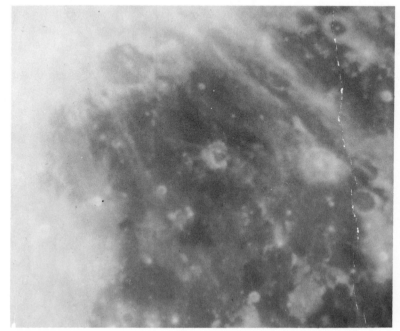

it and more gradually on the opposite side. It is as though they were all that remain of a nearly circular rampart 1120 kilometers across surrounding the great sea.

The lunar craters have nearly circular walls, steep and often shelving on the inside and sloping more gradually to the plain outside. Lofty peaks surmount some of the walls, and peaks also appear near the centers of many craters. Some craters have floors depressed 1000 meters or so below the plain. In others the floors are elevated; the inside of the crater Wargentin is nearly as high as the top of the wall itself. Some craters have rough, bright floors; Aristarchus is the brightest of these. Others, such as Plato in the lunar Alpine region and Tsiolkovsky on the far side, are as dark inside as the seas. While we use terms such as lofty and steep, the visual appearance of craters is quite deceiving. Relative to their diameters lunar craters are quite shallow. A small crater 2 kilometers in diameter will have a depth of only 300 meters. A great crater 100 kilometers in diameter will have a depth of only 4 kilometers and its central peak might reach a height of 2.5 kilometers.

Hundreds of thousands of craters are recognized on the moon. The far side is more cratered than the near side. Crater sizes range from a few centimeters in diameter to Clavis some 240 kilometers in diameter to Mare Orientale almost 1000 kilometers in diameter. Most of these craters are of impact origin, many originating from material gouged out of the moon when the great craters such as Copernicus and Tycho, were formed. Not all craters are of such origin, however.

Sets of craters called **chain craters,** must be of igneous origin (Fig. 6.11). Their form can be attributed to slumping into a lava tube or extensive out-gassing through a crack in the surface. Chains of craters are recognized up to a length of 150 kilometers.

The lunar **rays** are bright streaks often 10 to 20 kilometers wide and up to 250 kilometers long, which radiate from points near a few of the craters and pass over mountain and plain alike without much regard for the topography. The longest and most conspicuous ray system radiates from the crater Tycho near the moon's south pole; it is a prominent feature of the full moon as viewed with any telescope. A system of shorter and more crooked rays is centered near the crater Copernicus. These ray systems are associated with ejected material from the primary craters.

The lunar **rilles** are clefts as wide as 2 or 3 kilometers and sometimes 2 kilometers in depth. They probably result from slumping action analogous to graben here on earth. Some rilles are irregular, whereas others run nearly straight for many kilometers paying no

FIGURE 6.11
A striking Apollo photograph showing various types of craters, mountains, rays, rilles on the moon. (National Aeronautics and Space Administration photograph.)

attention to local features. Some run through craters and high walls maintaining their relatively flat, level bottoms.

Lunar **valleys** are deep grooves possibly of similar origin to the rilles. Lunar **faults** may be related to the rilles as well. The surface slips vertically and the classic example is the Straight Wall some 100 kilometers long and 300 meters high having a slope of almost 45°. Lunar ridges, sometimes called **wrinkle ridges,** remind one of lava flows in that they are long wide tongues of material overlying the general surface at that point. Lunar **domes** remind one of uneroded pingos found on earth. They are round, smooth, often 10 kilometers in diameter, and 100 meters or so high.

THE PHYSICS AND ORIGIN OF THE MOON

6.7 The atmosphere of the moon

The moon does not have an atmosphere in the sense that the earth does. This was well known a century before the Apollo Program, hence the elaborate precautions taken in protecting the astronauts from direct radiation and providing life-support systems for their spacecraft and spacesuits. The fact that the moon has no atmosphere is supported by many observations. For example, the occultation of a star by the moon is essentially instantaneous, and even planets can be occulted by the moon with no discernable distortion of their image. Polarization measures place an upper limit on density of the lunar atmosphere of 10^{-9} that of the earth. Careful measures of the occultation of radio sources (notably the Crab nebula, Section 13.23) placed this limit even lower giving 6×10^{-13} the density of the earth's atmosphere.

The reason for the lack of atmosphere is the escape velocity from the moon. Since the moon's gravitational acceleration is 1/6 that of the earth, the escape velocity is only 2.4 kilometers per second at the surface. Since the energies of all particles in a gas are essentially the same, light particles must move faster than heavy particles. When the velocity exceeds the velocity of escape by a factor of 5 these particles will slowly "boil" away. On the earth only hydrogen and helium escape in significant amounts; on the moon only the very heaviest gaseous elements can remain. Hence, any atmosphere must be confined to a thin (a few meters) layer close to the surface of the moon and be composed of elements such as xenon and krypton. Instruments placed on the moon by the Apollo astronauts give an upper limit to the pressure of 10^{-13} earth atmosphere with occasional significant increases above this pressure, which can be attributed to out-gassing of the nearby spacecraft landing module.

Whatever atmosphere the moon had when it reached its present size and mass, ½ of it would be gone in 3×10^8 years and essentially all of it would have disappeared in 10^9 years. The youngest rocks returned by the Apollo astronauts are 2×10^9 years old, therefore the moon has had no atmosphere for the previous 10^9 years because any gases released during the rock forming process would have disappeared.

6.8 Geology of the moon

Any complete discussion of the geology of the moon would now fill several textbooks. Although the prefix geo- implies the earth, we follow the lead of geologists studying the moon and use the term lunar geology rather than coin a new -ology for each planet studied. Seismic experiments are beginning to unravel the sub-surface structure of the moon. Ever since the first detailed lunar pictures revealed the tracks of large rocks that rolled down into craters, some seismic activity has been accepted. This activity could be the result of meteoritic impacts, moonquakes, or landslides released by the large thermal differences from moon day to night and back, or all three, which appears to be the case.

Seismic records reveal long, low-level surface rumbles that originate from land slides. Also, signals characteristic of impacts (calibrated from various space equipment impacts) are clearly observed and are apparently occurring all over the moon. Moonquakes are observed to originate about 800 kilometers below the surface. These moonquakes occur on a cyclic basis commensurate with the lunar month and, hence, are stresses released by tidal forces.

Based on these facts, it has been theorized that the moon has a crust perhaps a few kilometers thick and a mantle about 70 kilometers thick. The moon may have a core, but it is very small if it exists. Certainly where the moonquakes occur, halfway to the center, the material must be quite rigid so that the stresses being relieved could accumulate. It also must support the mascons that have larger mean densities than that of the lunar average.

The rocks returned by the Apollo astronauts resemble basaltic and granite rocks of the earth. In detail they are quite different in composition and structure. For example, aluminum seems much more abundant in all of the samples. The ages of the rocks range from 2.5×10^9 years to 4.6×10^9 years with the great majority being around 3.5×10^9 years old.

Detailed laboratory studies continue and are of great importance. We can never explore the whole moon. Indeed, the Apollo program came to an end on 19 December 1972, and now we must extrapolate the findings by using the laboratory results and our remote sensing techniques. Changes in the observed properties with temperature, during lunar eclipses, for example, will again be an important source of information.

6.9 The origin of the moon

From physical theory we know that the moon is receding from the earth at a rate of a few centimeters a year. We will soon measure the exact amount using the laser–rectoreflector technique. We can follow this analysis back in time about 4×10^9 years (an interesting coincidence) when the rotation of the earth must have been quite rapid—about 4.5 hours per day. The period of revolution of the moon was of the same order and this must be the distance at which the earth–moon system began. If the moon were closer it would have been destroyed by the tidal forces of the larger body, earth.

From some point in time the moon has gradually receded from the earth because the earth tides tend to brake or slow the earth, and reduce its angular momentum. Since angular momentum cannot be destroyed, it is transferred to the moon by moving the moon farther from the earth.

One theory developed to explain the moon's origin states that the moon was captured by the earth. In this case the physics of the problem is quite clear. Such a capture is virtually impossible in the absence of a third body with a special location and energy. The available third body is the sun, but any solution using the sun leads to the earth absorbing most of the energy of the capture with a resulting temperature of more than 3000°K, which would have melted the earth.

Even so, two independent theories deserve speculative mention: one is by A. G. W. Cameron and the other by F. Singer. Cameron believes that the moon originated by an accretion process near the orbit of Mercury. He based this upon the fact that the moon and Mercury have nearly identical albedos and similar surface compositions and histories. Then, in a gravitational tug of war, the moon moved into an earth orbit resulting in Mercury's orbit having a rather large eccentricity. Cameron does not consider how the moon became a satellite of the earth, but his theory does explain the moon's orbit being closer to the ecliptic than the earth's equator. Singer's theory begins here and assumes that the moon was captured into a retrograde orbit, which reduces the energy to be absorbed. Later, through tidal braking, its orbit has become direct and its rotation period synchronous with its revolution. The capture energy is only reduced a few percent, so we are left with a very hot earth at the time of capture. Interestingly enough, there are some reasons to believe that the earth went through a molten stage about 4×10^9 years ago which would agree with Singer's theory and help to explain formation of the oceans and the atmosphere of the earth.

Qualitative arguments can be advanced against both parts of this picture, but essential to both is the capture of the moon by the earth. A capture orbit tends to have a large eccentricity ($e > 0.15$), as shown by the known cases of satellite capture, and the energy to be absorbed by the earth remains quite large.

There is another theory in which the earth–moon system was born as it is, with each body reaching its present mass and size by accretion. The mean density of the rocks and surface material of the moon is 3.3 (about the same as the earth's surface rocks) and the mean density of the entire moon is about 3.34, so the moon's constitution must be about the same throughout. Despite this conclusion, we know that the moon is differentiated geologically as is the earth and Venus. All indications are that Mars is also, and we infer that Mercury is as well. Geological differentiation results from active vulcanism and there is evidence for some lunar volcanic activity eons ago, which ended about 3×10^9 years ago.

ECLIPSES OF THE MOON

Like any other opaque object in the sunshine, the earth casts a shadow in the direction away from the sun. The **umbra** of the shadow is the part from which the sunlight would be completely excluded if some light were not scattered into it by the earth's atmosphere (Fig. 6.12). The umbra is a long, thin cone reaching an average of 1,382,500 kilometers into space before it tapers to a point. This darker part of the shadow is often meant when we speak of the **shadow**. It is surrounded by the **penumbra,** from which the direct sunlight is only partly excluded.

Suppose that a large screen is held at right angles to the direction of the shadow, and that it is moved out into space in that direction. The umbra of the earth's shadow would fall on this screen as a dark circle growing smaller with increasing distance of the screen, until at the moon's distance the diameter of the shadow would be 9170 kilometers, or nearly 3 times the moon's diameter of 3476 kilometers.

Because the earth's shadow points away from the sun, it sweeps eastward around the ecliptic once in a year as the earth revolves. At intervals of a synodic month, the faster moving moon over-

FIGURE 6.12
The moon in the earth's shadow. Light is scattered into the umbra by the earth's atmosphere so that the moon has a deep copper tint during totality.

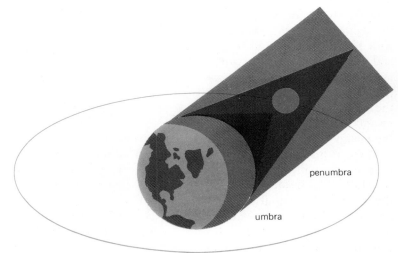

penumbra

umbra

moon's orbit

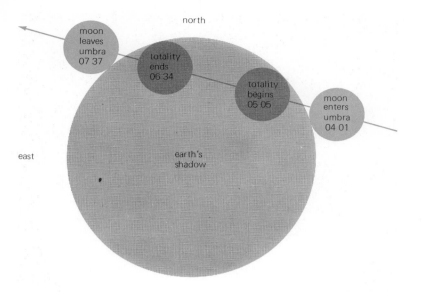

north

moon
leaves
umbra
07 37

totality
ends
06 34

totality
begins
05 05

moon
enters
umbra
04 01

east

earth's
shadow

FIGURE 6.13
The various phases of totality for the lunar eclipse of 25 May 1975. The times are given in Universal Time. This eclipse will be visible over most of North America.

takes the shadow and sometimes passes through it. Since the moon's orbit is tilted about 5°8′ with respect to the ecliptic, this occurs when the nodes of the moon's orbit are near the earth–sun line. The geometry of an eclipse is very simple and worth working through as an exercise. Since the line of the nodes must be near the earth–sun line, it is possible that no umbral eclipses will occur in certain years, as in 1976, but by the same token as many as three eclipses may occur, as in 1982.

Umbral eclipses occur when the moon passes through the umbra of the earth's shadow. The longest eclipses, when the moon goes centrally through the shadow, last about 3^h40^m; this duration is counted from the first contact with the umbra until the moon leaves the umbra completely. Total eclipse can last as long as 1^h40^m, preceded and followed by partial phases, each of about an hour's duration. Usually the eclipse is not central (Fig. 6.13), so that its duration is shorter. Often the moon passes so far from the center of the shadow that it is never completely immersed in the umbra, and the eclipse is partial throughout.

Penumbral eclipses occur when the moon passes through the penumbra of the earth's shadow without entering the umbra. The weakened light of the part of the moon that is in the penumbra is visible to the eye when the least distance of the edge of the moon from the umbra does not exceed one-third of the moon's diameter; it is detected in the photograph (Fig. 6.14) when the

shortest distance does not exceed two-thirds of the moon's diameter and by photometric means when the distance is still greater.

A lunar eclipse is visible wherever the moon is above the horizon during its occurrence, that is, over more than half the earth, counting the region that is rotated into view of the moon while the eclipse is going on. The ephemeris times of the circumstances of all umbral eclipses and of penumbral eclipses when they occur are published in advance in the *American Ephemeris and Nautical Almanac*.

Six umbral lunar eclipses are scheduled in the interval from 1974 to 1979 inclusive as visible, at least in part, in a considerable area of the United States and Canada (see Table 6.1). Five of these are total.

The first conspicuous effect of the lunar eclipse is seen soon after the moon enters the umbra. A dark notch appears at the eastern edge of the moon and slowly spreads over the disk. The shadow is so dark in comparison with the unshaded part (Fig. 6.15) that the moon might be expected to disappear in total

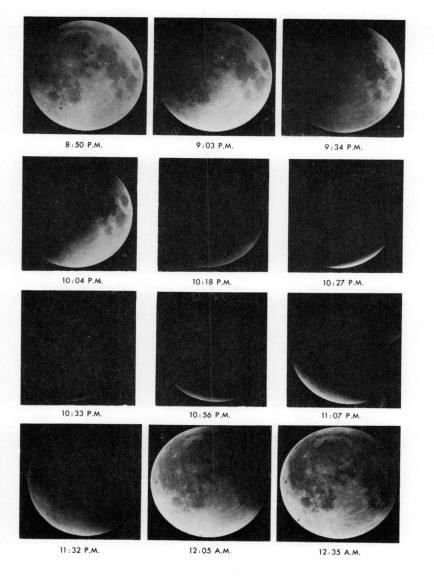

FIGURE 6.15
*A total lunar eclipse sequence. (Griffith
Observatory photograph.)*

8:50 P.M.

9:03 P.M.

9:34 P.M.

10:04 P.M.

10:18 P.M.

10:27 P.M.

10:33 P.M.

10:56 P.M.

11:07 P.M.

11:32 P.M.

12:05 A.M.

12:35 A.M.

eclipse. As totality comes on, however, the moon usually becomes
plainly visible.

The moon in total eclipse is still illuminated by sunlight that
filters through the earth's atmosphere around the base of the
shadow. The light is diffused by the air into the shadow and onto
the moon. It is redder than ordinary sunlight for the same reason

TABLE 6.1

Umbral lunar eclipses visible in the United States and Canada

Middle of eclipse

DATE	UNIVERSAL TIME	DURATION OF UMBRAL PHASE	DURATION OF TOTALITY
29 Nov. 1974	$15^h 14^m$	$3^h 30^m$	$1^h 17^m$
25 May 1975	5 49	3 36	1 29
18–19 Nov. 1975	22 24	3 30	0 42
4 Apr. 1977	4 19	1 36	—
24 Mar. 1978	16 23	3 39	1 31
6 Sept. 1979	10 55	3 12	0 46

SOURCE: R. Duncombe, U.S. Naval Observatory.

that the sunset is red, so that the totally eclipsed moon has an unfamiliar hue. The brightness of the moon then depends on the transparency of the atmosphere around the base of the shadow. Enough light usually sifts through to show the surface features clearly. On a dozen occasions from 1601 to 30 December 1963, the totally eclipsed moon became very dim or even invisible, generally because of volcanic dust in the atmosphere.

QUESTIONS

1 The Romans before 50 B.C. used polished stones, such as rubies, as aids to vision. Suppose they had arranged two such lenses and observed the moon; how do you think the course of astronomy would have changed?

2 What is meant by the statement "the moon is gibbous?" Where did the word originate, and what does it mean?

3 Why is earthlight bluer than sunlight?

4 What is the difference between the synodic month and the sidereal month?

5 What is the explanation of the harvest moon?

6 Explain the principal librations of the moon.

7 Why do we believe the maria are younger than the lunar highlands?

8 The energy of a gas particle is $\frac{1}{2}mv^2$, where m is the particle's mass and v is its velocity. If our statement that all particle energies in a gas have the same energy is true, approximately how much faster is hydrogen of relative weight 1 moving compared to krypton of relative weight 84?

9 What are the three sources of seismic activity on the moon?

10 Work through the geometry mentioned in Section 6.10 and determine the approximate limits between which a lunar umbral eclipse will occur.

FURTHER READINGS

BALDWIN, RALPH B., *The Face of the Moon*, University of Chicago Press, Chicago, 1949.

——— *The Measure of the Moon*, University of Chicago Press, Chicago, 1963.

——— *A Fundamental Survey of the Moon*, McGraw-Hill, New York, 1965.

CHERRINGTON, E. H., JR., *Exploring the Moon Through Binoculars*, McGraw-Hill, New York, 1969.

FIELDER, GILBERT, *Structure of the Moon's Surface*, Pergamon, London, 1961.

KOPAL, DZENEK, *The Moon*, 2nd ed., Academic Press, New York, 1964.

KUIPER, GERARD P., ED., *Photographic Lunar Atlas*, University of Chicago Press, Chicago, 1960.

OPPOLZER, THEODOR VON, *Canon of Eclipses*, Dover, New York, 1962.

PAGE, THORNTON, AND LOU WILLIAMS PAGE, EDS., *Wanderers in the Sky*, Macmillan, New York, 1965.

An early diagram showing a method of drawing the eccentric orbit of Mars. The diagram appears in Astronomia Nova.

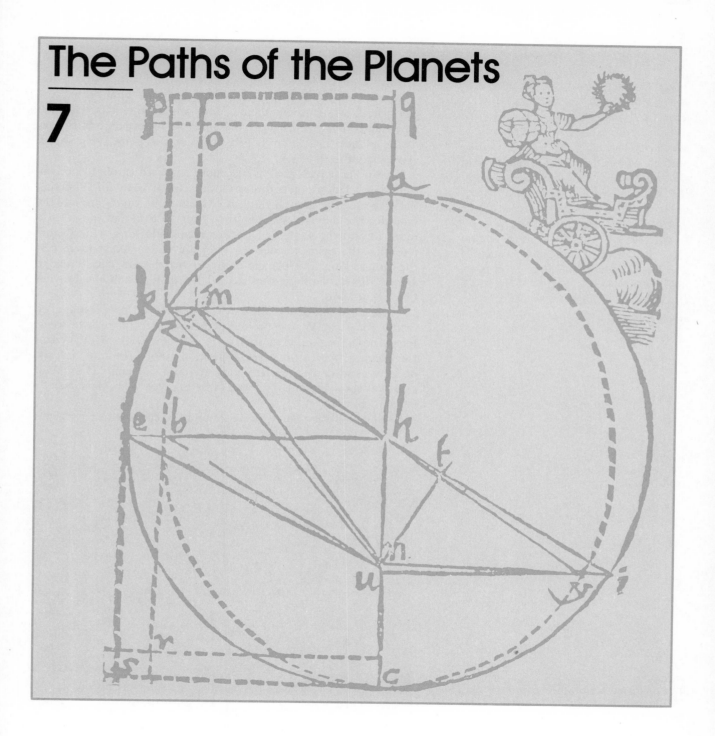

7

Seven bright celestial bodies move about among the "fixed stars" that form the constellations. They are the sun, the moon, and the five planets—Mercury, Venus, Mars, Jupiter, and Saturn, which have the appearance of stars to the unaided eye. These were the planeta, or "wanderers," of the ancients. These seven bodies are among our nearest neighbors in space. In the foreground of the starry scene they are conspicuous in our skies. Their brightness and their complex movements against the background of the stars have made them objects of special interest through the ages.

Anyone who has viewed the spectacle of the heavens that is displayed in one of the larger planetariums knows how the planets swing back and forth in their courses among the stars. Celestial movements of a year can be represented in a short time in the sky of the planetarium. The looped paths of the planets are shown very clearly there.

For the most part, the planets move eastward through the constellations. This is their **direct** motion, for it is in this direction that they revolve around the sun. At intervals, which are not the same for the different planets, they seem to turn and move backward, toward the west, that is, they **retrograde** for a while before resuming the eastward motion. They are said to be **stationary** at the turns. Thus the planets seem to march and countermarch among the stars, progressing toward the east around the heavens in series of loops.

These apparent movements of the planets are readily observed in the sky itself. Watch the red planet Mars from week to week, for example, beginning as soon as it rises at a convenient hour of the evening. Notice the planet's position among the stars on each occasion, and mark the place and date on a star map. The line of dots will show presently, as it does in Fig. 7.1, that Mars steers a devious course.

Just how do the planets move around the earth so as to proceed

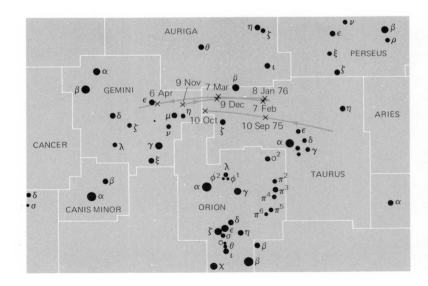

FIGURE 7.1
The predicted path of Mars during the 1975 opposition. Mars will be about as far north as it ever gets.

in loops among the constellations? By what combinations of uniform circular motions centered in the earth can their observed movements be represented? This was the problem the early scholars wished to solve. Intuitively they agreed that celestial motions ought to be uniform and in circles; then, too, the circle is an easy figure for calculations.

The globe of the earth was stationary at the center of their universe. The sphere of the stars turned around it daily. Within that sphere the seven wanderers shared its daily turning, and also moved eastward around the earth at various distances from it.

7.1 The geocentric system

The most enduring early plan for solving the problem of the planetary motions was developed by Ptolemy at Alexandria in the second century and is, accordingly, known as the **Ptolemaic system.** It was a plan of epicycles. In the simplest form of the system (Fig. 7.2) each planet was supposed to move on the circumference of a circle, the **epicycle,** while the center of that circle revolved around the earth on a second circle, the **deferent.** By such combinations the attempt was made to represent the observed movements of the planets. The Ptolemaic plan became more complex as time went on. During many centuries that intervened between the decline of Greek culture and the revival of learning in Europe, Arabian astronomers undertook to improve the system so that it would more nearly represent the planetary movements. Among other devices they tried to get a better fit by adding more epicycles. Each planet was eventually provided with from 40 to 60 epicycles turning one upon another. It was then that King Alphonso of Castile remarked that had he been present at the Creation he might have given excellent advice. The theory of the central earth had begun to seem unreasonable.

From the times of the early Greek scholars there was an undercurrent of opinion that the earth is not stationary. The followers of Pythagoras, who taught that the earth is a globe, supposed that it is moving. Aristarchus of Samos, in the third century B.C., is said to have been convinced that the earth rotates daily on its axis and revolves yearly around the sun. There were others who caught glimpses of the truth, such as Hipparchus who measured the moon's orbit and found it was oval (see Section 6.7). Such ideas, however, then seemed unbelievable to almost everyone and received little attention, simply because a turning globe would throw off all its occupants just as a spinning wheel does.

By the time Columbus sailed west on his famous voyage, there was growing dissatisfaction with the theory of the central, station-

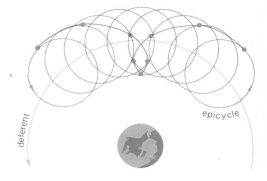

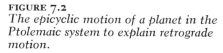

FIGURE 7.2
The epicyclic motion of a planet in the Ptolemaic system to explain retrograde motion.

ary earth. Before the companions of Magellan had returned from the first trip around the world, another European scholar had independently reached the conclusion that the earth is in motion. His name, as we say it, was Nicolaus Copernicus. His theory of the moving earth was published in 1543.

The **Copernican system** set the sun in the center instead of the earth, which now took its rightful place as one of the planets revolving around the sun. It is therefore referred to as a **heliocentric system**. His theory retained the original idea that the planets moved uniformly in circles, and accordingly retained a system of epicycles. Copernicus could offer no convincing proof of either the earth's revolution or daily rotation, which he also advocated.

7.2 Observations and interpretations

Tycho Brahe greatly improved the instruments and methods of observing the positions of the celestial bodies. He saw clearly that an improvement in the theory of the motions of the planets required more reliable data on their apparent movements among the stars. Born in 1546, the most fruitful years of his life were spent in his observatory on the formerly Danish island of Hven, 30 kilometers northeast of Copenhagen. He died in Prague in 1601.

Tycho's observations were made before the invention of the telescope; his chief instruments were large quadrants and sextants having plain sights. With these, he and his assistants observed the planets night after night and determined their right ascensions and declinations with a degree of accuracy never before attained. He gave special attention to the planet Mars, a fortunate choice because its orbit is not as nearly circular as are the orbits of some of the other bright planets.

Johannes Kepler was Tycho's assistant in his last year in Prague. He inherited the records of the positions of the planets, which his mentor had kept for many years, after an interesting conflict with Tycho's survivors. Kepler studied the records patiently in the hope of determining the actual motions of the planets. In 1609 he announced two important conclusions, and in 1618 he discovered the third. They are known to us as **Kepler's laws**:

1 **The planets move around the sun in ellipses having the sun at one of the foci.** Thus the planets do not go around the earth, and their orbits are not circles.

2 **Each planet revolves in such a way that the line joining it to the sun sweeps over equal areas in equal intervals of time.** This

is called the law of equal areas. The nearer the planet comes to the sun, the faster it moves, as we have already noticed (see Section 4.5) in the case of the earth.

3 The squares of the periods of revolution of any two planets are in the same ratio as the cubes of their mean distances from the sun. This useful relation is called the **harmonic law.**

Note that Kepler's third law as stated can be written symbolically as

$$\left(\frac{P_1}{P_2}\right)^2 = \left(\frac{a_1}{a_2}\right)^3$$

where the subscripts refer to planets 1 and 2, which can be the earth and Jupiter or any other planet. This relation can be rewritten in the following way:

$$\frac{a_1{}^3}{P_1{}^2} = \frac{a_2{}^3}{P_2{}^2}$$

and if we use years for the periods and astronomical units for the distances the ratio is unity. To check this statement we can use Jupiter, where $P = 11.86$ years and $a = 5.203$ astronomical units. Then $P^2 = 140.659$, $a^3 = 140.851$, and the ratio is 1.0014, which is very nearly unity. The reasons that it is not are that we are using approximate values and Kepler's relation is itself an approximation, a more precise form being given by Newton (see page 126).

Here ended the attempts to represent the movements of the planets by uniform circular motions centered around the earth. There was still no evidence, however, that the earth itself revolves around the sun.

THE LAW OF GRAVITATION

While Kepler was deriving his laws that describe the movement of the planets around the sun, Galileo Galilei, his contemporary in Italy, was laying the foundations of mechanics. Galileo questioned the traditional ideas about the motions of things and set out to determine for himself how they really move. It remained for Isaac Newton in England to formulate clearly the new laws of motion and to show that they apply not merely to objects immediately around us but to the celestial bodies as well. The principal feature of the new mechanics was the concept of an attractive force that operates under the same rules everywhere in the Universe.

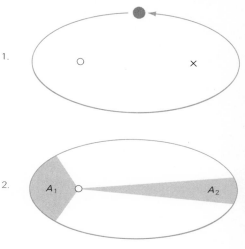

1.

2.

$A_1 = A_2$

7.3 The laws of motion

Before the time of Galileo, an undisturbed body was supposed to remain at rest. Hence, it seemed appropriate that the earth should be stationary. Anyone who asserted that the earth is moving might well be asked to explain by what process it is kept in motion.

Galileo's experiments led him to the new idea that uniform motion in a straight line is the natural state. An object will go on forever in the same direction with the same speed unless it is disturbed. Rest is the special case where the initial speed happens to be zero. Uniform motion in a straight line, therefore, demands no explanation. It is only when the motion is changing either in direction or in speed that an accounting is required. We say then that a force is acting on the body, and inquire where the force originates.

The strength of the **force**, F, is measured by its effect on the body on which it acts; it equals the **mass**, m, of the body multiplied by its acceleration, α, or the rate of change of its velocity (directed speed)

$$F = m\alpha$$

The acceleration may appear as increasing or diminishing speed, or changing direction, or both. A stone falling vertically faster and faster is accelerated. An object moving in a circle with constant speed is accelerated. In both cases a force is acting.

The laws of motion formulated by Newton in his *Principia* (1687) are substantially as follows:

1 Every body persists in its state of rest or of uniform motion in a straight line unless it is compelled to change that state by a force impressed upon it.

Where a force is applied:

2 The acceleration is directly proportional to the force and inversely proportional to the mass of the body, and it takes place in the direction of the straight line in which the force acts.

3 To every action there is always an equal and contrary reaction.

The second law defines force in the usual way. The first law states that there is no acceleration where no force is acting; the motion of the body remains unchanged. The third law asserts that the force between two bodies is the same in each direction. A bat exerts no greater force on the ball than the ball exerts on the bat; but the lighter ball experiences a greater acceleration than the heavier bat and batter combined.

Armed with these laws of motion, Newton succeeded in reducing Kepler's three laws of the planetary movements to a single universal law. It is said that the fall of an apple one day as Newton sat in his garden started the great mathematician to thinking of this problem. Does the attractive force that brings down the apple also control the moon's revolution around the earth? Does a similar force directed toward the sun cause the planets to revolve around it?

A force is continuously acting on the planets, because their courses around the sun are always curving. It is an attractive force directed toward the sun; this fact can be deduced from Kepler's law of equal areas. From further studies of Kepler's laws, Newton discovered the law of the sun's attraction. He found that the force between the sun and a planet is directly proportional to the product of their masses, and inversely proportional to the square of the distance between their centers.

Newton next calculated the law of the earth's attraction. An apple falls 4.9 meters in the first second. The moon, averaging 60 times as far from the earth's center, is drawn in from a straight-line course 0.13 centimeter a second, which is about 4.9 meters divided by the square of 60. Using more exact values than these, he showed that the force of the earth's attraction for objects around it is inversely proportional to the squares of their distances from its center. Although his studies could not extend beyond the planetary system, Newton concluded that he had discovered a universal law and so announced it in his **law of gravitation:**

Every particle of matter in the universe attracts every other particle with a force that varies directly as the product of their masses and inversely as the square of the distance between them.

$$F \propto \frac{m_1 m_2}{r^2}$$

7.4 Revolution of the planets

Using the earth as our example, let us try to understand its revolution following Newton's law. By Newton's law there is an attractive force between the earth and the sun, which is the same in the two directions. Started from rest they would eventually come together. The earth is moving, however, nearly at right angles to the sun's direction at the rate of 29.8 kilometers a second, and in one second it is attracted less than 0.3 centimeter toward the sun. It is this deviation from a straight-line course second after second through the year that causes the earth to revolve around the sun as in Fig. 7.3. At the position E, the earth, if undisturbed, would

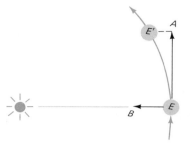

FIGURE 7.3
The earth's motion as explained by the laws of motion. The earth E is attracted toward the sun along the line EB. At the same time the earth would move along the line EA due to its orbital velocity. The net effect is that as the earth falls toward the sun it moves along a curved path EE.

continue on to A, by the first law of motion. It arrives at E′ instead, having been attracted in the meantime toward the sun the distance EB.

Properly speaking, the earth and sun mutually revolve around a point between their centers. The ancient problem of whether the sun or the earth revolves was not well stated. Both revolve. If the earth and sun were equally massive, the point around which they wheel yearly would be halfway between their centers. Because the sun is a third of a million times as massive as the earth, this center of mass is only 450 kilometers from the sun's center; it is not far from the center for all the other planets.

Thus the planets revolve around the sun, although not precisely around its center. The first law of motion explains their continued progress, and the force of gravitation causes them to revolve around the sun instead of going away into space.

The orbits of the planets relative to the sun are ellipses having the sun's center at one focus. These **relative orbits** are the ones we employ for the planets generally, and similarly for the satellites revolving around their planets. They are the same in form as the actual orbits and differ from them only slightly in size.

Newton showed from his law of gravitation that the orbits of revolving bodies in general may be any one of the three conic sections. These are the ellipse (see Section 3.4), parabola, and hyperbola (Fig. 7.4). The ellipse includes the circle, where the eccentricity is zero. The parabola, eccentricity 1, is open at one end, and the hyperbola is wider open. Evidently the permanent members of the sun's family have closed, elliptical orbits.

If the earth, now revolving in nearly a circle at the rate of 29.8 kilometers a second, could be speeded up, its orbit would become larger and more eccentric. At the speed of 42 kilometers a second the orbit would become a parabola, and the earth would depart from the sun. This is the velocity of escape from the sun at the earth's distance.

$$v_{escape} = \sqrt{2} \ v_{circle}$$

7.5 The masses of the planets

The law of gravitation views the physical universe as a scheme of masses and distances. It is, therefore, of considerable interest to inquire how the masses are measured. The **masses** of some planets (i.e., the quantities of material they contain) can be found by Kepler's harmonic law as restated more precisely by Newton:

The squares of the periods of **any two pairs** of mutually revolving celestial bodies, **each multiplied by the combined mass of the**

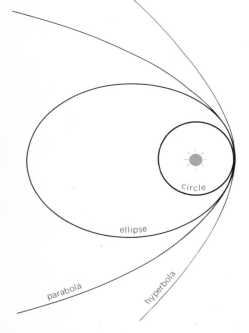

FIGURE 7.4
The various possible orbits from circular to hyperbolic. The tangential velocity and the distance from the sun determine the form of the orbit.

pair, are in the same proportion as the cubes of the mean distances that separate the pairs.

This is shown symbolically by

$$\frac{P_1{}^2 m_1}{P_2{}^2 m_2} = \frac{a_1{}^3}{a_2{}^3}$$

where m_1 represents the total mass of the first pair and m_2 that of the second pair.

Suppose that we wish to find the mass of the planet Saturn. We write this proportion, taking Saturn and one of its satellites as one pair and the earth and sun as the second pair. Let the unit of mass be the combined mass of the earth and sun, the unit of distance the mean distance between the earth and sun, and the unit of time the period of the earth's revolution around the sun. The relation becomes simply: The mass of Saturn and its satellite,* equals the cube of the mean distance of the satellite from Saturn divided by the square of its period of revolution around Saturn.

The masses of planets having satellites have been found in this way. It is more difficult to weigh planets, such as Mercury, Venus, and Pluto, which have no satellites; their masses are determined by their disturbing effects on the motions of neighboring bodies, such as orbiting or passing spacecraft or, as in former times, other planets.

7.6 Courses and forces

Early astronomers tried to represent the planetary movements by combinations of circular motions centered in the earth. Copernicus set the sun in the center instead of the earth. Kepler discovered that the planets revolve around the sun in ellipses instead of circles and epicycles. So far the interest was confined to the courses themselves.

Newton's law of gravitation directed the attention to mighty forces controlling the courses of the planets. This law has made possible the present accurate predictions of the planetary movements. It has promoted the discoveries of celestial bodies previously unknown, from their effects on the motions of known bodies. It applies equally well to mutually revolving stars.

For most purposes astronomers make their calculations on the basis of Newton's law of gravitation. Only rarely does the newer theory of relativity predict celestial events with appreciably greater accuracy. The advance of Mercury's perihelion around the sun at

* The mass of the satellite may be neglected.

a faster rate than is predicted by the law of gravitation is a well-known example. The apparent displacements of stars away from the sun's place in the sky at total solar eclipses is another example of the occasionally greater merit of relativity in explaining events on the astronomical scale. On the atomic and nuclear scale as well as on the cosmological scale we are forced to use the theory of relativity.

In 1905, Albert Einstein propounded the **principle of relativity,** which states as axiomatic that the laws of physics are the same as determined by one observer as by another observer. According to this principle the velocity of light is a constant, independent of time, direction, or position and the coordinate system in which it is being measured. A basic difference between Newtonian mechanics and relativistic mechanics appears in the concept of time. In Newtonian mechanics time is absolute, in relativistic mechanics time is relative—relative to the observer. As a result no single observer is favored or, to make a positive statement, all observers are equally favored.

Although the mathematics are beyond the limits of this text, certain general statements can be made. At low speeds (or low velocities) relativistic mechanics and Newtonian mechanics are identical. At high speeds they yield quite different results (see page 402). For example, for two masses at rest with respect to each other we can define the rest mass of one as m_0. The same mass moving at some high velocity (v) will have an apparent mass m which is greater than m_0 by the amount

$$m = \frac{m_0}{[1 - (v/c)]^2}$$

This simple relation tells us that as v gets very close to c (the velocity of light), m becomes very, very large.

The same thing happens for time. The moving object has a time given by

$$t = \frac{t_0}{[1 - (v/c)]^2}$$

Near the velocity of light the unit of time becomes very, very long in terms of the normal rest unit, t_0. This is known as time dilation. An example of this occurs in experimental cosmic-ray physics: When a cosmic ray strikes an atom in the upper atmosphere it may destroy the nucleus releasing particles called mesons. Independent mesons live only a brief period of time before they spontaneously (automatically) decay into either electrons or positrons (positive electrons). This decay time is well measured in the laboratory, and it is too short for the meson to make the journey from the upper atmosphere to the surface of the earth, whatever the meson's velocity. However, mesons as by-products of cosmic rays are observed at the surface of the earth and the only explanation is that they are traveling at velocities such that their time with respect to ours is dilated.

The examples of time and mass are derived from the **special theory of relativity.** The earlier examples of the excessive advance of the perhelion of Mercury and the apparent outward displacement of stars near the sun's limb are consequences of Einstein's **general theory of relativity** in which the principle of relativity is extended to systems accelerating with respect to one another. According to Newtonian theory (and relativistic theory) the major axis of the orbit of Mercury will precess (rotate) at a rate of about 532 seconds of arc per century resulting from effects of the sun and other planets, whereas it was well known that it was rotating at a rate of about 574 seconds. According to general relativity an accelerated body in the presence of a large mass should show an additional precessional advance of 43 seconds of arc per century and this accounts for the difference. In rounding off the observed values we are leaving the impression that there is a disagreement of 1 second of arc per century. The situation is much better than that, the actual difference being only 0.08 second of arc per century.

Mercury's orbit
2000   1900
1900   2000
perihelion of orbit

THE PLANETARY SYSTEM

The meaning of the word **planet** as a body revolving around the sun began with the acceptance of the Copernican system, which added the earth to the list of planets, subtracted the sun from the original list, and reduced the moon to its proper place as a satellite of the earth.

The known membership of the planetary system has increased greatly since Copernicus' time. Knowledge of satellites attending other planets began in 1610 with Galileo's discovery of the four bright satellites of Jupiter. The planet Uranus, barely visible to the naked eye, was discovered in 1781. Neptune, which is always too faint to be seen without the telescope, was found in 1846. The discovery of the still fainter and more remote Pluto in 1930 completed the list of the nine known **principal planets.** In 1801 Ceres was the first and largest of the **asteroids** or **minor planets** to be discovered.

The earth is one of the principal planets. The moon is one of thirty-two satellites that accompany six of these planets. Thousands of asteroids and great numbers of comets and meteor swarms are also members of this large family that, including the sun itself, is known as the **solar system.**

7.7 Aspects and phases of the planets

The names of the planets in order of mean distance from the sun are:

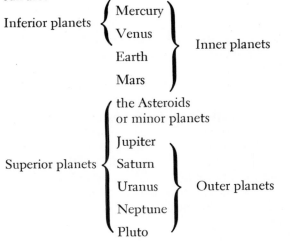

They are classified as inferior and superior planets, and also as inner and outer planets. The **inferior planets** are closer to the sun than the earth, and the **superior planets** revolve outside the earth's orbit. The **inner planets** revolve inside the main zone of the asteroids, whereas the outer planets have their orbits outside this zone. The four inner planets and Pluto as well are sometimes known as the **terrestrial planets,** because they are small as compared with the giant planets, Jupiter, Saturn, Uranus, and Neptune.

Not all the asteroids are confined to the main zone; some of them invade the regions of the principal planets. A part of the projected orbit of Pluto (Fig. 7.5) is nearer the sun than is part of Neptune's orbit.

Certain terms are used to describe the position of a planet in its orbit as viewed from the earth.

The **elongation** of a planet at a particular time is its angular distance from the sun as seen from the earth. Certain positions of the planet relative to the sun's place in the sky have distinctive names and are known as the **aspects** of the planet. The planet is

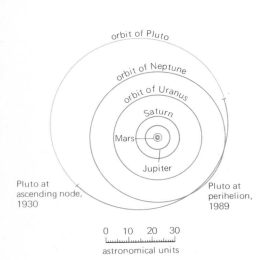

FIGURE 7.5
The orbits of the major planets. Only the orbits of Mercury and Pluto differ significantly from zero, or circular orbits. The part of Pluto's orbit south of the ecliptic is indicated by the tinted line.

in **conjunction** with the sun when the two bodies have the same celestial longitude, so that the planet's elongation is not far from 0°. It is in **quadrature** when the elongation is 90°, and is in **opposition** when its celestial longitude differs by 180° from the sun's, so that its elongation is near 180°.

The inferior planets, Mercury and Venus, have limited elongations; they appear to us to oscillate to the east and west of the sun's place. From superior conjunction, beyond the sun, they move out to **greatest eastern elongation,** which does not exceed 28° from the sun for Mercury and 48° for Venus. Here they turn westward relative to the sun, pass between the sun and the earth at **inferior conjunction,** then move out to **greatest western elongation,** and finally return toward the east behind the sun. As Fig. 7.6 shows, the inferior planets go through the complete cycle of phases, just as the moon does. Their phases are nearly full at superior conjunction, quarter in the average at the greatest elongations, and nearly new at inferior conjunction.

The superior planets, such as Mars and Jupiter, revolve around the sun in periods longer than a year. They accordingly move eastward through the constellations more slowly than the sun appears to do. With respect to the sun's place in the sky, they seem to move westward (clockwise in Fig. 7.7), and attain all values of elongation in that direction from 0° to 180°. At **conjunction** they pass behind the sun to subsequently appear in the east before sunrise. At **western quadrature** they are near the celestial meridian at sunrise. At **opposition** they rise around the time of sunset, and at **eastern quadrature** they are near the meridian at sunset.

From Fig. 7.7 we also see that the superior planets show the full or nearly full phase at all times to the earth. Mars near its quadrature appears conspicuously gibbous, because it is the nearest of these planets to the earth, so that its hemisphere turned toward the sun and earth are considerably different.

7.8 Retrograde motions of the planets

Consider again the loops in the planets' movements among the stars, which mystified the early astronomers and promoted the complex machinery of the Ptolemaic system. At intervals the planets seem to interrupt their motions toward the east, and retrograde, or move back toward the west, for a while. The retrograde motions occur because we are observing from a planet that is revolving at a different rate from the others.

A superior planet, such as Mars, retrogrades near the time of its opposition. The earth, which is moving faster, then overtakes it

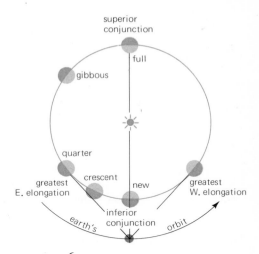

FIGURE 7.6
Aspects and phases of an inferior planet. The elongations are limited; the phases are like those of the moon.

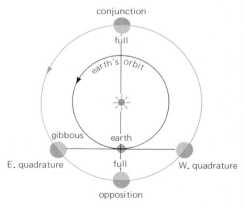

FIGURE 7.7
Aspects and phases of a superior planet. The aspects are similar to those of the moon. The only phases are full and gibbous.

and leaves it behind. On the other hand, Mars has its fastest direct motion near its conjunction with the sun, where its own motion and its displacement by ours are in the same direction. The inferior planets retrograde near inferior conjunction. In general, a planet retrogrades when it is nearest the earth.

7.9 Mean planetary motions of the planets

The distances of the planets from the sun are given in Table 7.1, where other information about the planets and their orbits appears as well. These are mean distances; the distances vary because the

TABLE 7.1 *Distances of planets from the sun*

Name	Symbol	Mean distance from sun AU	Mean distance from sun km × 10⁶	period of revolution Sidereal	period of revolution Synodic	Eccentricity	Orbital inclination to ecliptic
				DAYS	DAYS		
Mercury	☿	0.3871	57.91	87.969	115.88	0.206	7° 0′
Venus	♀	0.7233	108.20	224.701	583.92	0.007	3 24
Earth	⊕	1.0000	149.60	365.256	—	0.017	0 0
Mars	♂	1.5237	227.94	686.980	779.94	0.093	1 51
				YEARS			
Ceres	①	2.7673	413.98	4.604	466.60	0.077	10 37
Jupiter	♃	5.2028	778.33	11.862	398.88	0.048	1 18
Saturn	♄	9.5388	1429.99	29.458	378.09	0.056	2 29
Uranus	♅	19.1819	2869.57	84.013	369.66	0.047	0 46
Neptune	♆	30.0579	4496.60	164.794	367.49	0.009	1 46
Pluto	♇	39.5177	5911.77	248.430	366.74	0.249	17 9

Name	Equatorial diameter (km)	Mass (⊕ = 1)	Density (water = 1)	Period of rotation in days	Inclination of equator to orbit	Oblateness	Stellar magnitude	Albedo
Sun ☉	1,392,000	332,960	1.41	25.38	7° 10′	0	−26.8	
Moon ☾	3,476	0.012	3.34	27.322	6 41	0	−12.6	0.07
Mercury	4,868	0.05	5.49	58.65	0 0	0	− 1.9	0.06
Venus	12,112	0.82	5.26	244	3 18	0	− 4.4	0.76
Earth	12,756	1.00	5.52	0.997	23 27	0.003		0.36
Mars	6,787	0.11	3.94	1.026	24 59	0.005	− 2.8	0.16
Jupiter	142,830	317.9	1.33	0.410	3 4	0.067	− 2.5	0.73
Saturn	119,330	95.1	0.71	0.426	26 44	0.105	− 0.4	0.76
Uranus	47,200	14.5	1.70	0.451	97 53	0.071	+ 5.7	0.93
Neptune	51,000	17.5	1.63	0.658[a]	28 48	0.025	+ 7.6	0.84
Pluto	5,800	0.10[a]	5.5[a]	6.39	?	0.156[a]	+14.9	0.14[a]

[a] = approximate values

orbits are ellipses with the sun at one focus. To find the greatest distance that a planet departs from the mean, multiply the mean distance by the fraction representing the eccentricity of the planet's orbit. Thus the mean distance of Mercury from the sun is 57.9×10^6 kilometers, and the eccentricity of its orbit is 0.206; the greatest variation from the mean is therefore 11.9×10^6 kilometers. Mercury is 46×10^6 kilometers from the sun at perihelion and 69.8×10^6 kilometers at aphelion.

A relation known as the **Titius-Bode relation,** often called **Bode's law,** is an easy way to remember the relative distances from the sun of all except the most remote planets. Write in a line the numbers: 0, 3, 6, 12, and so on, doubling the number each time to obtain the next one. Add 4 to each number, and divide the sums by 10. The resulting series of numbers, 0.4, 0.7, 1.0, 1.6, 2.8, . . . , represents the mean distances (D) of the planets expressed in astronomical units. (The **astronomical unit,** AU, is the earth's mean distance from the sun.)

$$D = A + B \times 2^n$$

where $A = 0.4$, $B = 0.3$, and $n = -\infty, 0, 1, 2, ...$

Compare the distances found by this rule with the actual mean distances in astronomical units given in Table 7.1. The agreement is quite close except for Neptune and Pluto, although it would be less impressive for Mercury if the rule of doubling the number had been followed from the start.

When the distance of one planet from the sun is given, the distance of the others can be calculated from their periods of revolution by Kepler's harmonic law. The earth's mean distance from the sun is taken as the yardstick that sets the scale for the distances of planets from the sun, of satellites from their planets, and of stars as well. This is the reason for calling the yardstick the astronomical unit and for determining its value as accurately as possible.

We have seen in Chapter 6 that the moon's distance can be found by observing its parallax from two stations on the earth. The distance of the sun would be measured less reliably in this way, because the sun's parallax is much smaller and also because the stars are less available as reference points in the daytime. More dependable optical values of the astronomical unit have been derived by observing the larger parallaxes of the nearer planets, particularly of the asteroid Eros at its closest approaches to the earth. The value of the solar parallax adopted since 1896 in the astronomical almanacs by international agreement is 8″.80. This is the difference in the direction to the sun's center as it would be viewed on the equator from the center and edge of the earth when the earth is at its average distance from the sun. The corresponding mean distance of the earth from the sun is about 149.6×10^6 kilometers.

The timing of many radar echoes from the planet Venus near its inferior conjunction beginning in 1961, as reported by several radio observatories, gave remarkably consistent results among themselves. The calculated average value of the solar parallax is $8''.794$, which gives a value for the astronomical unit of about 149,597,896 kilometers. A quarter of the discrepancy between the optical and radio results is ascribed by M. C. Eckstein to the use of slightly different constants of the solar system in the two determinations. The remainder of the discrepancy is not explained. In any case, it is readily appreciated that the radar values are the more accurate and the averages of these values are now used to calculate tables of positions of the planets, stars, etc.

It should strike the reader as interesting that we really do not measure the parallax of the sun, but derive it. The methods used, whether optical or radar, are identical to those of Kepler. The relative orbits of the earth and Venus (in the radar case) are known very accurately in terms of the astronomical unit. Determining the distance in some measuring units (meters, kilometers, etc.) when the two planets are close together allows us to scale the orbits in our measuring units and hence determine the astronomical unit. Knowing the radius of the earth we can then derive the parallax of the sun.

The astronomical unit is often referred to as unit distance and unit distance is often given in terms of light travel time. The light travel time is 499.012 seconds. In later discussions we use two light travel time units, the distance light travels in one year, called the light year, and 206,265 times the astronomical unit distance, called the parsec (see page 253).

7.10 Regularities of the revolving planets

The revolutions of the planets around the sun and of the satellites around their planets exhibit some striking regularities, which apply more generally to the larger bodies. These regularities and the exceptions in the cases of the less massive members may provide important clues concerning the origin of the solar system.

1 **All the planets revolve around the sun from west to east,** as the earth does. This includes all the asteroids. Most satellites have direct revolutions around their planets, and this is also the favored direction of all rotations in the system excepting Venus and a few satellites and asteroids.

2 The orbits of the planets and satellites are nearly circles. However, the orbits of the smallest principal planets, Mercury and Pluto, and of some asteroids have greater eccentricities.

TABLE 7.2 *Planetary satellites*

Name	Discovery	Mean distance from primary (km)	Mean sidereal period of revolution	Diameter	Stellar magnitude
Moon		384,397	27.322 days	3476 km	−12.6
		SATELLITES OF MARS			
Phobos	Hall, 1877	9,350	0.319	13	+11.6
Deimos	Hall, 1877	23,480	1.261	10	+12.8
		SATELLITES OF JUPITER			
Fifth	Barnard, 1892	181,000	0.498	193	+13.0
I Io	Galileo, 1610	421,800	1.769	3220	4.8
II Europa	Galileo, 1610	671,100	3.551	2900	5.2
III Gaynmede	Galileo, 1610	1,070,000	7.155	5270	4.5
IV Callisto	Galileo, 1610	1,883,000	16.689	4500	5.5
Sixth	Perrine, 1904	11,478,000	250.583	96[a]	13.7
Seventh	Perrine, 1905	11,737,000	259.667	30[a]	16.0
Tenth	Nicholson, 1938	11,721,000	259.208	19[a]	18.6
Twelfth	Nicholson, 1951	21,243,000	631	19[a]	18.8
Eleventh	Nicholson, 1938	22,531,000	692	19[a]	18.1
Eighth	Melotte, 1908	23,496,000	737	19[a]	18.8
Ninth	Nicholson, 1914	23,657,000	758	19[a]	18.3
		SATELLITES OF SATURN			
Janus	Dollfus, 1967	157,700	0.749	300	+14.0
Mimas	Herschel, 1789	185,600	0.942	480	12.1
Enceladus	Herschel, 1789	238,200	1.370	560	11.8
Tethys	Cassini, 1684	294,800	1.888	960	10.3
Dione	Cassini, 1684	377,500	2.737	960	10.4
Rhea	Cassini, 1672	527,200	4.517	1287	9.8
Titan	Huygens, 1655	1,221,000	15.945	4830	8.4
Hyperion	Bond, 1848	1,483,000	21.276	480[a]	14.2
Iapetus	Cassini, 1671	3,560,000	79.331	1090[a]	11.0
Phoebe	Pickering, 1898	12,952,000	550.33	190[a]	16.5
		SATELLITES OF NEPTUNE			
Miranda	Kuiper, 1948	130,500	1.413	300	+16.5
Ariei	Lassell, 1851	191,800	2.520	800	14.4
Umbriel	Lassell, 1851	267,200	4.144	650	15.3
Titania	Herschel, 1787	438,400	8.706	1130	14.0
Oberon	Herschel, 1787	586,300	13.463	960	14.2
		SATELLITES OF URANUS			
Triton	Lassell, 1846	355,200	5.876	3700	+13.5
Nereid	Kuiper, 1949	5,562,000	359.875	320[a]	18.7

[a] = approximate values

3 The orbits of most planets and satellites lie nearly in the same plane. With the exception of Pluto's orbit, the orbits of the principal planets are inclined less than 8° to the ecliptic plane, so that these planets are observed always near the ecliptic, and mostly within the boundaries of the zodiac (see Section 3.7).

These three regularities do not apply to many comets and meteor swarms, especially to those having the longer periods of revolution.

The true periods of revolution of the planets, or their **sidereal periods,** increase with distance from the sun in accordance with Kepler's harmonic law, from 88 days for Mercury to nearly 250 years for Pluto. The synodic periods are also given in Table 7.1. They are the intervals between two successive conjunctions of the planet with the sun, as seen from the earth; for the inferior planets the conjunctions must both be either inferior or superior. In other words, the synodic period is the interval in which the inferior planets that are moving faster gain a lap on the earth, or in which the earth gains a lap on the slower superior planet. Mars and Venus have the longest synodic periods because they are nearest the earth and run it the closest race around the sun.

QUESTIONS

1 Arabian astronomers were the recipients and custodians of Greek astronomical knowledge from roughly A.D. 400 to A.D. 1500, yet they seem to have contributed no major advance. Speculate why this was so.
2 What do you think King Alphonso's advice would have been?
3 Rationalize a flat earth in the face of repeated observations that the top of the mast of a ship going to sea disappears from view last.
4 If the earth is spinning, why don't we fly off?
5 What is meant by the term "relative orbit?"
6 How far from the center of the sun is the center of mass of the Jupiter–Sun system?
7 Suppose the period of a planet is 8 years. How far would it be from the sun in astronomical units?
8 Show that the actual distance of Pluto at perihelion is greater than that of Neptune at aphelion.
9 What do we mean by the terms elongation, conjunction, quadrature, and opposition?
10 What are the dynamic regularities exhibited by the solar system?

FURTHER READINGS

BLANCO, V. M., AND S. W. MCCUSKEY, *Basic Physics of the Solar System*, Addison-Wesley, Reading, Mass., 1961.

BRANDT, JOHN C., AND PAUL W. HODGE, *Solar System Astrophysics*, McGraw-Hill, New York, 1964.

DREYER, J. L. E., *History of the Planetary System from Thales to Kepler*, Dover, New York, 1953.

PAGE, THORNTON, AND LOU WILLIAMS PAGE, EDS., *Wanderers in the Sky*, Macmillan, New York, 1965.

ROTH, G. D., *Handbook for Planet Observers*, Van Nostrand, New York, 1966.

VAN DE KAMP, PETER, *Elements of Astromechanics*, Freeman, San Francisco, 1964.

Drawing of a sun-centered solar system by Kepler following Galileo's theories. (N.Y. Public Library, Rare Books Division.)

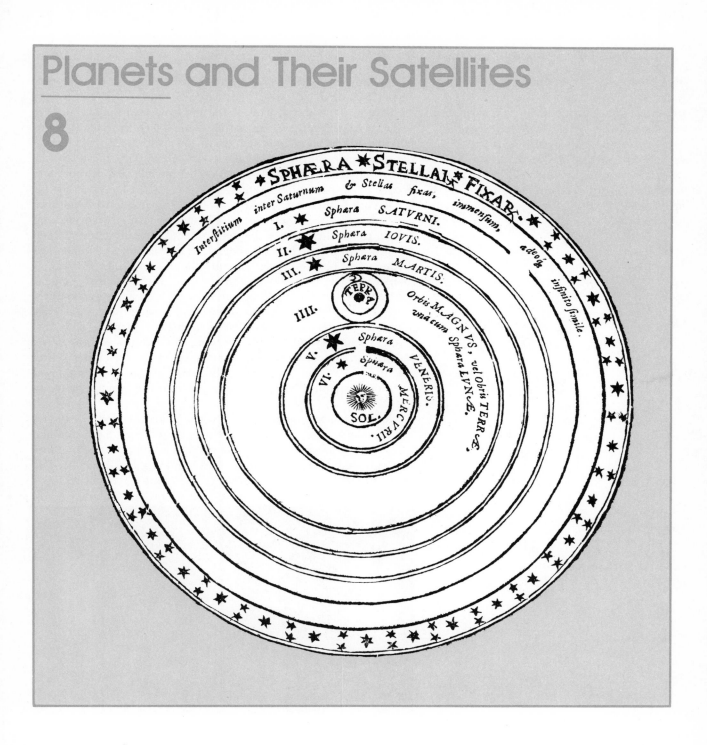

8

The planets have been objects of detailed study since the application of the telescope to astronomy. First the emphasis was on the visible features, then on the physical features of the surfaces and atmospheres. In this chapter we consider each planet in some detail and in a later chapter take up the question of the origin of the solar system.

These two planets revolve around the sun inside the earth's orbit, Mercury once in 88 days, Venus in 225 days. They accordingly oscillate to the east and west of the sun's place in the sky and are never very far from it. At times they come out in the west at nightfall as evening stars; at other times they rise before the sun as morning stars. Both planets, as we view them with the telescope, show the whole cycle of phases (see Section 7.7) just as the moon does. Neither planet has a satellite.

Mercury is the nearest to the sun and the smallest of the principal planets. Its diameter of 4886 kilometers is less than one and a half times the moon's diameter. Venus, the brightest planet, outshines all other celestial bodies except the sun and moon. Its diameter, 12,110 kilometers, is only slightly less than the earth's diameter. Our nearest neighbor among the principal planets, Venus comes within an average of 42×10^6 kilometers of the earth. Since 1962 various spacecraft, radio, and radar observations have added to our detailed knowledge of both these planets.

The terms *evening star* and *morning star* are applied most often to the appearances of the inferior planets in the west after sunset and in the east before sunrise. The terms also may be employed for superior planets when they are visible in the evening or morning sky.

Mercury is occasionally visible to the naked eye for a few days near the times of its greatest elongations (Fig. 8.1). It then appears in the twilight near the horizon as a bright star, sometimes even a little brighter than Sirius, and twinkling like a star because of its small disk and low altitude. The elongations occur about

FIGURE 8.1
Mercury near elongation in 1934. The phase effect is obvious but no surface detail can be seen. (Lowell Observatory photograph.)

7 June 1934   11 June 1934   12 June 1934

22 days before and after inferior conjunction. Because the synodic period is only 116 days, several greatest elongations occur in the course of a year; they are, however, not equally favorable.

Mercury's altitude above the horizon at sunset and sunrise varies considerably on these occasions in our latitudes. The altitude is greatest, and the planet is therefore most easily visible, when the ecliptic is most inclined to the horizon (see Section 3.4). Therefore, the most favorable times to see Mercury as evening star are at its greatest eastern elongations in the early spring, and as morning star at its greatest western elongations in the early autumn. Such occasions are especially favorable at that time when the planet is near its greatest distance from the sun. In its rather eccentric orbit Mercury's distance from the sun's place in the sky at its greatest elongations varies from 28° at its aphelion to as little as 18° at its perihelion.

The best views of Mercury with the telescope are obtained in the daytime when the planet is well above the horizon. In addition to the phases, some dark markings are glimpsed and are also

TABLE 8.1
Elongations of Mercury

		Eastern				Western	
1974	9	Feb.	18°.2	23	Mar.	27°.8	
	4	June	23 .5	22	July	20 .2	
	1	Oct.	25 .9	10	Nov.	19 .1	
1975	23	Jan.	18 .6	6	Mar.	27 .2	
	16	May	22 .0	4	July	21 .6	
	13	Sept.	26 .8	25	Oct.	18 .4	
1976	7	Jan.	19 .2	16	Feb.	26 .2	
	28	Apr.	20 .4	15	June	23 .3	
	26	Aug.	27 .2	7	Oct.	17 .9	
	20	Dec.	20 .1				
1977				28	Jan.	24 .9	
	10	Apr.	19 .5	27	May	25 .0	
	8	Aug.	27 .4	21	Sept.	17 .9	
	3	Dec.	21 .3				
1978				11	Jan.	23 .4	
	24	Mar.	18 .7	9	May	26 .4	
	22	July	26 .9	4	Sept.	18 .1	
	16	Nov.	22 .6	24	Dec.	22 .0	
1979	8	Mar.	18 .2	21	Apr.	27 .4	
	3	July	25 .9	19	Aug.	18 .6	
	29	Oct.	23 .9	7	Dec.	20 .7	
1980	19	Feb.	18 .1	2	Apr.	27 .8	
	14	June	24 .5	1	Aug.	19 .5	
	11	Oct.	25 .2	19	Nov.	19 .6	

recorded in photographs, which are reminiscent of the lunar seas. The great increase in the planet's brightness from the quarter to the full phase, as the shadows become shorter, indicates that its surface is as mountainous as the surface of the moon. Also, its albedo of only 7 percent is similar to the moon. Like the moon, too, Mercury has no atmosphere, and would not be expected to have any because of its small size and low surface gravity.

It was long believed that Mercury rotated once in its orbital path around the sun. Visual and photographic observations, which are always difficult, seemed to confirm this belief. Under such conditions the permanently dark portion of the planet would have a temperature not far above absolute zero. However, accurate radiometer measures indicated that the temperature of the dark side was very nearly equal to that of the sunlit side. There are two simple explanations for this contradiction: either the planet has an atmosphere or its period of rotation is different from its orbital period. The low surface gravity of the planet and high temperature facing the sun preclude Mercury's having much of an atmosphere (see Section 6.7). Therefore a test was proposed to see whether the planet's rotation period was different from that of its revolution.

Radar telescopes had successfully determined the rotation period of Venus as explained below. In the case of Mercury the existing radar systems using conventional pulse methods, such as those used in weather determinations and aviation, could not be used. The pulses would be so weak that even with the largest telescope they might not be detected. The Arecibo radio telescope was equipped with a CW (continuous-wave) radar system and its 1000-foot paraboloid was used to beam the signal toward Mercury and receive the faint returning signal. After many observations over several years Mercury has been found to have a period of rotation of 58.646 days. This explains the apparently fallacious visual and photographic observations. The rotation period is almost half the synodic period—thus we were observing the same hemisphere at successive conjunctions.

Venus emerges slowly from superior conjunction, behind the sun, to appear as an evening star requiring 220 days to reach greatest eastern elongation. Then in only 72 days it moves back to inferior conjunction, this time between us and the sun, to become a morning star. In another 72 days it reaches greatest western elongation, where it turns again to begin the 220-day return to superior conjunction. Thus the entire synodic period is 584 days.

The greatest brilliancy of Venus as evening and morning star occurs about 36 days before and after inferior conjunction. On these occasions it appears 15 times as bright as Sirius, the brightest star, and 6 times as bright as the planet Jupiter. Then Venus appears through the telescope in the crescent phase, which is between 5 and 6 times as great from horn to horn as the apparent diameter of the fully lighted disk it shows when it is beyond the sun. Figure 8.2 shows the changing phase and apparent size as viewed with a telescope. Around the times of greatest brilliancy, Venus becomes visible to the naked eye in full daylight, like a star in the blue sky.

Venus, because of its great brilliance and rapid motion, is often mistaken for some unnatural or transient phenomenon in the sky. When such a phenomenon is reported, the student should consult tables of planetary elongations (for example, Table 8.2) before drawing hasty conclusions. Jupiter, Saturn, and Mars can be quite spectacular morning and evening objects as they approach and depart from conjunction and have also given rise to misinterpretation.

Optical observations from the earth have given limited information about Venus. The permanently cloudy atmosphere allows us to sample only its uppermost layers and tells us relatively nothing about the surface. Spectrograms indicate the presence of carbon dioxide in abundance and give marginal evidence for some water vapor and free oxygen. Markings in the clouds are occasionally glimpsed with the telescope, and are more clearly seen in photographs taken with a blue filter (Fig. 8.3). We may reason that if these markings are parallel to the equator of Venus, they would indicate a rather short period of rotation. Nothing definitive can be said. Fortunately, this issue has also been resolved by using CW and conventional pulsed radar techniques.

Assume there is no rotation and consider that Venus is overtaking and passing the earth on an essentially straight line (Fig.

FIGURE 8.2
The phases of Venus in its orbit. The full phase effect in Fig. 8.3 (27 Sept. 1910) proves the Copernican view of the solar system.

TABLE 8.2
Elongations and conjunctions of Venus

| Year | Elongations | | | | Conjunctions | | | |
	Eastern		Western		Superior		Inferior	
1974			Apr. 4	46°.4	Nov.	6ᵈ13ʰ	Jan.	23ᵈ21ʰ
1975	Jun. 18	45°.4	Nov. 7	46°.6			Aug.	27 13
1976					Jun.	18 4		
1977	Jan. 24	47°.0	Jun. 15	45°.8			Apr.	6 6
1978	Aug. 29	46°.2			Jan.	22 6	Nov.	7 22
1979			Jan. 18	47°.0	Aug.	25 13		
1980	Apr. 5	45°.9	Aug. 24	45°.9			Jun.	15 7

FIGURE 8.3
The phases of Venus as photographed at the telescope. Note the apparent change in diameter caused by the different distances. The complete halo at inferior conjunction proves that Venus has an atmosphere. (Lowell Observatory photograph.)

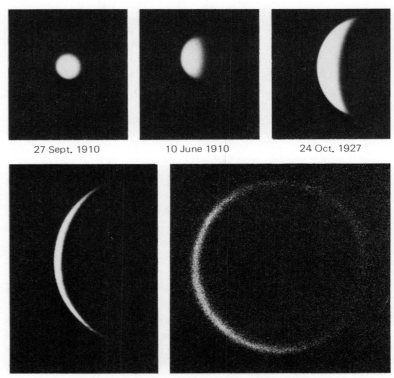

27 Sept. 1910   10 June 1910   24 Oct. 1927

25 Sept. 1919   19 June 1964

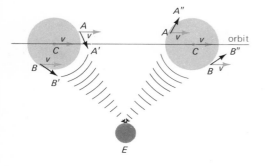

FIGURE 8.4
The apparent forward rotation of a non-rotating inferior planet as it passes the earth. The left edge appears to be approaching BB' more rapidly than the right edge AA'. After passing the right edge it appears to be receding more rapidly BB'' than the left edge AA''.

8.4). Because the planet has a perceptible disk, the projection of the orbital velocity v at A upon AE will be AA' and at B it will be BB'. Later after conjunction the projections are AA'' and BB''. In the first case BB' is larger than AA' and in the second case BB'' is larger than AA'', so the planet seems to have an apparent forward rotation. We call the rotation ω_1. Let the true rotation of the planet be ω_2; then the resultant rotation will be $\omega = \omega_1 + \omega_2$ ($\omega \to 0$ as CE increases). Now if ω_1 and ω_2 have the same sign algebraically, that is, ω_2 is also direct rotation, the spread in the returned signal, given in cycles per second, will have a maximum at conjunction. If, however, ω_2 is opposite in sign to ω_1, we will have a minimum at conjunction. The observational results are shown in Fig. 8.5 illustrating that the period of Venus is 243 days (i.e., it is **retrograde**).

In 1962, when the spacecraft Mariner 2 passed 35,000 kilometers above the surface of Venus, it was able to assemble and to transmit to the earth more definite information about the planet and

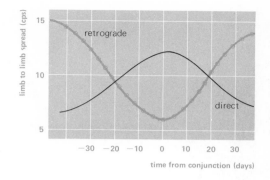

FIGURE 8.5
The observed points differentiate between direct and retrograde rotation and, along with the transmitting frequency and the size of the planet, assign the period as well.

its surroundings. Viewed at that distance the disk of Venus subtended an angle of 16°. Analysis of the radiometer data assigned to both the planet's surface for the bright and dark hemispheres a temperature of 720°K, which is hot enough to melt lead. Similarly, a temperature of 300°K was indicated at the top of the clouds on both the day and the night sides of the planet. Water is evidently not present on the surface; the amount of water vapor in the atmosphere was shown to be less than a thousandth of that in the earth's atmosphere. Venus did not seem to have an appreciable magnetic field or a trapped radiation belt around it. The planet's rotation should be very slow to explain the observations, and, as we have seen above, this is indeed the case. Finally, the pressure at the surface seemed to be more than 80 times that of the earth.

Unfortunately, a series of Venera spacecraft gave information that seemed to contradict the above, so the basic facts remained in controversy for a decade. During this period, radar observations yielded a consistent value for the diameter of the planet (12,110 kilometers) and a rough map of the surface as well. While the details of the mapping technique are beyond the scope of our text, the results show a typically terrestrial planet (Mercury, earth, moon, Mars) with several very high mountains. The mountains could be tracked by radar and easily confirmed the retrograde rotation of Venus. Cratering, typical of Mars, the moon, and presumably Mercury is apparent. The radar mapping of Venus is a tribute to man's ingenuity.

In 1972 Venera 7 and 8 landed on Venus and operated just long enough to confirm and improve the Mariner 2 observations. Specifically, the surface temperature was nearly 750°K on both the day and night side, the surface pressure was almost 90 times that of the earth's atmosphere and the surface rocks resembled that of granite and gave evidence of geological differentiation. On the descent into the atmosphere the spacecraft measured winds of 180 kilometers per hour in the upper atmosphere, 6 kilometers per hour in the lower atmosphere, and while direct measures were not made to test the atmosphere on the surface, we can infer that only a light breeze exists. All of this is in keeping with our knowledge of the physics of gaseous atmospheres and a slowly rotating planet. In addition, Venera 8 sampled the atmosphere confirming our knowledge of the upper atmosphere; it showed that there was only a trace of water vapor (H_2O), and detected a trace of ammonia. The latter observation is significant since ammonia is present in the gaseous outer planets and in interstellar space.

8.1 Transits of Mercury and Venus

The inferior planets occasionally **transit,** or cross directly in front of the sun at inferior conjunction. They then appear as dark dots against the sun's disk.

About 13 transits of Mercury occur in the course of a century; they are possible only within 3 days before or after 8 May, and also within 5 days of 10 November, when the sun passes the nodes of the planet's path. Transits are scheduled for the remainder of the century on 13 November 1986, 6 November 1993, and 15 November 1999, which will be a grazing transit. Transits of Mercury are not visible without the telescope.

These transits, which can be timed rather accurately, have been useful for improving our knowledge of the planet's motions. Since Mercury's diameter is well known it can be used to accurately scale the size of sunspots and other solar features as it makes its transit (Fig. 8.6). The mathematician Leverrier discovered from records of many transits that the perihelion of Mercury's orbit is advancing faster than would be predicted by the law of gravitation. The major axis of the planet's orbit is turning eastward, mainly because of the attractions of other planets. The observed excess (43" per century) in its turning has been explained by the theory of relativity as we have noted in Section 7.6.

Transits of Venus are less frequent; they are possible only when the planet arrives at inferior conjunction within about 2 days before or after 7 June or 9 December, the dates when the sun passes the nodes of the planet's path. They are now coming in pairs having a separation of 8 years. The latest transits occurred in 1874 and 1882; the next ones are scheduled for 8 June 2004, and 6 June 2012. Transits of Venus are visible without a telescope, however, the normal precautions for observing the sun must be observed (see Section 10.10).

MARS, THE RED PLANET

Next in order beyond the earth, Mars revolves once in 687 days at the average distance of 227.9×10^6 kilometers from the sun, and rotates on its axis once in 24^h37^m. Its diameter is 6798 kilometers, or slightly more than half the earth's diameter. Its atmosphere is not sufficiently extensive and clouded to hide its surface, which exhibits a variety of markings. Here, again, just as with Venus a series of highly successful spacecraft, notably Mariner 4, 6, 7, and 9 flights have dramatically advanced our knowledge in a short period of time. The persistent idea that Mars contains

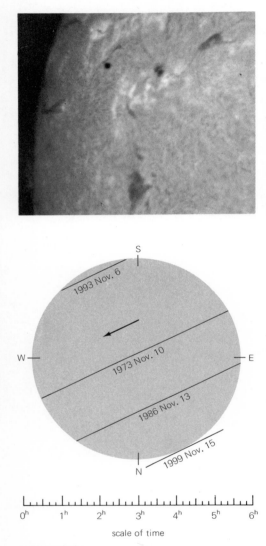

FIGURE 8.6
The transits of Mercury remaining in this century are shown on the bottom. The 10 Nov. 1973 transit is shown on the top. Such transits are still useful to scale the size of sunspots.

The earth as a celestial object. The earth appears in the moon's sky as a colorful ball. (National Aeronautics and Space Administration Photograph.)

Astronomical experiment on the moon. Astronaut J. Young is shown behind and to the right of an ultraviolet telescope. The telescope was used to study the Large Magellanic cloud at ultraviolet wavelengths absorbed by the earth's atmosphere. (National Aeronautics and Space Administration Photograph.)

The planet Jupiter. Note the colorful band structure and the whorl-like notches in the bands. The great red spot is at the upper left. (Photograph from the Hale Observatories.)

The planet Saturn. The very smooth banded structure on the ball of the planet is striking. Cassini's division in the rings stands out. (Lick Observatory Photograph.)

The Hourglass nebula (right). Another example of an emission nebula. (Lick Observatory Photograph.)

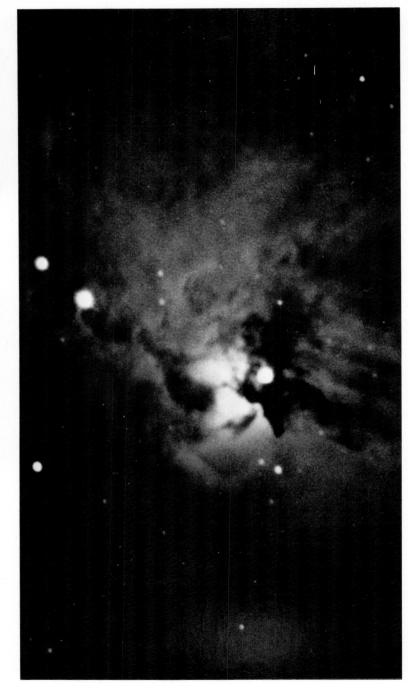

The Pleiades. The nebulae around the bright stars of this famous galactic cluster are excellent examples of reflection nebulae. (Photograph from the Hale Observatories.)

The Trifid nebula (left). This fine nebula in Sagittarius is a good example of an emission nebula. (Photograph from the Hale Observatories.)

The planet Mars. The Martian south pole is up and a cloud is visible along the lower left edge. (Lick Observatory Photograph.)

The Ring nebula (right). This famous nebula, located in Lyra, is an example of the ideal planetary nebula. The nebula is actually a shell, the ring appearance resulting from projection effects. (Photograph from the Hale Observatories.)

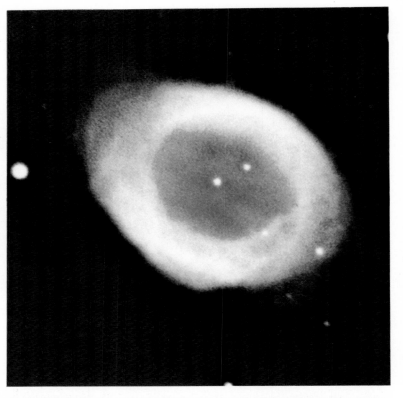

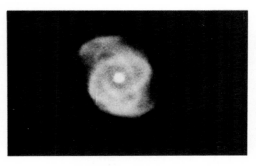

NGC 7293. Note the distorted corkscrew appearance of this planetary nebula in Aquarius. (Photograph from the Hale Observatories.)

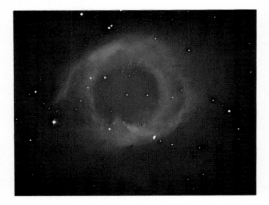

NGC 6543. A corkscrew appearance is evident in this planetary nebula and it is even more distorted than that of NGC 7293. (Lick Observatory photograph.)

The Crab nebula (right). This famous nebula originated as a violent supernova. (Photograph from the Hale Observatories.)

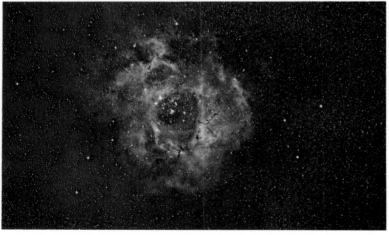

The Veil nebula (above). The beautiful,
graceful arc of this nebula in Cygnus is
actually a part of a large wreath-like structure;
it is the remnant of a supernova event.
(Photograph from the Hale Observatories.)

The large Rosette nebula is the apparent
remains of star formation as the beautiful
open cluster readily attests. (Photograph
from the Hale Observatories.)

The Great Andromeda galaxy. *Faintly visible to the naked eye, the Andromeda galaxy is quite spectacular in this photograph, along with its two brightest companion galaxies. (Photograph from the Hale Observatories.)*

Spiral galaxy NGC 598. A spiral galaxy seen almost face on. Note the intense point source in the center and the small nucleus typical of an Sc galaxy. (Photograph from the Hale Observatories.)

Spiral galaxy NGC 7331 is an Sb-type spiral galaxy located in Pegasus. This galaxy is similar to the Milky Way galaxy. (Photograph from the Hale Observatories.)

NGC 3034 is an irregular galaxy. A super violent event has occurred in the center of this galaxy and is tearing it apart. (Photograph from the Hale Observatories.)

Three-element radio interferometer telescope. Three 27-meter paraboloids form the interferometer in a valley at Green Bank, West Virginia. The two nearest telescopes are movable and the greatest spacing from the farthest telescope is 3.8 kilometers. (National Radio Astronomy Observatory Photograph.)

Skylab in orbit (right). Orbiting Skylab has a solar telescope and several small stellar ultraviolet telescopes as part of its observing equipment. (National Aeronautics and Space Administration Photograph.)

The great collecting mirror at the top of the R. R. McMath solar telescope at the Kitt Peak National Observatory. The tunnel slopes downward to the right.

certain forms of life has made this planet an object of special interest, particularly at its closest approaches to the earth.

8.2 Mars through the telescope

Mars is best observed telescopically at its oppositions. Oppositions of Mars with the sun recur at average intervals of 780 days, or about 50 days longer than 2 years. They accordingly come 50 days later in successive alternate years. Owing to the eccentricity of its orbit, Mars varies considerably in its distance from the earth at the different oppositions, from less than 56×10^6 kilometers at perihelion to more than 100×10^6 kilometers at aphelion.

Favorable oppositions occur when the planet is also near its perihelion. At these unusually close approaches, which are always in the late summer, Mars becomes the most brilliant starlike object in the heavens, with the single exception of Venus, and also attracts attention then because of its red color. With a telescope magnifying only 75 times, its disk appears as large as does the moon's disk to the unaided eye. The dates of oppositions of Mars from the latest to the next favorable oppositions are given in Table 8.3 and Fig. 8.7. A pair about equally favorable will occur in 1986 and 1988.

Mars presents its poles alternately to the sun just as the earth does; its seasons resemble ours except that they are about twice as long. The winter solstice occurs near the time of perihelion, as in the case of the earth. Thus, as with us (see Fig. 3.10), the summer in the southern hemisphere begins when the planet is nearest the sun, and in the northern hemisphere when it is farthest from the sun. On the earth, as we have seen, the seasons are not caused by distance from the sun. When the northern and southern hemispheres are compared, there are no appreciable differences in their range of winter and summer temperatures. The variation of our distance from the sun is relatively small, and our southern hemisphere has more water to modify extreme temperatures.

The temperature difference is noticeable on Mars because it has a more eccentric orbit, where the greatest distance from the sun exceeds the least by 19 percent, or 42×10^6 kilometers. The temperatures of the seasons are appreciably the more extreme in the Martian southern hemisphere. The snow cap around its south pole becomes larger in the winter season than does the north polar cap; it completely disappears in the summer, which the northern one has not been observed to do. It is the south polar cap that is pointed toward the earth at the favorable oppositions, and this is the one that more often appears in the photographs.

White caps, which appear alternately around the poles of Mars,

TABLE 8.3

Oppositions of Mars

Date of opposition	Distance ($\times 10^6$ km)	Magnitude
25 Oct. 1973	67	−2.2
15 Dec. 1975	87	−1.5
22 Jan. 1978	100	−1.1
25 Feb. 1980	103	−0.8

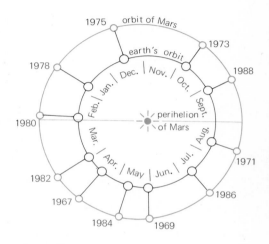

FIGURE 8.7
The varying distance of Mars from the earth at oppositions between 1967 and 1988. Mars is at perihelion late in August so the most favorable oppositions always occur then.

are the most conspicuous features of the view with the telescope. Each snow cap expands rapidly as winter progresses in that hemisphere, and shrinks with the approach of summer. The southern cap has attained a diameter of 5900 kilometers, so that it then extends more than halfway from the pole to the equator. As the cap retreats toward the pole, the main body may leave a small white spot behind, presumably on the summit or cooler slope of a hill.

The polar caps are more conspicuous in photographs in violet light, which reveal the areas of haze above them. The haze may be caused by ice crystals similar to those in our cirrus clouds. Patches of haze also appear in the cooler air near the sunrise and sunset lines. Yellowish spots like clouds of dust are visible at times on various parts of the disk. Often the haze is widespread over the disk, making the view of the surface more difficult, and then it clears quite suddenly over large areas.

Spectroscopic observations lead us to assign an upper limit of only 0.025 the pressure of the earth's atmosphere. We should note that above 70 kilometers in general the density of the Martian atmosphere is greater than that of the earth, resulting from the lower gravity of Mars.

Radio observations indicate a high-temperature halo about Mars at the proper height to be interpreted as a radiation belt similar to the earth's Van Allen belts (see Section 1.2). Such observations are interpreted to mean that Mars has a magnetic field analogous to the earth's. However, spacecraft magnetometers have failed to detect a measurable magnetic field 10,000 kilometers above the planet.

Many years ago Baldwin advanced the idea that Mars should be splattered with craters, that generally the craters would be larger than the moon's because of the proximity to the asteroid belt and the planet's thin atmosphere, and that the seasonal variations may result from the depositing and melting of frost at high elevations. This theory was at odds with many earlier ideas, but it has been confirmed. The various markings have led to a considerable speculation about the nature of the surface of Mars. Various dark markings appear in a variety of shapes and sizes against a generally reddish background that imparts its color to the reflected light of the planet. The reddish areas themselves are often called "deserts."

The large dark areas are given names of bodies of water that have survived from early maps, just as the maria on the moon have survived. The prominent dark areas are Syrtis Major (Great Bog) and Solis Lacus (Lake of the Sun). With the shrinking of each polar cap the dark markings of that hemisphere are more distinct; they become blue-green with the Martian spring and then fade and turn brown as winter approaches. Thus they have been interpreted as behaving in a manner similar to areas of vegetation. An alternate explanation is that in the Martian winter they are covered with a dirty frost and in the spring this melts away. Another explanation is offered by D. B. McLaughlin who thought they were volcanic ash deposited by prevailing winds, thus explaining the darkening in the spring. The desert sands similarly deposited with the changing wind patterns in the autumn would account for the return to lighter hues. This may be close to actuality.

A complex network of fine dark lines known as the **canals** of Mars was first reported by the astronomer Schiaparelli, who observed it with a small telescope and described the lines as like "the finest thread of spider's web drawn across the disk." This careful observer found the canals less difficult to see in the hemisphere where the snow cap was melting, and was tempted to suppose that vegetation bordering them might add to the width of the strips.

Many observers have viewed the network since then, often with telescopes of moderate size. It seems that the canals appear only during rare instants of exceptionally steady seeing. Thus a continued succession of photographs with the very short exposures permitted by the largest telescopes might be expected to reveal the network eventually if it exists. Yet many other experienced observers have been unable to detect long, narrow canals even visually and have doubted their existence as such.

8.3 Mars close up

Thanks to a series of remarkably successful spacecraft the study of Mars telescopically is relegated to the past, except for meteorological and spectroscopic purposes. Mariner 4 photographed a small portion of Mars on a fly-by and revealed the cratered surface for the first time. This put to rest the theories of canals, intelligent beings, etc., but did not exclude the possibility of "life" per se. Mariners 6 and 7 added to our information, and then the most successful unmanned spacecraft to date, Mariner 9, photographed almost all of the surface of Mars and its moons as well.

Mariner 9, designed to last 3 months while in Martian orbit, arrived at Mars on 13 November 1971 in the midst of a global dust storm that began in June 1971 and returned valuable pictures of the event plus the satellites. Then, in late January 1972, after the great storm subsided, Mariner 9 began the process of detailed

FIGURE 8.8
Mariner 9 photographs of Mars showing a vast chasm with eroding branching canyons. Much of the erosion is due to wind which, in the low density atmosphere, reach great velocities. (National Aeronautics and Space Administration photographs.)

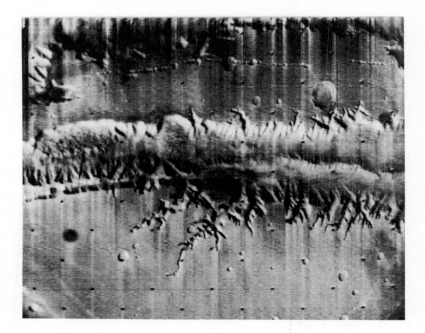

photography of the surface (Fig. 8.8) of the planet and continued through 27 October 1972 when it was turned off.

The Mariner 9 pictures and remote sensors show a surface pressure of only 0.006 that of the earth's atmosphere, dust clouds rising to a height of 15 kilometers, and occasional frost clouds at 60 kilometers. The photographs show large volcanos and canyons of tectonic (seismic) origin. The surface is not homogeneous; it is heavily cratered but also has several very large, smooth desert areas. The northern hemisphere is different from the southern hemisphere. This is interesting because this is the case for the earth and the moon as well. Geologic faulting is obvious and one great canyon (Fig. 8.9) coincides with a persistant "canal" on all of the earlier visual maps of Mars.

Most of the craters do not have sharp rims, which indicates erosion processes have occurred. In fact the entire surface shows clear effects of "weathering," and some of the great canyons and gulleys appear to be very similar to arroyos. The latter are certainly quite different in origin from lunar rilles. In addition they extend 400 kilometers and are often 5 kilometers wide. They are very similar to the Colorado River in Arizona as it appears in satellite pictures of the earth. This seems to indicate that Mars has, in its past, been eroded by water. The fact that even "new" craters are

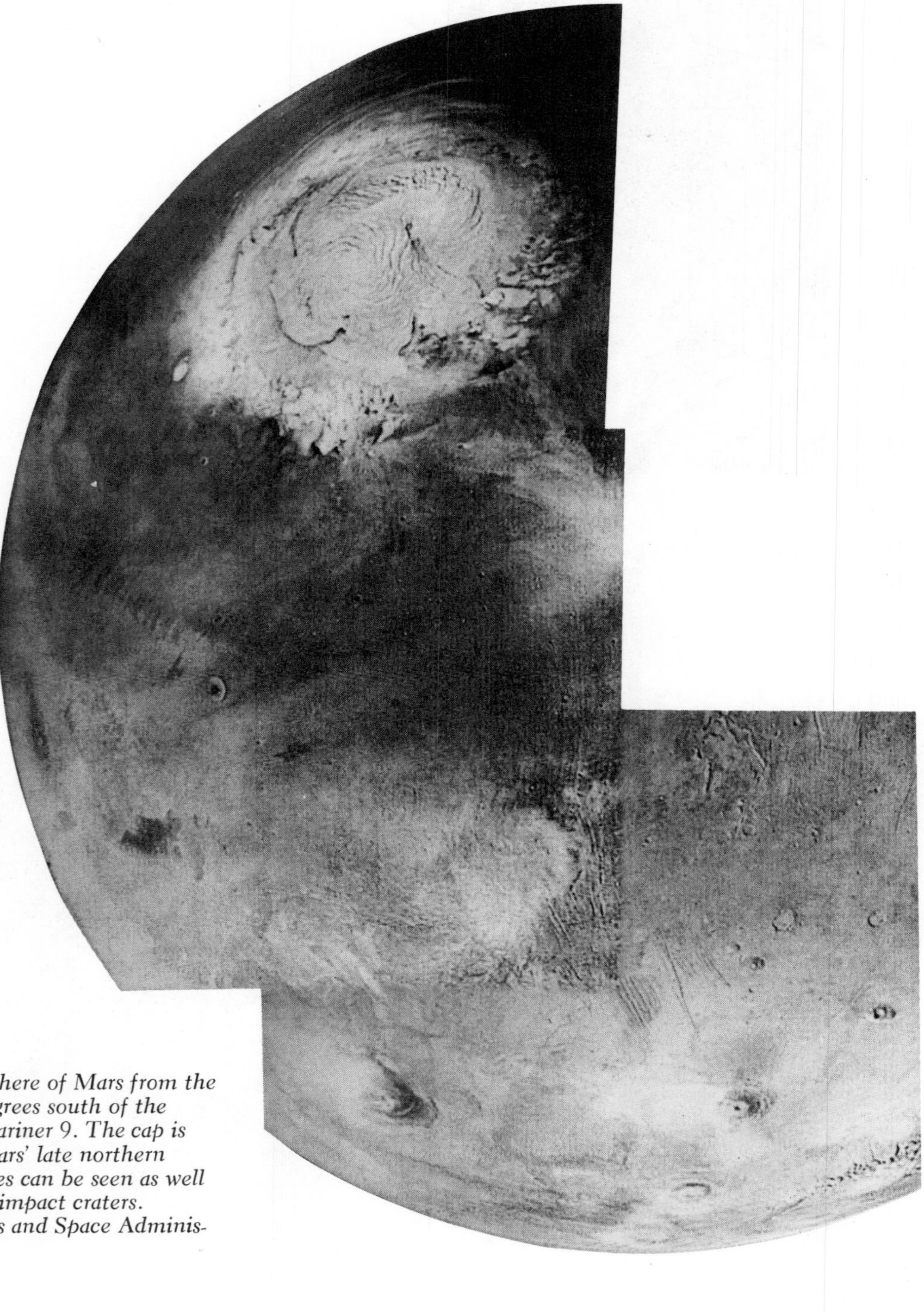

FIGURE 8.9
The northern hemisphere of Mars from the polar cap to a few degrees south of the equator as seen by Mariner 9. The cap is shrinking as this is Mars' late northern spring. Huge volcanoes can be seen as well as some of the major impact craters. (National Aeronautics and Space Administration photograph.)

eroded shows that wind erosion is the present mechanism for erosion; sand particles can achieve extremely high velocities in the low-density atmosphere and hence have great momenta.

At the same time as the Mariner 9 flight, two spacecraft, Mars 2 and 3, arrived at Mars to discharge landing capsules. Unfortunately the capsule from Mars 2 failed completely and that from Mars 3 transmitted for only 20 seconds to its mother craft before it failed.

8.4 The climate of Mars

The atmosphere of Mars is rarer than ours and is deficient in certain ingredients necessary for animal life. Free oxygen is not detected in the spectrum analysis of sunlight reflected from the planet; it is only 0.6 percent as abundant there as in the earth's atmosphere. The chief constituent must be nitrogen, which is not revealed by this means. There is only 0.1 percent as much water vapor as is found on earth, and twice as much carbon dioxide.

We have already noted that the Martian atmosphere is less dense than the earth's at the planetary surface, but more dense in the upper layers; thus, Martian clouds of vapor content are found far above its surface. The corresponding layers, such as the Mesopause, are located much farther from the surface of Mars than they are in our atmosphere. We may theorize that the upper Martian atmosphere has an ionosphere similar to ours, but again much farther from the surface.

The average temperature of the surface of Mars is $-40°C$ as compared with $+16°C$ for the earth. The climate of Mars seems to be too severe for the flourishing of even the hardiest known vegetation in such profusion as to color the dark areas. We should note, however, that the equatorial regions frequently are well above freezing.

Two Viking spacecraft are scheduled to land on Mars in late May 1976 and are designed to answer many questions about the climate and geology of Mars as well as provide added information as to the possibility of primitive organisms existing on the surface.

8.5 The two satellites of Mars

Mars has two satellites, Phobos and Deimos. Both are quite small; Phobos is about 13 kilometers in diameter and Deimos is about 10 kilometers in diameter. The satellites are actually rather irregular, have very low albedos (0.05), and show craters (Fig. 8.10). They are located so close to the planet that they are invisible except in rather large telescopes and hence were not discovered

FIGURE 8.10
Deimos (left) and Phobos (right) as seen by Mariner 9. The impact cratering is evident. These remarkable photographs put to rest many erroneous notions concerning the moons of Mars. (National Aeronautics and Space Administration photographs.)

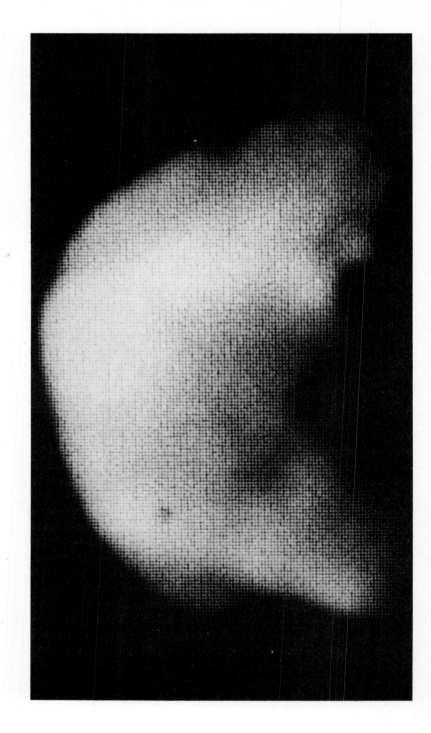

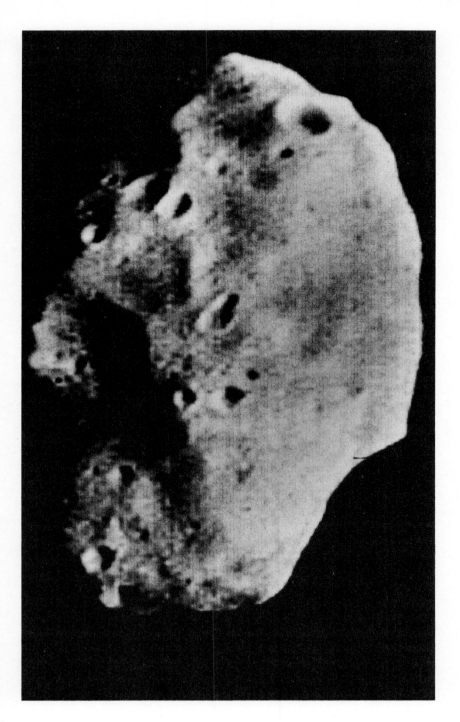

until 1877. They were theorized by Johnathan Swift in his satire *Gulliver's Travels* some 150 years prior to their actual discovery.

Phobos has a direct revolution at the distance of only 9×10^3 kilometers from the surface of Mars, once around in 7^h39^m, or less than one third the period of the planet's rotation in the same direction. Viewed from Mars, therefore, Phobos rises in the west and sets in the east. No other known natural satellite in the solar system revolves in a shorter interval of time than the rotation period of its primary.

Deimos, the outer satellite, revolves eastward around Mars once in 30^h18^m at a distance of 23×10^3 kilometers. It is smaller than Phobos and only a third as bright. This satellite rises in the east in the Martian sky, but drops behind the rotating planet so slowly that it goes through its whole cycle of phases for an observer there before it sets.

THE ASTEROIDS

The **asteroids**, or **minor planets**, revolve around the sun mainly between the orbits of Mars and Jupiter. Invisible to the naked eye, with the occasional exception of Vesta, they are "starlike" in the sense that they appear as point sources of light instead of disks even through large telescopes. The majority have periods of revolution between 3.5 and 6 years.

8.6 The orbits of asteroids

Toward the close of the eighteenth century, the astronomer Bode invited his colleagues to share in a search for a planet between the orbits of Mars and Jupiter. He explained that a series of numbers, which later came to be known as the Titius-Bode relation (see Section 7.9), represented the relative distances of the known planets from the sun with a single exception. No planet had been found corresponding to the number 2.8.

While the search was being organized, the missing planet was discovered incidentally by Piazzi in Sicily on 1 January 1801, because of its motion among the stars. The mean distance of the new planet, which Piazzi named Ceres, proved to be almost exactly 2.8 times the earth's distance from the sun. There was greater surprise when other minor planets were discovered later at about the same distance as Ceres.

Many thousand asteroids have been detected by the generally short trails they leave in the photographs by their motions against the background of the stars (Fig. 8.11) during the exposures. More

FIGURE 8.11
*Trails of three asteroids. (An early photo-
graph by M. Wolf, Königstuhl-Heidelberg.)*

than 1650 have had their orbits determined and have accordingly
received permanent running numbers and names in the catalogs.
Many of these, however, could readily become hopelessly lost
because of considerable and rapid perturbations of their orbits by
the attraction of Jupiter. As an example of the concerted effort
to prevent their being lost, F. K. Edmondson and his associates
at Indiana University have recovered a number of the fainter ones
by moving the telescope to follow their expected motions during
the exposures. Thus the light of the asteroid is concentrated in a
small image during the exposure while the stars appear as trails
and very faint stars are not recorded at all.

The asteroids are small when compared with the principal
planets. Ceres, the largest, is 767 kilometers in diameter. Pallas,
Juno, and Vesta are also among the larger asteroids. Only 34
asteroids are larger than 100 kilometers. The majority are less than
100 kilometers, and some are known to be scarcely a kilometer in
diameter. Their albedos run from 0.03, less than that of Mercury,
to 0.3 which is nearly that of the earth.

The motions of asteroids depart considerably from the regularities we have noted (see Section 7.10) in the movements of the
larger planets. Although they all have direct revolutions, some
have orbits so much inclined to the ecliptic that they venture far
outside the zodiac. Some have rather highly eccentric orbits; one
asteroid, Hidalgo, has its aphelion as far away as Saturn, and

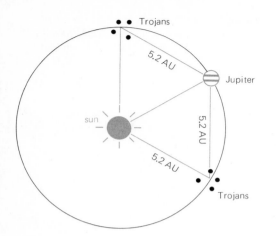

another, Icarus, comes at its perihelion nearer the sun than the orbit of Mercury.

The asteroids are not distributed at random through the zone they mainly frequent between the orbits of Mars and Jupiter. They avoid distances from the sun where the periods of revolution would be simple fractions, particularly one-third, two-fifths, and one-half, of Jupiter's period. There they would be subject to frequent recurrences of disturbances by Jupiter. These gaps in the asteroid distribution are known as **Kirkwood's gaps.** Where the periods are equal to Jupiter's, however, there are two regions in which asteroids congregate. The **Trojan asteroids** oscillate around two points near Jupiter's orbit, which are equidistant from that planet and the sun. They are named Achilles, Agamemnon, and so on after the Homeric heroes. More than a dozen are recognized.

Several known asteroids come within the orbit of Mars and make closer approaches to the earth than do any of the principal planets. Eros, for example, can come within 22×10^6 kilometers of the earth, at which time this 24-kilometer wide object appears as bright as a star of the 7th magnitude. These favorable oppositions, when Eros is also near perihelion, occur rather infrequently; the latest one was in 1931, and the next one is scheduled for 1975. The large parallax on such occasions permits accurate measurements of their distances, which have been valuable for verifying the scale of the solar system (see Section 7.9).

Examples of asteroids that come even closer are Apollo and Adonis (Fig. 8.12). The perihelion distance of Adonis is only slightly greater than Mercury's mean distance from the sun; this asteroid passes a little more than 1.6×10^6 kilometers from the earth's orbit and about the same distance from the orbits of Venus and Mars. Another neighboring asteroid, Icarus, attracted W. Baade's attention in 1949 when it left a long trail on a photograph taken with a 48-inch Schmidt telescope. At perihelion its distance from the sun is less than 30×10^6 kilometers.

Asteroids which have been observed within a few million kilometers of the earth are about a kilometer in diameter. At closest approach they appear as faint stars moving so swiftly across the heavens that they can easily be missed. Most of them have vanished before their orbits could be reliably determined.

8.7 Asteroids as fragments

The erratic orbits of some asteroids suggest that these are fragments that have been propelled in various directions by the collisions of larger bodies. This theory may account not only for the smaller asteroids that pass near us but also for the meteorites that

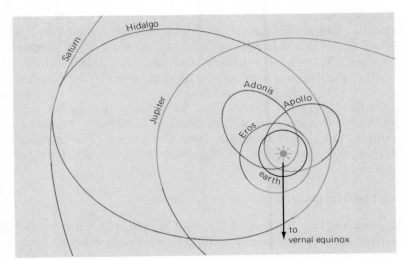

FIGURE 8.12
Orbits of four unusual asteroids. Broken lines represent parts of the orbits below the plane of the earth's orbit. (Adapted from a diagram by D. Brouwer.)

fall on the earth and are occasionally large enough to blast out meteorite craters. If this theory is correct we should expect a considerable amount of dust and very small fragments throughout the asteroid belt. While the inner portion of the belt does have somewhat more dust than the average in the ecliptic plane in the vicinity of the earth, the outer portion of the belt does not.

Many asteroids fluctuate periodically in brightness, as would be expected of rotating fragments having irregular shapes. Eros is an example. In its rotation once in 5^h16^m, this asteroid, shaped roughly like a brick, presents its larger sides and smaller ends to us in turn. Thus it becomes brighter and fainter out there in the sunlight twice in each period. The light variation is greatest when the equator is presented edgewise to us, and it becomes less in other parts of the orbit when a polar region is turned more nearly in our direction.

An extensive photoelectric study of representative asteroids by G. P. Kuiper and associates has shown that over 90 percent vary periodically in brightness, having two maxima and minima in each period of rotation. The periods range from 2^h52^m to about 20 hours. The rotation axes have random orientations. In the case of the asteroid Eunomia the direction of rotation is definitely retrograde.

JUPITER, THE GIANT PLANET

Jupiter is the largest planet in the solar system. Its equatorial diameter, 142,800 kilometers, is 11 times as great as the earth's diameter. Its mass exceeds the combined mass of all the other planets. With the exceptions of Venus and occasionally Mars, this planet is the brightest starlike object in our skies. Even a small telescope shows its 4 bright satellites and cloud belts clearly. Jupiter has 12 known satellites, the greatest number attending any planet.

At the distance of almost 778×10^6 kilometers from the sun, Jupiter revolves around the sun once in 11.89 years, so that it advances one sign of the zodiac from year to year. The period of its rotation, about $9^h 50^m$, is the shortest among the principal planets.

8.8 The structure and atmosphere of Jupiter

The markings on the disk of the giant planet, which run parallel to its equator, are features of its atmosphere. Bright **zones** alternate with dark **belts.** The broad equatorial zone is bordered by the north and south tropical belts. Then come the north and south tropical zones, and beyond them a succession of dark and bright divisions extending to the polar regions. Bright and dark spots appear as well; they often change in form quite rapidly, as atmospheric markings might be expected to do. Yet some of them are of surprisingly long duration. The Great Red Spot (Fig. 8.13) is an extreme example; this oval spot 48,000 kilometers long, has been observed for at least a century. It behaves like a floating solid. The various markings go around in the rotation at different rates, owing to the unequal horizontal movements of the clouds themselves.

The atmosphere consists mainly of hydrogen and helium. Its refraction effect on the light of a star was recently observed by W. A. Baum and A. D. Code as the planet began to occult a star, and was interpreted by them to mean that Jupiter's outer atmosphere is very dense. Methane and ammonia contaminate the atmosphere, as is indicated by the presence of their bands in the spectrum. At the temperature of $-130°C$, methane is still gaseous, whereas ammonia is mainly frozen into crystals. The observed temperature is somewhat higher than that expected from absorption and reradiation of solar energy.

Radiation from Jupiter in the radio wavelengths is of three observed types:

1 Thermal radiation consistent with the atmospheric temperature already stated.

$$\text{intensity} \propto \nu^2$$

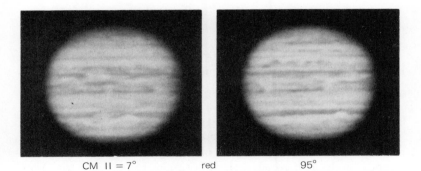

CM II = 7°   red   95°

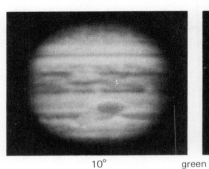

10°   green   97°

11°   blue   100°

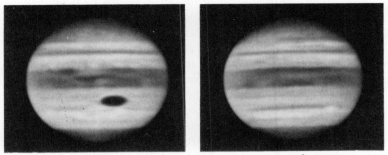

16°   UV   105°

FIGURE 8.13
Jupiter photographed in four different colors. Note in the left sequence how the red spot is dark in UV, blue, and green light because it absorbs those colors but reflects red. The right sequence was taken after approximately 90° rotation of the planet. (Lowell Observatory photographs.)

2 Nonthermal emission that issues in short blasts from the planet's ionosphere.

3 Nonthermal emission of a different type, which is weak and is recorded only at microwave frequencies.

intensity $\propto \nu^{-a}$

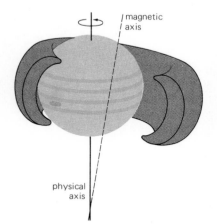

The short blasts are analogous to static from lighting discharges. The other nonthermal component is synchrotron radiation originating from a concentration of energetic charged particles, trapped by the Jovian magnetic field similar to our terrestrial Van Allen belts. Jupiter's magnetic poles are inclined to its rotational poles similar to the earth's, but unlike the terrestrial situation, the magnetic axis does not pass through the center of Jupiter.

The bulging of Jupiter's equator, which is clearly shown in the figure, provides one clue about conditions in the interior, which is hidden beneath the clouds. With its swift rotation the planet would be even more oblate if its mass were not highly concentrated toward its center. The temperature and the low average density of the whole planet (about 1.3 times the density of water), which requires very light material in the outer parts are also used to speculate about Jupiter's structure.

Several theoretical models have been created in an attempt to satisfy the limited clues. At one extreme R. Wildt of Yale University, about 30 years ago, designed a model in which the atmosphere is mainly hydrogen and has a depth of 13,000 kilometers. Beneath it there is a layer of ice 27,000 kilometers thick around a rocky core 61,000 kilometers in diameter. The core of the heavier elements is 6 times as dense as water, or somewhat more than the earth's average density. Hydrogen contributes 50 percent of the mass of this model, an abundance now considered too low.

At the opposite extreme, Wildt, DeMarcus, and others have proposed a newer model of the planet consisting entirely of hydrogen. Anything resembling an atmosphere in this model is only a few kilometers thick. Solid hydrogen begins at a depth of 3000 kilometers. Metallic hydrogen extends from a depth of 13,000 kilometers to the center; it becomes highly compressed under the very great pressure that increases to 32 million atmospheres at the center. Intermediate models have been discussed by these and other investigators. Whatever the choice may be, we conclude that Jupiter has little resemblance to the earth.

Jupiter, because it dominates the solar system's planets, is of extreme importance to our eventual understanding of the origin and evolution of the solar system. At this writing a spacecraft,

Pioneer F, is on a long journey to Jupiter. It encountered Jupiter on 3 December 1973 passing some 14,000 kilometers above the clouds. Its trajectory is designed to allow it to study the atmosphere of one of Jupiter's satellites, Io, and the great red spot as well as other parameters of the planet and its atmosphere. During its voyage it has seen several large meteoroids passing by and has been hit by a number of very small ones.

After passing Jupiter, it will swing partly around the planet and be accelerated; this additional energy will allow it to leave the solar system on an unobstructed path. At a distance of 30 astronomical units its transmissions will become too weak even for our largest radio telescopes. It will take Pioneer F about 80,000 years to cover a distance equal to that of the nearest star. In the unlikely event that it is intercepted by other humanoids (intelligent life, roughly like humans) it carries an appropriate pictogram (Fig. 8.14).

8.9 Jupiter's twelve satellites

Next to the moon, the four bright satellites of Jupiter are the most conspicuous satellites in our skies. They could be glimpsed with the naked eye if they were farther removed from the glare of the planet. They were discovered by Galileo early in 1610 and are referred to as the Galilean satellites.

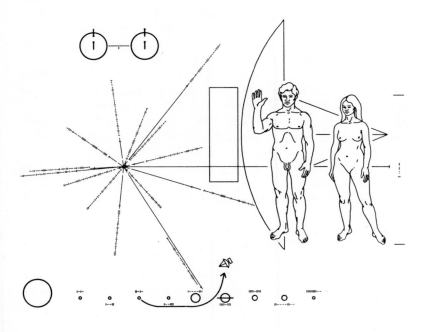

FIGURE 8.14
The pictograph carried on a metal plaque aboard Pioneer F. Pioneer F was accelerated by Jupiter as it flew by on 3 December 1973 and is now on its way out of the solar system. Should the spacecraft ever be found intact by other beings, hopefully they will be able to decipher the plaque. (National Aeronautics and Space Administration photograph.)

The bright satellites are numbered in order of distance from the planet; they have personal names as well, although these are not so often used. The 1st and 2nd satellites are about the size of the moon. The 3rd and 4th satellites are half again as great; they are the largest of all satellites and even surpass Mercury in diameter. Like the moon they have equal periods of rotation and revolution. Io, the innermost Galilean satellite, appears to have an atmosphere and is orange. We should expect that the other Galilean satellites have atmospheres, but this is not apparent in the observations that we are able to make.

These satellites have direct revolutions in nearly circular orbits, which are almost in the plane of the planet's disk or into its shadow. At other times they pass in front of the planet; their transits across its disk are visible with telescopes of moderate size, and the shadows they cast upon it are visible as well. The timing of these phenomena enabled O. Roemer in 1675 to determine the velocity of light (c).

The other satellites of Jupiter are too small and faint to be easily seen through the telescope. They are designated by numbers from 5 to 12 in order of their discovery. The 5th satellite is the nearest of all to the planet; in the small eccentricity and inclination of its orbit and in its direct revolution it resembles the bright satellites. The seven outer satellites have orbits of considerable eccentricity and inclination to the ecliptic. They fall into two groups. The 6th, 7th, and 10th satellites have direct revolutions at mean distances of a little more than 11×10^6 kilometers from the planet. The 8th, 9th, 11th, and 12th satellites have retrograde revolutions at distances about twice as great.

The data on the outer satellites appear in Table 7.2. The order of distance of the outermost four has little significance, because disturbances of their motions by the sun's attraction may change that order in a few years. The diameters of the satellites that do not appear as disks are estimated from the brightness of the satellites.

SATURN, THE RINGED PLANET

Saturn is the most remote of the bright planets and is, therefore, the most leisurely in its movement along the constellations. It revolves once in 29.5 years at the mean distance of 1428×10^6 kilometers from the sun. It appears as a bright yellow star in our skies, and it ranges in brightness from equality with Altair to twice that of Capella. This planet ranks second to Jupiter in size and mass; its equatorial diameter is 120,800 kilometers. It has the lowest average density, 0.7 times the density of water, and the most prominent bulge at the equator of any of the planets. Its unique system of rings makes it one of the most impressive celestial sights with the telescope.

8.10 Saturn and its rings

Saturn resembles Jupiter except for its smaller size, mass, and density. Methane bands are stronger in its spectrum, and there is less evidence of ammonia, presumably because this ingredient is frozen out of its atmosphere, which is $-150°$C.

From its broad, yellow equatorial zone to its bluish polar caps, the cloud markings show less detail than do those of Jupiter. Although enduring spots are rarely seen on Saturn, those that have been observed seem to show that the planet's rotation period is longer in the higher latitudes. Thus a period of 10^h14^m was derived for a bright spot in the equatorial zone in 1933 (Fig. 8.15) and for another in 1960, whereas a period of about 10^h40^m was found for a group of spots around latitude $+60°$ in 1960. T. A. Cragg at the Hale Observatories concludes that Saturn rotates in these two basic periods instead of having a steady increase of period with increasing latitude, as in the case of the sun.

Saturn's rings are invisible to the naked eye and were, therefore, unknown until after the invention of the telescope. When Galileo began observing Saturn in 1610, he glimpsed what seemed to be two smaller bodies in contact with the planet on opposite sides. The supposed appendages disappeared two years later and subsequently reappeared. This changing appearance of the planet remained a mystery until about half a century later, when Huygens, with a larger telescope, concluded that Saturn is encircled by a broad flat ring. In still later times we have found that there are four concentric rings instead of a single one.

The entire ring system is 275,000 kilometers across, but is scarcely more than 15 kilometers thick. The width of the **outer ring**, or A ring, is 16,000 kilometers. The **bright ring**, or B ring, is 26,000 kilometers wide. It is separated from the outer ring by the 4800 kilometer **Cassini division**, named after its discoverer; this is the most obvious division in the rings. The third or **crape ring**, referred to as the C ring, is actually continuous with the bright ring and is about 19,000 kilometers wide. The crape ring is quite faint and often not very clearly shown in pictures but nevertheless rather easily visible with telescopes of moderate size, although it was not discovered until 1850. In 1970 the **inner ring**, or D ring, was discovered by P. Guerin and it was noted that there

FIGURE 8.15
Several views of Saturn from a good view of the north of the rings (top left) through the rings' edge on (bottom left and top right) to the south of the rings (bottom right). (Lowell Observatory photograph.)

is a clear division between it and the crape ring analogous to the Cassini division. The inner division is about 1600 kilometers wide and the D ring is on the order of 19,000 kilometers wide. The rings show some remarkable changes in overall reflectivity, occasionally being quite dull for a given opening.

Saturn's rings are inclined 27° to the plane of the planet's orbit, and they keep the same direction during its revolution. They accordingly present their northern and southern faces alternately to the sun, and also to the earth which is never more than 6° from the sun as viewed from Saturn. Twice during the sidereal period of 29.5 years the plane of the rings passes through the sun's position (Fig. 8.16), requiring nearly a year each time to sweep across the earth's orbit. In that interval our own revolution brings the rings edgewise to us from 1 to 3 times, when they disappear if viewed through small telescopes and are only very narrow bright lines through larger telescopes.

The latest edgewise presentation occurred from 1965 to 1966. The widest opening of the southern face occurred in 1974 and the next widest opening of the northern face will be in 1989. When the rings are open widest, their apparent breadth is 45 percent of their greatest diameter and one sixth greater than the planet's polar diameter. On these occasions Saturn appears brighter than usual, because the rings at this angle reflect 1.7 times as much sunlight as does the planet's disk.

Saturn's rings consist of solid particles that revolve like satellites around the planet in nearly circular orbits in the plane of its equator and in the direction of its rotation. They are mainly icy particles, as Kuiper's study of sunlight reflected from the rings suggests. The light from the separate pieces runs together at the great distance of Saturn to give the appearance of a continuous surface. We have noted already the changing reflectivity.

If the rings were really continuous, all parts would rotate in the same period, and the outside, having farther to go, would go around faster than the inside. The spectrum shows (Fig. 8.17), however, that the inside has the faster motion, as it should have in accordance with Kepler's harmonic law if the rings are composed of separate pieces.

The outer edge of the outer ring has the longest period of rotation, 14^h27^m. The inner edge of the bright ring rotates once in 7^h46^m, and the material of the crape ring must go around in still shorter time. Meanwhile the planet itself rotates in a period of about 10 hours. Thus the outer parts of the ring system move from east to west across the sky of Saturn, whereas the inner parts seem to go around from west to east, like Phobos in the sky of Mars.

Further evidence that the rings are not continuous is given on the rare occasions when a bright star passes behind the rings. Bright stars can be seen, almost undiminished in brightness, as they apparently move behind the rings.

8.11 The satellites of Saturn

Titan, the largest and brightest of Saturn's 10 known satellites, is

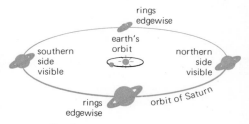

FIGURE 8.16
Diagrammatic explanation of the ring system of Saturn from above and below. Twice in the course of Saturn's revolution the rings are presented edgewise to the sun. Each time the plane of the rings requires about a year to sweep across the earth's orbit.

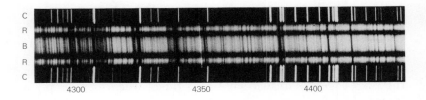

FIGURE 8.17
Spectrum of ball and rings of Saturn. In the spectrum of the ball of the planet, in the middle, the lines slant because of the planet's rotation. In the spectrum of the rings, above and below, the lines have the opposite slant. This shows that the rings revolve more rapidly at their inner edges, proving their discrete nature. (Lowell Observatory photograph.)

Oppositions of Saturn

6	Jan.	1975
20	Jan.	1976
2	Feb.	1977
16	Feb.	1978
1	Mar.	1979
13	Mar.	1980

considerably larger than the moon. It is one of the few satellites known to have an atmosphere (Io and Triton are the others). Titan's spectrum shows methane bands in Kuiper's photographs at McDonald Observatory. This satellite resembles Mars and Io in its reddish color, probably because of similar action of the atmosphere on the surface rocks. Titan has hydrogen in its atmosphere and at the surface its atmosphere is 10 times more dense than that of Mars. Its temperature, 93°K, is lower than predicted indicating that the greenhouse effect operates. Even against the low surface gravity, hydrogen is present indicating there must be a source to replenish the hydrogen lost to interplanetary space.

Four or five other satellites can be seen with telescopes of moderate size, appearing as faint stars in the vicinity of the ringed planet. All the satellites have direct revolutions around the planet with the exception of Phoebe, the most distant and the faintest one; Phoebe has retrograde revolution like Jupiter's outer group of satellites. Some of the satellites vary in brightness in the periods of their revolutions; they evidently rotate and revolve in the same periods, and are either irregular in form or have surfaces of uneven reflecting power. The very high reflection from the inner satellites suggests that they have icy surfaces, and their low densities may mean that they are composed mainly of ice. Presumably, the new satellite Janus, discovered recently by A. Dollfus, will exhibit the same characteristics as the other inner satellites.

URANUS AND NEPTUNE

The discovery of Uranus, in 1781, was accidental and unexpected. Herschel was examining a region in the constellation Gemini when he noticed a greenish object which seemed to him somewhat larger than a star. The object eventually proved to be a planet more remote than Saturn, and it received the name Uranus. Forty years later, when this planet had gone nearly halfway around the sun, its orbit was calculated from many observed positions, with

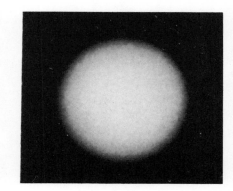

allowance for the disturbing effects of other known planets. After that, however, the new planet did not follow precisely the course it was expected to pursue. Astronomers finally concluded that its motion in the heavens was being altered by the attraction of another planet, Neptune, which was still more remote and as yet unseen.

Uranus is nearly 51,000 kilometers in diameter; it revolves once in 84 years at 19 times the earth's distance from the sun. It rotates once in less than 11 hours, having its equator inclined nearly at right angles to the ecliptic. Barely visible to the unaided eye, Uranus shows a small greenish disk through the telescope, on which the markings are not clearly discernible (Fig. 8.18). The temperature of the visible disk is 118°K. The spectrum includes a dark band in the infrared region observed by Kuiper and identified by Herzberg showing the presence of molecular hydrogen. This was the first direct evidence of molecular hydrogen in the atmospheres of the major planets. Bands of the contaminating methane appear in the spectra of both Uranus and Neptune. A haze appears in the upper atmosphere, especially around the poles.

Five satellites revolve around Uranus in nearly circular orbits in the plane of its equator and are therefore inclined nearly at right angles to the ecliptic. The orbits were presented edgewise to us in 1966 and will next appear flatwise in 1987. The 5th satellite, discovered by Kuiper at McDonald Observatory in 1948, is the faintest and nearest to the planet.

The existence of Neptune was predicted by Leverrier in 1846. By comparing observed positions of Uranus during the preceding quarter of a century with the predicted ones, he was able to calculate the place at that time of the unseen disturber in the sky. An astronomer at the Berlin Observatory, where an accurate star

FIGURE 8.18
Uranus as seen from Stratoscope II. The left picture is a composite of 17 frames. No markings are visible but a limb darkening effect is obvious. (Princeton University photograph.)

map was available, directed the telescope toward the specified region in the constellation Aquarius and soon found Neptune within a degree of the place assigned it by Leverrier. The discovery was acclaimed as a triumph for the law of gravitation, on which the calculation was based.

Neptune, being about 47,000 kilometers in diameter, is almost the same size as Uranus and revolves once in 165 years. It has a direct rotation taking 15.8 hours to complete, according to earlier spectroscopic measures. Using other means, however, O. Gruenther has recently reported a period of 12.7 hours. Although Neptune is invisible to the naked eye, with the telescope it appears as bright as a star of the 8th magnitude and shows a small greenish disk on which very weak markings have been seen. It seems to closely resemble Uranus.

Neptune has two known satellites. The first, Triton, is somewhat larger than the moon and is slightly nearer the planet than the moon's distance from the earth; it is nearly twice as massive as the moon and probably has an atmosphere. Triton has a retrograde revolution, contrary to the direction of the planet's rotation. The second satellite, discovered by Kuiper in 1949, is much smaller and more distant from the planet. It has a direct revolution in an orbit having an eccentricity of 0.76, the greatest for any known satellite.

PLUTO, THE MOST REMOTE PLANET

The discovery of Pluto was announced by Lowell Observatory on 13 March 1930, as the successful result of a long search at that observatory for a planet beyond Neptune. The planet was discovered by C. Tombaugh in his photographs taken in January of that year. The search had been instituted by Percival Lowell, who had calculated the orbit of a trans-Neptunian planet from slight discrepancies between the observed and predicted movements of Uranus, which seemed to remain after the discovery of Neptune.

Pluto is visible with the telescope appearing like a star of the 15th visual magnitude (Fig. 8.19). Its diameter is 5800 kilometers according to calculations by Kuiper with a disk meter on the 200-inch telescope, which is not contradicted by an upper limit of 10,700 kilometers assigned by I. Halliday from a near occultation of a star by Pluto. Unless the density is greater than would be expected, its mass cannot exceed a tenth of the earth's mass. Pluto is thought to have a gritty snow-covered surface of nitrogen on the morning quarter and an atmosphere that is considerably rarer than

ours. At Pluto's distance from the sun, the sun subtends less than
1 minute of arc, hence the surface temperature on the day side
does not exceed −210°C. Its period of rotation is 6.390 days, as
determined by M. F. Walker and R. H. Hardie from periodic
fluctuations in its brightness.

At its average distance the planet is 39.5 astronomical units, or
5910×10^6 kilometers, from the sun. It revolves once in 248 years,
which is half again as long as the period of Neptune's revolution.
Its orbit is inclined 17° to the ecliptic, the highest inclination for
any principal planet, so that Pluto ventures at times well beyond
the borders of the zodiac.

The eccentricity, 0.25, of Pluto's orbit is the greatest for any

FIGURE 8.19
*Small sections of the astrographic plates
used to discover Pluto. Note the significant
motion of the planet after only six days.
(Lowell Observatory photograph.)*

principal planet. On this account and because of the great size of its orbit, its distance from the sun varies enormously. At aphelion it is 2870×10^6 kilometers beyond Neptune's distance from the sun in its projected orbit, whereas at perihelion it comes 48×10^6 kilometers nearer the sun than the orbit of Neptune; yet in their present orbits the two planets cannot approach each other more closely than 385×10^6 kilometers. At the time of its discovery Pluto was near its ascending node (Figs. 7.5 and 8.19) and also near its mean distance from the sun. The distance will diminish until the planet reaches its perihelion in the year 1989. In Fig. 7.5 the part of the orbit south of the ecliptic is indicated by the broken line.

QUESTIONS

1 Explain how Mercury's elongations can differ by as much as 10°.
2 How do the phases of Venus and Mercury prove the heliocentric theory of the solar system?
3 One of the early applications of the telescope showed that at crescent phase, the crescent of Venus extended well around the edge of the planet. What can be deduced from this observation?
4 If the surface pressure of Mars is only 6/1000 that of the earth, how can the atmospheric pressure at 70 kilometers altitude exceed the earth's atmospheric pressure at that altitude above the earth?
5 What are the two types of radio noise from Jupiter? How do they originate?
6 Briefly describe how Roemer could obtain the velocity of light from the timing of eclipses of Jupiter's satellites.
7 Which three satellites have atmospheres? Two or three others might have atmospheres, which are they?
8 Describe the rings of Saturn. What observations show that the rings are not solid?
9 Show that an occultation of a star by Neptune can last almost 2 hours and 17 minutes. Show that a similar occultation by Pluto can last a little over 28 minutes.
10 From the information given on p. 170, calculate the circular velocity at Pluto's mean distance from the sun. What is the escape velocity from the solar system at this distance?

FURTHER READINGS

BRANDT, J. C. AND PAUL W. HODGE, *Solar System Astrophysics*, McGraw-Hill, New York, 1964.

KRAUS, J. D., *Radio Astronomy*, McGraw-Hill, New York, 1966.

MEHLIN, T. G., *Astronomy and the Origin of the Earth*, Brown, Dubuque, Iowa, 1971.

PAGE, THORNTON AND LOU WILLIAMS PAGE, EDS., *Wanderers in the Sky*, Macmillan, New York, 1965.

SLIPHER, E. C., *Mars*, Northland Press, Flagstaff, Arizona, 1962.

———, *The Brighter Planets*, John Hall, Ed., Northland Press, Flagstaff, Arizona, 1964.

WHIPPLE, FRED L., *Earth, Moon, and Planets*, 3rd ed., Harvard University Press, Cambridge, Mass., 1968.

9

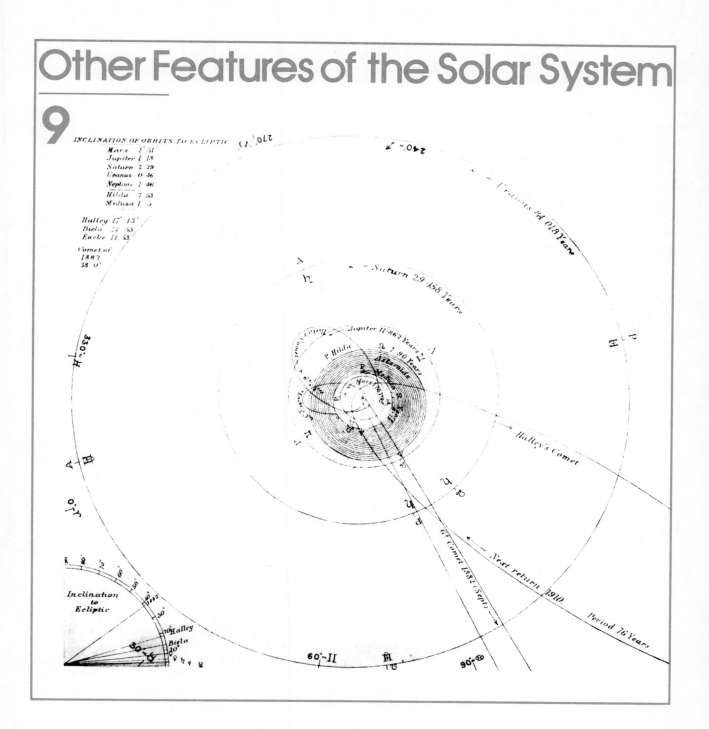

INCLINATION OF ORBITS TO ECLIPTIC

Mars	1° 51'
Jupiter	1 18
Saturn	2 29
Uranus	0 46
Neptune	1 46
Hilda	7 53
Medusa	1 5

Halley	17° 43'
Biela	12 33
Encke	12 53

Comet of
1882
38° 0'

Inclination
to
Ecliptic

9

Comets and meteors revolve around the sun in orbits that are generally more eccentric than those of the planets. Meteors are products of the disintegration of comets; meteor streams are associated with the orbits of comets. Meteorites, which are allied more closely with asteroids, come through the atmosphere to the earth, and very large ones produce meteorite craters. Comets, meteors, asteroids, the planets, and the sun originated from a great cloud of gas and dust.

A conspicuous comet visible to the naked eye has a head and a tail. The head consists of a hazy, globular **coma**, sometimes having a brighter **nucleus** near its center. The luminous **tail** is directed generally away from the sun and occasionally extends a considerable distance across the heavens. Many comets, however, are almost featureless telescopic objects. Spectacular comets to the unaided eye, such as Halley's comet, are infrequent.

9.1 Discovery of comets

Comets are likely to be discovered either in photographs of regions of the heavens often taken for other purposes or in visual searches with small telescopes. They are often discovered by interested amateurs, as was the case with the recent comet Ikeya-Seki (Fig. 9.1). The western sky after nightfall or the eastern sky before dawn are most promising for the search. A comet generally appears as a small hazy spot, and its gradual movement among the stars shows decisively that it is not a faint star cluster or nebula. Having found a comet, the observer should report its position, direction of motion, and brightness to the Central Bureau for Astronomical Telegrams, Smithsonian Astrophysical Observatory, Cambridge, Massachusetts, which serves as a receiving and distributing station in this country for such astronomical news.

As soon as three positions of the comet (its right ascension and declination) have been observed at appropriate intervals, a preliminary orbit is calculated. Then it is usually possible to decide from the records whether it is a new comet or an identified comet returning. A catalog of cometary orbits prepared by B. Marsden of the Smithsonian Astrophysical Observatory in 1972 includes 600 different comets. An average of five or six comets are picked up each year, and generally three or four of them have not been previously recorded.

Comets are provisionally designated by the year of their discovery followed by a small letter in the order in which the discovery is announced. An example is Comet Ikeya-Seki 1967n which was the fourteenth comet found in 1967. After the orbit is determined, the permanent designation is the year (not always the year of discovery) followed by a Roman numeral in order of perihelion passage during that year. Thus Comet Ikeya-Seki cited above became Comet 1968 I, the first comet to pass its perihelion in 1968. Many comets are also known by the name of the discoverer, or discoverers, such as comet Kohoutek (1973f) or Halley's comet.

FIGURE 9.1
Comet Ikeya-Seki photographed by H. Giclas in the morning twilight. Note the spiral like tail structure. (Lowell Observatory photograph.)

FIGURE 9.2 [BELOW AND NEXT PAGE]
*Two pictures of Halley's comet in 1910.
Note the motion of the comet and the
change in tail structure. (Lick Observatory
photograph.)*

Halley's comet (Fig. 9.2) is named in honor of Edmund Halley, a contemporary of Isaac Newton, who predicted its return. Halley calculated the orbit of a bright comet of 1682 to be a parabola and noted its resemblance to the orbits of comets of 1531 and 1607, which he had also determined from records of their observed places in the sky. Deciding that these were three appearances of the same comet, which must therefore be revolving in an ellipse, he recalculated the orbit and predicted its return to the sun's vicinity "about the year 1758." The comet was sighted on Christmas night of that year and reached perihelion early in 1759. It came around to perihelion again in 1835 and in 1910, its latest appearance.

Twenty-eight returns of this comet are identified from the records as far back as 240 B.C. It was Halley's comet that appeared in the year 1066 at the time of the Norman conquest of England and is depicted on the Bayeux tapestry. The intervals between returns to perihelion have averaged 77 years, varying a few years because of disturbing effects of the planets.

Halley's comet is not only the first periodic comet to be recognized, but it is also the only conspicuous one of the many periodic comets known today that return to perihelion more often than once in a century. It has a retrograde revolution around the sun in an elongated orbit (Fig. 9.3) at distances from the sun that range from half to more than 35 times the earth's distance. Invisible near its aphelion, which is more remote than Neptune's distance, it is now located between Uranus and Neptune and will return to the sun's vicinity in 1986.

9.2 The orbits of comets

The orbits of comets depart from the regularities we have noted (see Section 7.10) in the case of the principal planets. These orbits are generally of high eccentricity and are often much inclined to the ecliptic. With respect to their motions the comets are divided into two groups by a somewhat indefinite dividing line.

1 Comets having nearly parabolic orbits. In this more numerous group the orbits are so eccentric that they are not readily distinguished from parabolas in the small portions near the sun where the comets are visible. For example, the eccentricity of the orbit of comet Kohoutek was 0.999. The orbits extend far beyond the region of the planets, and the undetermined periods are all so long that only one appearance of each comet is likely to be found in the records. In this sense the comets are "nonperiodic." About half the revolutions are direct, that is, west to east (comet Ben-

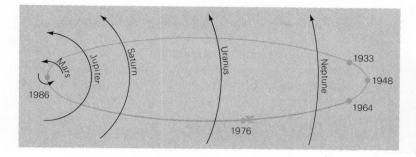

nett, 1969i), and the other half are retrograde (comet Abe, 1970g). Comet Bennett's period is about 1700 years and its aphelion distance is about 300 AU.

2 Comets having definitely elliptical orbits. These "periodic comets" revolve in periods not exceeding a few hundred years. The orbits are allied more closely with the planetary orbits. Although most of their orbits have a large eccentricity, they are frequently more moderately inclined to the ecliptic, and the revolutions are mainly direct. The retrograde revolution of Halley's comet is one of the exceptions, however.

Comet Schwassmann-Wachmann 1 (1925 II) is unusual in having a nearly circular orbit. This comet revolves around the sun once in about 16 years in an orbit somewhat larger than that of Jupiter. It is also unusual in its occasional surprisingly great and rapid flare-ups, which have been attributed to expansions of a dusty coma; an increase of 100 times in brightness has occurred within less than a day.

Two dozen or more comets revolve around the sun in periods averaging 6 years, or half of Jupiter's period. Their aphelions and one node of each orbit are not far from Jupiter's orbit, so that these comets can come close to the planet itself. They constitute **Jupiter's family of comets** (Fig. 9.4). Their direct revolutions and the low inclinations of their orbits to the ecliptic suggest that Jupiter has assembled the family by capture of comets passing by in originally larger orbits. At successive encounters the planet's attraction has progressively reduced the orbits to their present sizes. The membership is unstable; further approaches of these comets to the planet may occur so that some of them will be removed from the family much as Pioneer 10 has been ejected from the solar system (see Section 8.4). Three members of Jupiter's family have been especially noteworthy.

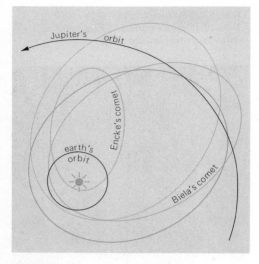

FIGURE 9.4
Orbits of four comets of Jupiter's family. Encke's comet has the smallest orbit of all comets.

Originally discovered in 1786, comet Encke was the first member of Jupiter's family to be recognized in 1819. Its period of revolution, 3.3 years, is the shortest for any known comet. Its aphelion is a whole astronomical unit inside Jupiter's orbit, having gradually been drawn in by this amount. The 49th appearance of the comet around perihelion was recorded in April 1974. Like other members of the family, comet Encke never becomes more than faintly visible to the unaided eye.

Biela's comet, having a period of 6.5 years, came to an end in a spectacular way. At its return in 1846 it was divided into two separate comets traveling side by side, and at the next return the separation had increased 2.4×10^6 kilometers. The comet was never seen again, but a stream of meteors in its orbit, the Andromedids or Bielids, gave fine showers when it encountered the earth in 1872 and 1885. The Giacobini-Zinner comet, having a similar period, is associated with a meteor stream that in 1933 and 1946 provided the most abundant showers of this century.

9.3 The nature of a comet

A comet's nucleus is a conglomerate of frozen material, mainly methane, ammonia, and water, having meteoric particles and dust embedded in it. In this theory, proposed by F. L. Whipple in 1950, the nucleus of the average comet does not exceed a kilometer or two in diameter. It is known that the mass of a comet is too small to disturb appreciably the motions of planets and satellites at close approaches, whereas the comet's orbit may be greatly altered by the attractions of these bodies. Another theory by C. R. O'Dell postulates that small porous particles in the outer reaches of the solar system are coated by gases frozen out of the original solar nebula. These particles are not much more than 150 microns in diameter and collect to form the nucleus of a comet.

Some of the ices evaporate at each approach of the comet to the sun. The gases issue explosively into the coma; they are then swept out through the tail by radiations from the sun and dispersed into space. By action of sunlight these gases are transformed in the coma to carbon, methane, hydroxyl, ammonia radicals, and cyanogen, as the spectra of the comets show, and are soon converted in the tail to more durable (stable) molecules, such as carbon monoxide, carbon dioxide, and nitrogen. As the surface of the ices in the nucleus is made increasingly gritty by the meteoric particles that remain, evaporation is retarded; thus the comet may return to the sun many times before its material is completely dissipated.

The tails of comets are of three types:

1 Gaseous tails that glow with blue fluorescence stimulated by the sun's radiations. These are directed almost straight away from the sun by the pressure of its radiations and are generally more conspicuous for comets relatively near the sun.

2 Dusty tails that glow by reflected sunlight. These are curved and are likely to depart more from the direction opposite the sun's position as the comets are farther away.

3 Material fanning out behind the revolving comets (Fig. 9.5) consisting of meteoric particles originally embedded in the ices of the nuclei, which have been released by evaporation of the ices and are being scattered along the comets' orbits as meteor streams.

Most comets have tails of both types 1 and 2, but it is the dust tail that is most prominent because it appears as a long smooth arc. For example, comet Bennett's dust tail was fully 20° long. The gas tail tends to twist and sometimes gives a corkscrew appearance.

The brightness and other observed features depend on the comets' distances from the sun and earth, and on their masses and compositions as well. Earlier physical knowledge of comets was derived mainly from visual observations of relatively nearby objects. Present information is also based on surveys of more remote comets photographed with large telescopes. Some of the more distant comets have characteristics that might not have been expected previously, as Elizabeth Roemer points out. Thus the photographs of comet Humason (Fig. 9.6) at the heliocentric distance of 2.5 AU show that the comet was remarkably active structurally and that its light was blue. When the comet reached 1.2 AU another abrupt change in brightness and activity occurred.

Comets on parabolic orbits presumably are remnant material from the original solar system nebula according to J. Oort. This theory states that comets with definitely elliptical orbits have been "caught" on their journey to the sun, and hence show fewer and fewer gaseous emissions, as is observed. If this is so, then the parabolic comets come from the primordial material, and we can learn a lot about the early nebula by "catching" a comet. According to Oort's theory there is a considerable residue of nebular material at a distance of about 5×10^4 AU where the nuclei of comets form. The temperature there is only about 30°K, thus allowing the ices and frosts required by the theories of Whipple and O'Dell. Passing stars disturb the orbits of the nuclei gravitationally and cause their long journeys into the inner solar system.

FIGURE 9.5
Comet Arend-Roland (1956h). The meteoric material fanning out below the comet's
head gives the appearance of a spike. (Lowell
Observatory photograph.)

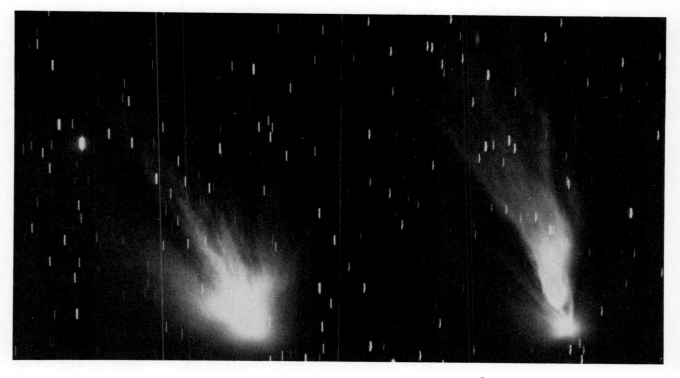

FIGURE 9.6
Comet Humason in 1962. Note activity in tail in one day at more than 2 AU from sun. This remarkable photograph also shows the daily motion of the comet even at that great distance. (Official U.S. Navy photograph.)

9.4 Meteors and their orbits

The trails of "shooting stars" across the starlit sky inform us of the flights of meteors through the air. Meteors are stony and metallic objects revolving around the sun. The majority have the size of dust particles or grains of sand. Those that are conspicuous to the unaided eye may be considerably larger. Meteors can be seen separately only when they happen to plunge through our atmosphere, where they are heated intensely by collision with air molecules. During the fraction of a second in which they are consumed, meteors produce luminous **trails.** The brighter meteors sometimes leave **dust trains** that remain visible from a few seconds to generally not longer than half an hour, and they may become twisted by air currents. Unusually bright meteors are known as **fireballs.**

The total number of meteor trails that would be visible to the unaided eye over all the earth's surface is of the order of 100 million a day. The number visible to a single observer, however, is usually very small. One must then add to this number many millions of meteors invisible to the eye, but revealed by radar and radio techniques that "look" at the ionized wake these meteors leave. It is estimated that the incoming meteors all together add 10^4 metric tons a day to the mass of the earth.

More meteors are observed between midnight and sunrise, since we are then on the forward side of the revolving earth. In the evening we are on the rear side and only those meteors able to overtake the earth can be seen.

When the same meteor trail is observed from two stations several miles apart, it is possible to determine the distance of the meteor from the stations. From the observed angular speed and direction of the meteor's flight the linear velocity is then obtained, providing enough data for calculating the meteor's orbit around the sun. For this purpose earlier visual methods have been replaced by the more precise photographic and radar techniques.

The first effective photography of meteor trails from two stations employing cameras provided with rotating shutters for interrupting the images of the trails at regular intervals to facilitate the timing was initiated by F. L. Whipple and associates at Harvard Observatory in 1936. The later extension of this program there and elsewhere with pairs of "super-Schmidt" cameras have produced double photographs of thousands of meteor trails and the calculation of many orbits of meteors that were members of the solar system and did not come from outside the system.

The meteor's speed alone is enough to determine the status of the meteor in this respect. If the speed on entering the atmosphere, with allowance for the motion and attraction of the earth, exceeds 11.2 kilometers per second, the meteor may have come from outside the solar system. This is the speed a body falling from infinity would have when it hit the earth, which means also that this is the speed needed to escape the gravitational pull of the earth. If the speed is less than this critical value, the meteor is a member of the system. The speeds of more than 10,000 meteors down to the 8th visual magnitude were determined by D. W. R. McKinley at Ottawa from records of radio beams returned from the trails. In not a single case was the speed great enough to suggest definitely that a meteor came from outside the system.

9.5 Meteor streams and showers

A **meteor stream** consists of many meteors revolving around the sun in about the same orbit, at least when they are in the earth's vicinity. A **meteor shower** occurs where the orbit of the stream

FIGURE 9.7
The radiant of a meteor shower. The drawing of a meteor shower shows how parallel moving objects appear to radiate from a point called the radiant. The 1966 Leonid meteor shower is shown in the photograph. The presence of nonshower meteors convincingly demonstrates the shower. Stars in the picture can be identified with the aid of star maps in Chapter 11. (Photographs by Dennis McLean.)

crosses the earth's orbit at either one or two points and whenever part of the stream and the earth arrive together at an intersection. The shower occurs around the same date, either annually if the stream is far extended or at longer intervals for a short stream. Only rarely have the showers been spectacular enough to attract the attention of people who are not watching for them.

The trails of the meteors in a shower are directed away from a small area of the sky, the center of which is the **radiant** of the shower. Because the trails are nearly parallel, they spread out over the sky from the radiant (Fig. 9.7), just as the parallel rails of a track seem to diverge from a distant point. Showers of meteors and the streams that produce them are named from the positions of the radiants among the constellations at the heights of the displays. Examples are the Perseids and the delta Aquarids. The place of the radiant is shifted during the progress of a shower as the earth changes the direction of its revolution.

The more abundant meteor showers are listed in Table 9.1, which is taken principally from more extended data by F. L. Whipple and G. S. Hawkins in *Handbuch der Physik*, Vol. 52, 1959. In each case the table gives the date of maximum display in universal time, the position of the radiant, so that it may be located in the star maps, and the name of the parent comet.

In August the Perseids provide the most familiar of the annual showers; the display from this wide stream extends through two or three weeks and the trails are frequently rather bright. The showers of the Orionids and Geminids are also among the most faithful of the annual showers. The radiant given for the delta Aquarids in late July is the mean of two radiants several degrees apart, and the same is true for the Taurids of November. Both streams make second crossings to produce the daytime showers of the Arietids and Beta Taurids, respectively.

The Leonids and Andromedids, which produced fine showers at intervals in the nineteenth century, have since been less conspicuous at the predicted times, although in 1966 the Leonids put on a fine show over the western portion of North America and the Pacific. Showers of Draconids in the evenings of 9 October in 1933 and 1946 were the most impressive of the present century, and an afternoon display on that date was recorded by radar at the Jodrell Bank Experimental Station in England in 1952. Three previously unknown showers, listed in the table, were also recorded in the daytime at this station.

TABLE 9.1

Meteor showers and associated comets

Shower	Maximum display (UT date)	Radiant at max. (equinox of 1950) R.A.	Decl.	Associated comet
Quadrantids	3 Jan.	15^h 20^m	$+48°$	
Lyrids	21 Apr.	18 0	$+33$	1861 I
η Aquarids	4 May	22 24	0	Halley (?)
Arietids[a]	8 June	2 56	$+23$	(=δ Aquarids)
ζ Perseids[a]	9 June	4 8	$+23$	
β Taurids[a]	30 June	5 44	$+19$	Encke
δ Aquarids	30 July	22 36	-11	(two streams)
α Capricornids	1 Aug.	20 36	-10	1948 n
Perseids	12 Aug.	3 4	$+58$	1862 III
Draconids	10 Oct.	17 36	$+54$	Giacobini-Zinner
Orionids	22 Oct.	6 16	$+16$	Halley (?)
Taurids	1 Nov.	3 28	$+17$	Encke (two streams)
Andromedids	14 Nov.	1 28	$+27$	Biela
Leonids	17 Nov.	10 8	$+22$	Temple
Geminids	14 Dec.	7 32	$+32$	
Ursids (Ursa Minor)	22 Dec.	13 44	$+80$	Tuttle

[a] Shower in daytime.

9.6 The zodiacal light

The triangular glow of the **zodiacal light** can be seen extending up from the west horizon after nightfall in the spring and from the east horizon before dawn in the autumn in our northern latitudes. Broadest and brightest near the horizon it tapers upward, leaning toward the south (Fig. 9.8). The glow is nearly symmetrical with the ecliptic and is most conspicuous when the ecliptic is nearly vertical.

Near the equator, where the ecliptic is nearly perpendicular to the horizon, the zodiacal light can be observed all year round. Here it is said to have been seen extended as a faint, narrow band encircling the sky. The light is mainly sunlight scattered by meteoric dust, which forms a ring around the sun in the plane of the earth's orbit, and by disintegration of comets. Near the sun the light is identified by a replica of sunlight in the spectrum of the corona.

The **gegenschein,** or **counterglow,** is a faint, roughly elliptical glow in the sky extending about 20° along the ecliptic and centered nearly opposite the sun's position. Barely visible to the unaided eye in favorable conditions, this glow is observed more clearly when photographed with very wide angle cameras or when recorded with the photoelectric cell. It is believed to be somewhat

FIGURE 9.8
The zodiacal light and comet Ikeya-Seki, 31 October 1965. The bright star at the top is Regulus in the constellation of Leo. Note how much brighter the comet's tail is as compared to the zodiacal light. Use the star maps in Chapter 11 to locate the ecliptic. (Photograph by H. Gordon Solberg, Jr., Las Cruces, New Mexico.)

variable in form and position and can be easily observed from above the earth's atmosphere. Among several proposed interpretations, the gegenschein has been tentatively attributed to sunlight scattered by a dust tail of the earth.

METEORITES AND METEORITE CRATERS

9.7 Falls of meteorites

Near noon one day in November 1492, a number of stones came down in a field near Ensisheim, Alsace. The largest one, weighing 118 kilograms, was placed in a church in that town; a smaller stone is exhibited in the Chicago Natural History Museum. This is the oldest observed fall of meteorites on record of which samples are still preserved.

The idea that stones fall from the sky goes back to very early times. There were stones preserved in some of the ancient temples that were doubtless of celestial origin, and these "stones from heaven" were objects of veneration. In later times, however, all reports of stones falling from the sky came to be regarded with suspicion. The stones seemed to choose remote places where there were no reliable observers. It may be, too, that the accounts of terrified spectators of some of the falls were so exaggerated that no one could believe them. Finally, in April 1803, a shower of two or three thousand stones fell at Laigle, France, and it was reliably reported. Yet the news spread so slowly that when 136 kilograms of meteorites came down near Weston, Connecticut, in December 1807, the first observed fall on record in the United States, many people were reluctant to believe that it was true.

Meteorites are masses of stony or metallic material, or both, which survive their flights through the air and fall to the ground. They arrive either singly or in many pieces. Several thousand individual pieces have come down in one fall, and in such cases they are likely to be distributed over an elliptical area having a major axis in the direction of the flight that is several kilometers long. Although meteors are products of the disintegration of comets, many meteorites are believed to be fragments of shattered asteroids (see Section 8.7).

Their speeds greatly reduced by air resistance, most meteorites cool before they reach the ground. In their brief flights through the air the heat has not gone far into their cold interiors, and the melted material has been swept away in droplets from their surfaces. They are usually cool enough to be handled comfortably when they are picked up immediately after landing and they do not penetrate far into the ground. Larger meteorites are less impeded by the air; some very massive ones have struck at such high speeds that they have blasted out large craters in the earth's surface (see Section 9.9).

9.8 Composition of meteorites

The meteorites themselves are essentially of two kinds, the stones (aerolites) and the irons (siderites). There are gradations between them (siderolites) from stones containing flecks of nickel–iron to sponges of metal with stony fillings. Inside their smooth, varnishlike fusion crusts the **stony meteorites** are often grayish, having a characteristic granular structure that serves to establish their celestial origin. The rounded granules are crystalline, chiefly silicates similar to those in our native igneous rocks. The largest known example, weighing at least a metric ton, fell on 18 February 1948, in Furnas County, Nebraska.

Iron meteorites are silvery under their blackened exteriors. They are composed mainly of alloys of iron and nickel, which are affected by acid in various degrees. When they occur in crystal forms, a characteristic pattern of intersecting bands parallel to the faces of an octahedron may be etched with dilute nitric acid on a polished section.

Individual meteorites from about 1600 falls have been recovered. They are generally named after the locality in which they were found; examples are the Canyon Diablo, Arizona, meteorites and the Willamette, Oregon, meteorite. Collections are exhibited in the Chicago Natural History Museum, the American Museum of Natural History in New York City, the Arizona State University at Tempe, and many other places.

Micrometeorites are particles of meteoritic dust, so small that they are not much altered when they fall through the atmosphere. In addition to the ones found on the ground, many micrometeorites have been collected in the opened nose cones of rockets and have been recovered for examination. These samples range from rather compact spheres to irregular dustballs.

Anyone finding a newly fallen meteorite is urged to place it into an inert atmosphere if possible and notify the Smithsonian Institution immediately. These objects have been sampling interplanetary space for eons and, hence, are excellent probes of regions our spacecraft will visit only infrequently.

About 35 individual meteorites weighing more than a metric ton are listed in F. C. Leonard's catalog. With the exceptions of the Furnas County stone, two stony irons, and a 1.8 metric ton iron individual from a Siberian fall in 1947, their falls were not observed. All the others are irons; the two largest are the Hoba and the Ahnighito meteorites (Fig. 9.9).

The Hoba meteorite lies partly buried in the ground in the Grootfontein district, Southwest Africa. Its rectangular upper surface measures 2.7 × 3.05 meters, its greatest thickness exceeds 1 meter, and its weight is unknown. The Ahnighito meteorite is the largest of three that the explorer R. E. Peary found near Cape York, Greenland, in 1894, and brought back to New York City. This meteorite measures about 3.35 × 2.13 × 1.83 meters and weighs a little more than 30.9 metric tons; it is exhibited in the American Museum–Hayden Planetarium. The Willamette meteorite, also in the Hayden Planetarium, weighs 13.6 metric tons. The largest meteorite found in the United States, this conical mass of nickel–iron was discovered in 1902, 16 kilometers south of Portland, Oregon. It evidently kept the same orientation in its flight and was fashioned by the rush of hot air. Some meteorites turned over and over as they fell and were rounded in the air.

Three large iron meteorites, each weighing more than 9.1 metric tons, were found in Mexico. They are the Bacubirito (26.3 metric tons), the Chupaderos (19 metric tons, in two pieces that fit together), and the Morito (10 metric tons). The last two are exhibited in the School of Mines in Mexico City.

9.9 Meteorite craters

The Barringer meteorite crater, near Canyon Diablo in northeastern Arizona, is a circular depression, 1.3 kilometers across and 174 meters deep (Fig. 9.10). Its rim, which rises 40 meters above the surrounding plain, is composed of debris thrown out of the pit from fine rock dust to blocks of limestone and sandstone weighing up to 6.3×10^3 metric tons apiece.

This crater is a scar left by the fall of a great meteorite probably not less than 50,000 years ago. The meteorite is estimated, at the minimum, to have had a diameter of 61 meters and a weight of 9×10^5 metric tons. It was only slightly retarded by the air because it was so massive, striking the earth with a mighty blow. The intense heat of the collision partly fused the meteorite and the rocks in contact with it; the gases expanded explosively, scattering what was left of the meteorite over the surrounding country and blasting out the crater. Within a radius of 9.6 kilometers around the crater, 27 metric tons of meteoritic iron have been picked up. The largest individual piece, weighing more than 640 kilograms, is exhibited in the museum at the north rim of the crater. Samplings indicate that the total amount of crushed meteoritic material around the crater is 11,000 metric tons.

FIGURE 9.9
The Ahnighito meteorite displayed on special scales designed by the Toledo Company to weigh this massive meteorite, 68,085 pounds. (American Museum-Hayden Planetarium photograph.)

FIGURE 9.10
Barringer Meteorite Crater near Winslow, Arizona. (Photograph by Meteor Crater Society, Winslow, Arizona.)

The Wolf Creek crater in West Australia, having a diameter of 0.85 kilometer at the bottom and a depth of 49 meters, is second in size among the readily recognized meteorite craters. These craters were produced by impacts of meteorites within the last million years. A number of "fossil craters" of suspected meteoritic origin have been detected in aerial photographs of Canada. Examples are the Brent and Holleford craters in Ontario; they are a few kilometers in diameter and their ages are estimated as 500 million years. It is not inconceivable that during the earth's history it has been hit by innumerable large meteorites. If it were not for the weathering and erosive action of wind and water, the land areas might well look like the moon and Mars. As we learn more about the moon's impact craters we will learn more about our own.

The geologist R. S. Dietz has pointed out that the shock generated by a meteorite in producing a large terrestrial crater transcends any volcanic or other earthly explosion. Where the surface structure of a region has become inconclusive with age and where no meteoritic remnants are reported, Dietz suggests that the work of such superintense shock waves may be recognized by the presence of a fracture pattern in the rocks known as shatter cones, and by a form of silica named coesite that is also produced under extremely high pressure. Examples of regions having one or both of these features are the Giant Kettle in southern Germany, the Wabar craters of Arabia, and the Vredefort Ring in the Transvaal of South Africa. The latter formation is 210 kilometers in diameter, comparable in size with the largest lunar craters.

THE ORIGIN OF THE SOLAR SYSTEM

Theories of the origin of the sun's planetary system are related to the problem of the origin of the sun itself and of stars in general. According to current opinion, which is described in Chapter 15, the stars evolve from contracting masses of cosmic gas and dust. Planetary systems are believed to develop around many stars. The problem of the origin of these systems refers specifically to the solar system, because this is the only known system of its kind.

A number of theories, some of them also known as hypotheses, have been proposed during the past two centuries as solutions to the problem. The simpler accounts of earlier times did not completely represent the more complex system that is recognized today. First we shall review the features that these theories try to explain. As examples of the earlier and later theories we consider particularly the nebular hypothesis of Laplace, the protoplanet hypothesis of Kuiper, and a newer hypothesis by A. G. W. Cameron and M. R. Pine.

We have noted previously the principal features of the solar system (see Section 7.10): the orbits lie in a plane, the orbits are nearly circular, the revolutions of the planets are in the same direction as the rotation of the sun, the sun rotates very slowly with its equator in the plane of the planets (Chapter 10), and the rotations of the planets with one exception are in the same direction. We note that the satellites lie generally in the plane of rotation of their planets and revolve in the direction of their planet's rotation, although there are a number of exceptions to this point.

Other features that may be important are the curious Titius-Bode relation (see Section 7.9), the terrestrial and gaseous planet division, the random distribution of the nonperiodic comets, and various tidal locks between the planets. We note also that the planets carry 98 percent of the angular momentum of the system. Another interesting feature common to the terrestrial planets including the moon is that the northern hemispheres are different from the southern hemispherse. We must admit however, that at the present stage of our knowledge we do not know what is or is not important, so theories proposed must try to explain as many features as possible.

9.10 The nebular hypothesis

The nebular hypothesis of the origin of the solar system was presented in 1796 by the mathematician P. S. Laplace. It was the most famous although not the first of the early theories. The philosopher Kant had thought of a nebular origin of the system half a century earlier.

Laplace's account began with a gaseous envelope surrounding the primitive sun and in slow direct rotation around it. As the envelope contracted toward the center, it rotated faster and accordingly bulged more at its equator. At length a critical stage was reached when the rotation became fast enough to make the centrifugal effect at the equator as great as the attraction toward the center. An equatorial ring of gas was then abandoned by the contracting envelope. Smaller rings were left behind successively as often as the critical stage was repeated. Each ring assembled gradually (Fig. 9.11) into a gaseous globe having its circular orbit around the sun the same as the ring from which it was formed. Some of the globes developed satellites in a similar manner as they condensed into planets. The mass of gas remaining at the

FIGURE 9.11
*Conceptual drawing of the condensing solar
nebula.*

center and containing most of the original material became the
sun.

Because of the simplicity of the account and the authority of
Laplace concerning the mechanics of the solar system, the nebular
hypothesis held a leading place among the scientific theories of
the nineteenth century. As its deficiencies were gradually recognized, they gave warning of situations to be avoided in later
theories. The objections to the nebular hypothesis are mainly as
follows:

1 The tendency of gases to disperse would scarcely have permitted the rings to assemble into planets.

2 The system developed by the theory would have been unlike
the present solar system in the distribution of **angular momentum**.
For each member this value is the product of its mass, the square
of its distance from the center of its revolution or the axis of its
rotation, and the rate of its angular motion. A principle of mechanics states that the sum of these products for all members of
an isolated system must always remain the same. According to
this principle, the nebular hypothesis required that the greater
part of the spin of the solar system should appear in the sun's
rotation, where only 2 percent is actually found.

3 The motions of smaller members of the solar system discovered since Laplace's time have shown many exceptions to the

regularities required by his theory. Moreover, the compositions of the planets and their atmospheres were not considered in the theory, because they were unknown at that early time.

9.11 Some later theories

About the year 1900, the geologist T. C. Chamberlin and the astronomer F. R. Moulton proposed the planetesimal hypothesis as a substitute for the nebular hypothesis. They supposed that another star passed close enough to cause much gas to emerge from the sun. The gaseous envelope thus produced around the sun condensed into small solid particles, called planetesimals, which eventually assembled to form members of the planetary system. When this idea and an alternate tidal hypothesis were considered less attractive, the interest of scientists reverted to the gaseous nebula already surrounding the primitive sun as a more promising approach to the evolution of the system.

In 1945, the physicist C. F. von Weizsäcker proposed a revised version of the nebular hypothesis. The solar nebula was assigned a composition mainly of the light chemical elements, like the composition of the sun, and an original mass equal to a tenth of the sun's mass. The division of the nebula was achieved not by shedding of successive rings, as in Laplace's theory, but by turbulent vortices produced by rotation of the nebula. Similar features are included in Kuiper's protoplanet hypothesis presented at about the same time.

9.12 The protoplanet hypothesis

The protoplanet hypothesis was proposed by G. P. Kuiper about 1950. Kuiper pointed out that the solar system might almost be regarded as a degenerate double star, in which the smaller mass was spread out to form a planetary system instead of condensing into a single star. Kuiper determined that the **protoplanets,** or fragments of the solar nebula, would be free from tidal disruption by the sun if the original nebula had at least 100 times the combined mass of the present planetary system. He also showed that the distances of the protoplanets from the center of the system would conform to Titius-Bode relation (see Section 7.9).

According to this theory, the heavier material of the protoplanets settled gradually to form solid cores. The lighter material, mainly hydrogen and helium, formed extensive atmospheres around the cores. At first, the primitive planets and their atmospheres contained excessive amounts of material. The giant planets had from 10 to 100 times their present masses. The protoearth was 1000 times as massive as the present earth. In the meantime the sun, which was originally large and cold at its surface, had become smaller and very hot. Its powerful radiations drove the remnants of the nebula and much of the protoplanets' atmospheres out of the system, just as tails of comets are repelled from the sun today. Jupiter and Saturn with their strong attractions were able to retain a considerable part of their hydrogen. Uranus and Neptune retained more limited amounts. The inner planets were stripped to little more than their cores.

In this theory we satisfy the conditions that the principal planets have direct revolutions around the sun, the rotation of the nebula is in the original direction, the planetary orbits are nearly circles and nearly in the plane of the flattened nebula's equator. At the start, these planets were forced by solar tides raised in them to rotate in the same direction and in the same periods as those of their revolutions around the sun. Later, as they contracted further, all except Mercury and Venus slipped away from the tidal brakes and began to rotate in shorter periods but generally in the same direction as before.

The satellites, as explained by the protoplanet hypothesis, developed by a repetition of the planet-forming process. In its contraction each protoplanet of sufficient mass left behind a flat disk of gas, like a small-scale solar nebula rotating around it. Protosatellites were formed by the breaking up of each nebula. Influenced by the rotations of their planets, many of the resulting satellites conform to the regularities of the planets themselves. These "regular" satellites have never succeeded in slipping away from the tide brakes because they are relatively near their planets. Thus they rotate in the same period in which they revolve around their primary, keeping one face toward their planet, as the moon does around the earth.

Although the moon in its motions is one of the regular satellites, it is unique among the satellites in its near equality in mass with its primary. The earth and the moon are more like a double planet and may have grown up together out of the solar nebula. Lunar expeditions have shed some light on this possibility.

A dozen satellites are "irregular." Their orbits are more inclined to the ecliptic or are more eccentric, or both, than the others. Half of them have direct and half have retrograde revolutions. Kuiper explains that these and others withdrew from the control of their planets of diminishing masses and began to move independently in orbits like those of their planets. The 12 irregular satellites were later recaptured with the aid of the nebula remaining around the planets. Other escaped satellites were not recaptured, but eventually returned close enough to be diverted into different orbits. The suggestion is made that Hidalgo and the Trojan asteroids were originally satellites of Jupiter, and that Pluto formerly belonged to Neptune's family.

Certain aspects of the protoplanet hypothesis make it less attractive now than it was 25 years ago. Our ideas about the origin and evolution of stars have changed, and we now know much more about the interstellar medium.

A new general, comprehensive theory has been proposed by A. G. W. Cameron and M. R. Pine that retains some of the general features of the protoplanet hypothesis but removes many of the objections. Cameron and Pine immediately take the separation of the terrestrial and gaseous planets as significant and explain this occurrence in a protostar nebula. The boundary of the primitive sun extended out as far as the asteroid belt. All of the planets developed through a fortuitous random congregation of rocky and nebular material, which within the dense nebula can only occur at certain intervals depending, in a complicated way, upon the density of the nebula, gravitational interactions, the temperature gradient in the nebula, and many other factors. This explains the Titius-Bode relation.

Cameron and Pine explain the formation of regular satellites of the major planets by a similar process in the protoplanet nebula. They point out that these satellites have regular spacing obeying a relation analogous to the Titius-Bode relation, but bearing no resemblance. Such relations are highly complex and depend upon the local conditions so that no general law can be stated that will predict the proper relation in advance. It is interesting that in this theory the terrestrial planets are unlikely to have regular satellites. They consider the moon and the Martian satellites to be irregular satellites, that is, captured bodies.

The Cameron-Pine hypothesis is so new it is hard to be objective in reviewing it. It is based much more strongly upon mathematics and the details of thermodynamics, geophysics, cosmochemistry, and prior theories, therefore promising to be a major stepping stone toward our total understanding of the origin and evolution of the solar system.

QUESTIONS

1 Explain how you would search for comets with some hope for success.
2 What are the two groups of comets?
3 When do we expect perihelion passage of the 49th return of Encke's comet?

4 Explain Whipple's theory of the nature of a comet. How does Oort's theory of the origin of comets support this?
5 Why are radar and radio techniques such effective tools in meteor astronomy?
6 Why do the authors urge anyone finding a fresh meteorite to notify the appropriate authorities immediately?
7 Why are large meteors seldom destroyed by their passage through the earth's atmosphere?
8 We believe meteorite impacts were much more frequent during the early history of the earth and moon. Why?
9 Describe the protoplanet hypothesis.
10 There are several details of the protoplanet hypothesis that do not agree with facts given in Chapters 6 through 9. List at least three.

FURTHER READINGS

BERLAGE, H. P., *The Origin of the Solar System*, Pergamon, Elmsford, N.Y., 1968.

HARTMANN, W. K., *Moons and Planets*, Bogden & Quigley, Tarrytown, N.Y., 1972.

HAWKINS, G. S., *Meteors, Comets, and Meteorites*, McGraw-Hill, New York, 1964.

MCKINLEY, D. W. R., *Meteor Science and Engineering*, McGraw-Hill, New York, 1961.

PAGE, THORNTON, AND LOU WILLIAMS PAGE, EDS., *The Origin of the Solar System*, Macmillan, New York, 1966.

SASLAW, W. C. AND K. C. JACOBS, EDS., *The Emerging Universe*, Univ. of Virginia Press, Charlottesville, 1972, Chap. 2.

WOOD, J. A., *Meteorites and the Origin of the Planets*, McGraw-Hill, New York, 1968.

The sun as depicted by Father Kircher in 1682. Note that some liberty with the facts was taken in the placement of sunspots.

The Sun with Its Spots

10

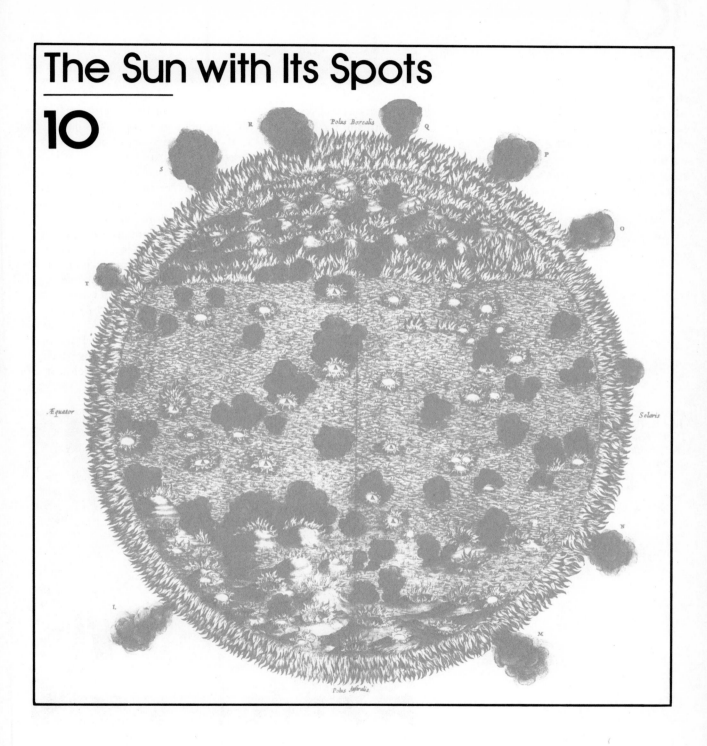

10

The sun is the dominant member of the solar system. It is also the only star near enough to us for its features to be examined in detail. Our account of the sun is accordingly associated both with the preceding descriptions of the solar system and with those of the stars in the following chapters.

On a clear day the sun is too bright to be safely observed without protection of the eye from its glare. Viewed through a dark glass it appears as a disk about as large as the full moon, and perfectly blank to the unaided eye. With the telescope the disk is enlarged and its features are revealed. It would be unwise, of course, to look directly through the telescope at the sun without a device for diverting most of the light and heat that it concentrates at the focus. A convenient procedure is to let the telescope project the image on a smooth cardboard screen held back of the eyepiece. In this way many people can observe at the same time.

Photographic records of the sun's surface and its surroundings are obtained in a variety of ways. Sunlight can be directed into fixed telescopes by coelostats. These permit the use of long-focus objectives that form large images of the sun. Examples of vertical solar telescopes are the 45.7 meter and the 18.3 meter towers of the Mount Wilson Observatory and the 21.3 meter and 15.2 meter towers of the McMath-Hulbert Observatory.

The McMath solar telescope, Fig. 10.1, of the Kitt Peak National Observatory has a sloping tube 152.4 meters long that is parallel to the earth's axis; three-fifths of the tube is below the

FIGURE 10.1
The R. R. McMath Solar telescope at Kitt Peak. The tunnel slopes down to the right. (Kitt Peak National Observatory photograph.)

ground. A heliostat, a rotating plane mirror 2 meters in diameter, reflects the sunlight down the tube to a 1.52 meter concave mirror near the bottom. This mirror, having a focal length of 91.4 meters, forms an image of the sun averaging 0.85 meter in diameter in the observing room at the ground level.

Photographs of the sun and its spectrum from high-altitude stations, balloons, rockets, and satellites are giving new information, and radio telescopes are providing data about the sun that are not available with optical instruments.

In the ordinary view the sun is a gaseous globe 13.92×10^5 kilometers in diameter, or 109 times the earth's diameter. The sun has 1.3 million times the volume of the earth. Because its mass is a third of a million times as great, it averages one-fourth the earth's density, or 1.4 times the density of water. The sun's temperature increases from 5800°K at its lowest visible level to about 7.4×10^6 degrees Kelvin at the center.

The above facts stated rather coldly, deserve elaboration because some points are not immediately obvious. The sun subtends about 31 minutes of arc, but a single point on the sun's surface has a parallax of almost 8.8 seconds of arc. (We define the solar parallax as the angle as viewed on the mean radius of the earth.) Knowing the radius of the earth we can compute the distance to the sun and, hence, its diameter and volume. Knowing the earth's mass and applying Newton's law, we can use the fact that the earth orbits around the sun to determine the sun's weight and then find its mean density. The mean density of the sun is about 1.4 grams per cubic centimeter. The surface density of the sun is only 10^{-7} gram per cubic centimeter so we should expect a very high central density.

Measuring the sun's surface temperature, while simple in theory, is not easy in practice. We can obtain the shape of the sun's radiation curve and compare it with an ideal glowing ball of various temperatures or we can measure the color of the sun and compare it with the color of the glowing ball. We can measure the energy received per square centimeter at the earth and, after correcting for the effects of the earth's atmosphere, determine the true value of the energy, which is referred to as the **solar constant** (1.374×10^6 ergs per square centimeter per second). By simple geometry we can determine the energy leaving each square centimeter on the sun and by the radiation laws we can then assign a temperature to the surface of the sun. Whatever method we use we arrive at a temperature of about 5750°K which is often rounded off to 5800°K or even 6000°K for convenience of memory. Now how do we convert this information into a value for

8″.8

the central temperature, knowing that the sun is gaseous throughout?

Simple calculus gives an approximate relation for the central pressure (P_0) in terms of the quantities we already know

$$P_0 = \frac{G\rho M}{2R}$$

For the sun, the radius $R = 6.96 \times 10^{10}$ centimeters, G (the universal constant of gravity) $= 6.67 \times 10^{-8}$ dyne centimeter^2 per gram^2, and we calculate that $P_0 = 1.3 \times 10^{15}$ dynes per centimeter^2! One earth atmosphere is 1.01×10^6 dynes per centimeter^2 so we can say that the pressure inside the sun is at least a billion earth atmospheres. No material can remain solid or even liquid at these pressures so we are dealing with a completely gaseous sphere as stated. From the physics of gases we know that pressure is related to temperature

$$T = \frac{mP}{k\rho}$$

where $m = 8 \times 10^{-25}$ gram, $k = 1.38 \times 10^{-6}$ erg per degree Kelvin. Substituting these values we find that $T = 5.4 \times 10^6$ °K as a lower value which substantiates the more accurate value quoted above.

We return to considering the inner structure of the sun, energy sources, etc., in Chapter 15. Although we consider the sun an average, well behaved star, we will find that the sun presents us with an unusual and scientifically fascinating mystery about its sources of energy.

THE PHOTOSPHERE

The **photosphere**, the visible surface of the sun, is mottled with brighter granulations and faculae and is often marked with darker sunspots (see Fig. 10.2). Strong magnetic fields are associated with the sunspots. The photosphere is the region as far into the sun as we can see. Here, where the pressure is only a hundredth of our air pressure at sea level, the gas becomes opaque. From this level the sunlight emerges, distributing energy equivalent to 3.9×10^{26} watts to illuminate and heat the members of the planetary system. Each square meter contributes 6.4×10^7 watts. The sun has been pouring out energy at this great rate for more than 5 billion years, during all the geological ages, and is expected to continue to do so for billions of years in the future.

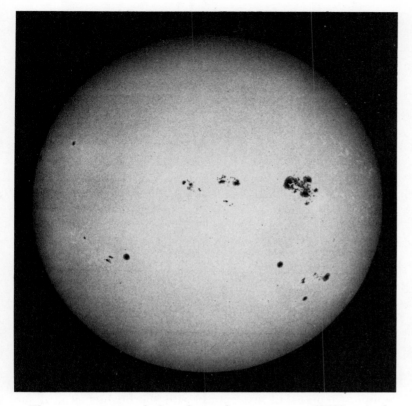

The temperature of the photosphere averages 5750°, on the absolute temperature scale, or less than 10,000°F.* It is somewhat higher near the center of the disk, where we look in directly, and is reduced to 5000°K near the edge, where our slanting view is obstructed at higher and cooler levels. Thus the sunlight from the edge is less bright and is redder than that from the center of the disk. This is referred to as **limb darkening** (see Fig. 10.4).

Through the telescope the photosphere presents a mottled appearance. Bright **granules** (Fig. 10.3) a few hundred kilometers in diameter cover a considerable part of the surface; they are hotter spots in the seething furnace formed by gases coming from below. Each granule lasts only a few minutes before it cools to the temperature of its surroundings. Larger bright spots, the **faculae**, are often conspicuous against the less luminous background near the edge of the disk. Dark spots on the sun have held greater interest.

* °F = ⅘(°K − 273) + 32

FIGURE 10.3
Solar granulation around a sunspot. This remarkably detailed photograph was taken by Stratoscope I from above 80,000 feet. (Photograph by Princeton University Observatory, ONR, NSF, NASA.)

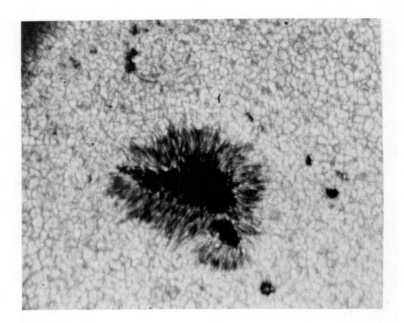

SUNSPOTS

Sunspots appear dark in contrast with the brighter general surface of the sun. They range in size from specks scarcely distinguishable from the spaces between the bright granules to the great spots visible without the telescope. They usually consist of two distinct parts: the **umbra**, the inner, darker part which is often divided, and the lighter **penumbra** around it.

Sunspots occur in groups; where a single spot is seen, it is likely to be a survivor of a group or the precursor of a new group. A normal group develops in about a week and then begins to decline. Two **principal spots** grow larger than the others, which form mostly between them. The **preceding spot** in the direction of the sun's rotation frequently becomes the larger of the two. The **following spot** is the largest of the spots in the rear. It subdivides and vanishes along with the smaller spots, until only the preceding spot is left to shrink and disappear. There are exceptions to this pattern.

One of the largest groups ever recorded appeared early in 1946 and lasted more than 3 months, an exceptionally long duration. The group attained the length of 320,000 kilometers and covered an area of 14.8×10^9 square kilometers. Its largest spot, in this case the following spot, measured 145,000 by 96,000 kilometers. A slightly larger group (Fig. 10.4) appeared in 1947, but it did not last as long as the one in 1946.

The sun's rotation is in the same direction as the earth's rotation and revolution; this is revealed by the gradual drift of sunspots across its disk (Fig. 10.5). Spots that come into view at the eastern edge disappear about two weeks later at the western edge, if they last that long. Because the sun's equator is inclined 7° to the plane of the earth's orbit, the paths of the spots across the disk are generally curved; the curve is greatest early in March, when the sun's south pole is in the earth's direction and again early in September, when its north pole is toward the earth.

Unlike the earth, which rotates with a set period in all latitudes, the rotation period of the gaseous sun is longer as the distance from the equator increases. Long-lived sunspots take 25 days to go completely around at the equator, although at first sight they

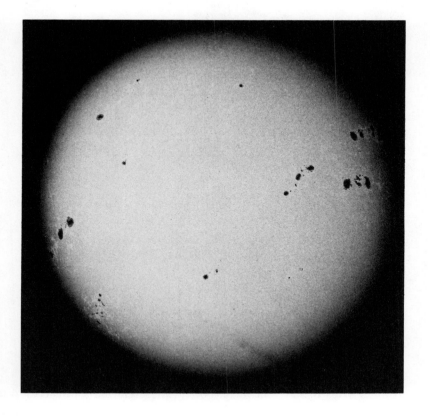

FIGURE 10.4
The sun on 15 September 1957. Note the large number of spots and also the limb-darkening effect. (Hale Observatories photograph.)

FIGURE 10.5
Sunspots show the sun's rotation. This large group of 1947 lasted more than three months. (Hale Observatories photograph.)

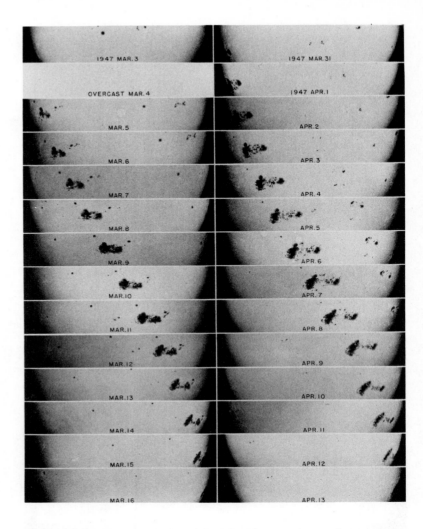

seem to take about 4 weeks. This discrepancy occurs because the earth has moved in its orbit at the same time the sun has completed its rotation, and the spots must move further around than 360°. At a latitude of 35° the true period is about 27 days. Spots seldom occur farther from the equator than this and the rotation period must be obtained from the Doppler effect at the limb of the rotating sun. At a latitude of 75° the period is 33 days. Both methods for determining the periods of rotation yield essentially identical results.

In some years the sun's disk is seldom free from spots, whereas in other years it may remain unspotted for several days in succession. Sunspot groups vary in number in a roughly periodic manner, a variation first announced in 1843 by H. Schwabe. The intervals between the times of maximum spottedness have averaged 11.1 years, but for the past half century have been more nearly 10 years. The numbers of groups at the different maxima are not the same. The rise to maximum is faster than the decline.

The maximum that occurred in 1958 was the highest on record in number of groups, but the groups and individual spots were generally smaller in size than at the maximum around 1948. The latest minimum occurred in 1964. Thus a period of geophysical cooperation, the International Year of the Quiet Sun, began in 1964 and extended through 1965. The most recent maximum occurred in the period 1968–1970 (Fig. 10.6). The average number was about the same for each year and was not as high as predicted. It is the broadest maximum on record. Records maintained since the late 1700s indicate that there may be a long-term cycle superposed upon the 11-year period and this long-term period is very nearly 88 years.

Sunspots occur mainly between latitudes 5° and 30° north and south of the sun's equator; very few have as yet been reported beyond 45°. At any particular time they are likely to appear in two rather narrow zones equidistant from the equator. As spots vanish and others appear, the zones shift toward the equator in cycles that parallel the sunspot number cycles.

About a year before sunspot minimum, small spots break out in the higher latitudes. Thereafter, the two zones of spot activity draw in toward the equator, and when the next minimum is reached, a few surviving members of the fading cycle are seen'

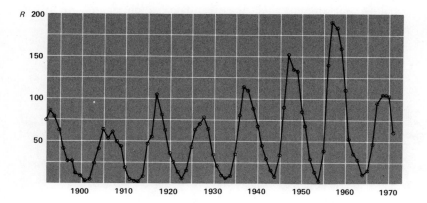

FIGURE 10.6
The average daily number of sunspots (R) plotted by the year.

around latitudes 5°. Meanwhile some spots of the new cycle have already become visible in the higher latitudes. Thus at the minimum of 1954 the Mount Wilson observers recorded 15 groups near the equator and 31 groups of the new cycle.

10.1 Magnetic fields

The sun's general magnetic field is beautifully revealed by the polar patterns in the corona during a total eclipse (Fig. 10.24), especially when the eclipse occurs during the sunspot minimum. In the picture the alignment of the electrons and other charged particles along the magnetic lines is quite clear. Magnetic field strengths are usually given in units called **gauss.** The general field strength at the sun's surface is about 2 gauss. However, in sunspot regions the local field strength may attain values of 1000 gauss or more.

When the image of a sunspot is focused on the slit of a spectroscope and passed through an analyzer, the dark lines of the solar spectrum appear split into two or more parts (Fig. 10.7). This **Zeeman effect,** named after the physicist who discovered it in the laboratory in 1896, is the splitting of the spectrum lines where the source of the light is in a strong magnetic field. The effect in the sunspot spectrum shows that the spot is magnetic and also reveals its **polarity**—whether the positive or negative pole of the magnet is toward us.

Most sunspot groups are **bipolar,** that is, their two principal spots have opposite polarities that conform to the following rule: During a particular cycle the leading spots in the sun's northern hemisphere have their positive poles toward us, and the following spots their negative poles. In the southern hemisphere the preceding spots present their negative magnetic poles, and the following spots their positive poles.

A remarkable feature of sunspot magnetism is the complete

FIGURE 10.7
Zeeman splitting in the sunspot spectrum at 5250Å.

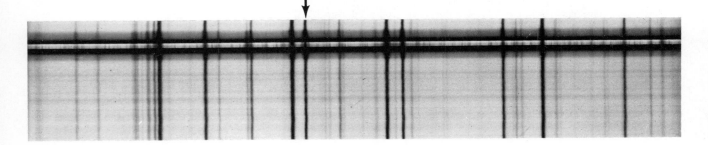

reversal of the pattern with the appearance of the groups of the next cycle (Fig. 10.8). The preceding spots in the northern hemisphere now present their negative poles, and so on. First reported by Mount Wilson astronomers around the sunspot minimum of 1913, this reversal of polarities has been observed at each succeeding minimum.

The Zeeman effect over the entire disk of the sun has been recorded repeatedly by the Babcocks and their associates using a scanning magnetometer. The general magnetic field may be regarded as having two components. (1) Polar fields of opposite polarity are observed around the two poles; these reversed signs at the sunspot number maximum in 1958. (2) Ring-shaped fields, parallel to the sun's equator and about equally distant from it to the north and south, have opposite field direction in the two hemispheres and migrate toward the equator in the sunspot cycle. These are replaced near sunspot minimum by a new pair of rings in the higher latitudes that have the reverse directions of the preceding magnetic fields. The equatorial rings are mainly submerged below the solar surface. As H. W. Babcock pictures them, they undulate like sea serpents. Where loops break out through the surface, they produce oppositely polarized regions in which sunspot groups may form. The magnetic fields are a basis for interpretations of the optically observed features at and above the solar surface.

THE SUN'S ATMOSPHERE

The sun's atmosphere consists of the gases above the photosphere reaching a height of 7000 kilometers (see Fig. 10.9). The **chromosphere**, extending to the height of several thousand kilometers, is so named because of its color, which is imparted chiefly by the red glow of hydrogen. It is normally the region where the spectacular solar flares are observed. The red **prominences** appear above the chromosphere, at times attaining heights of many hundred thousand kilometers. They are visible during total solar eclipses, and together with the inner corona are studied effectively with special devices at other times. The **corona**, the outermost solar envelope, appears as a filmy halo of intricate structure extending from about 7×10^3 kilometers to more than 7×10^6 kilometers.

10.2 The spectrum of sunlight

The spectrum of sunlight is an array of colors from violet to red, which is interrupted by thousands of dark lines. The lines are not seen in the rainbow or in the spectrum formed by a prism alone.

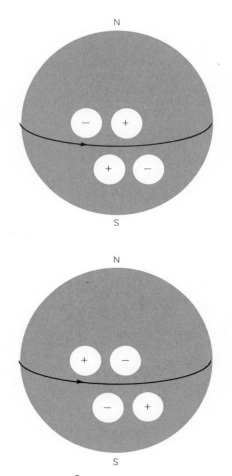

FIGURE 10.8
Reversal of polarities of sunspots with the beginning of a new cycle. The circles represent preceding and following spots of groups in the two hemispheres.

FIGURE 10.9
Schematic drawing of the sun's interior and exterior. The various surface and coronal features are indicated. (Modified from J. M. Pasachoff, "The Solar Corona." Copyright © 1973 by Scientific American, Inc. All rights reserved.)

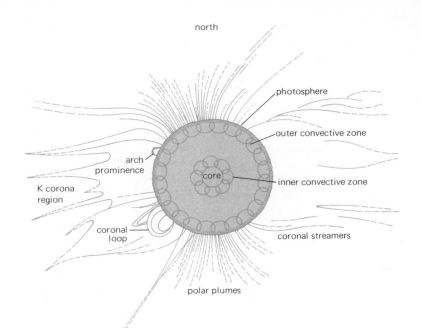

They require the selectivity given by the narrow slit before the dispersing element and, therefore, were not discovered until the slit was used. J. von Fraunhofer in 1814 was the first to see them clearly; he mapped several hundred dark lines and labeled them with letters beginning at the red end of the spectrum. The lines are still known by the letters that he assigned them.

The dark lines are images of the slit of the spectroscope at the wavelengths that are darkened in the sunlight; these are abstracted from the continuous light of the photosphere mainly by gases of the different chemical elements in the sun's atmosphere. The strongest solar lines are the Fraunhofer H and K lines of calcium near the violet end of the visible spectrum, which von Fraunhofer never saw himself, but are continuations of his lettering in his honor. The letters I and J are not used because they can be confused when written. There are also **telluric bands**—groups of lines in the sunlight resulting from molecules in the earth's atmosphere. The B band formed by terrestrial oxygen molecules is prominent in Fig. 10.10. The separate lines of the solar spectrum itself are produced by atoms of the different elements in the sun's atmosphere, which cannot generally combine into molecules at that high temperature.

The wavelengths (λ) of the lines are expressed in **angstroms**, abbreviated Å.* The visible region of the solar spectrum is be-

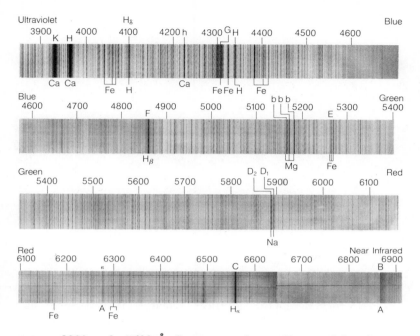

FIGURE 10.10
The visible solar spectrum from the ultraviolet to the near infrared. The colors and the Fraunhofer lines, B through K, are shown above the individual strips which have been cut from the original spectrogram. (Hale Observatories photograph.)

tween λ3900 and λ7600 Å. Some conspicuous lines and bands are listed in Table 10.1.

The ultraviolet region of the solar spectrum can be photographed from the ground only as far as λ2900 Å, beyond which it is absorbed by ozone and other constituents of our atmosphere. Features of the extreme ultraviolet and X-ray regions are being studied in photographs of spectra from rockets taken above the absorbing atmospheric levels. Below about λ1700 Å (Fig. 10.11) the continuous spectrum becomes so dim that the dark lines are no longer visible. Bright lines from various levels in the chromosphere and corona are prominent; they are generally lines of ionized atoms of the lighter chemical elements, such as oxygen, nitrogen, carbon, and magnesium. The brightest line of all is the alpha line of the Lyman series of hydrogen at λ1215.7 Å (section 12.10).

The infrared region has been mapped from the ground by photography and other means as far as λ200,000 Å. Here the solar lines are frequently obscured by broad telluric bands. Above λ1 millimeter (10,000,000 Å) the sun is mapped by radio techniques.

* 1 Å = 10⁻⁸ cm
      = 10⁻⁷ mm

TABLE 10.1
Fraunhofer lines

Fraunhofer letter	Wavelength (Å)	Identification
A	7594	oxygen (telluric)
B	6867	oxygen (telluric)
C	6563	hydrogen
D	5893	sodium (double)
E	5270	iron
F	4861	hydrogen
G	4310	composite blend
H	3968	calcium
K	3934	calcium
L	3820	iron
M	3735	iron
N	3581	iron

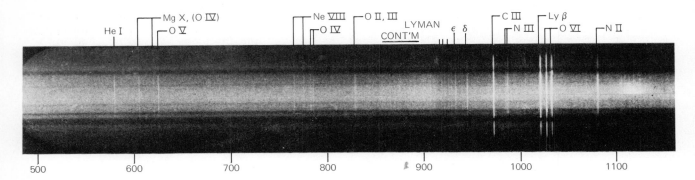

10.3 Chemical elements in the sun

More than 60 chemical elements are recognized in the sun's atmosphere. They have been identified by comparing laboratory spectra with the lines in the solar spectrum. Some of the remaining elements would not be expected to make much impression because they have not had their laboratory spectra determined well enough for dependable comparisons. We can conclude that practically all the 92 natural chemical elements found on earth are present in the sun.

The hot gases of the sun are generally composed of atoms. The molecules, or combinations of atoms, of 18 compounds are recognized as well by their dark bands in the spectrum. Examples are titanium oxide and the hydrides of calcium and magnesium, which occur in the cooler areas of sunspots. Only a few hardy compounds hold together above the unspotted regions.

Hydrogen is the most abundant element in the sun's atmosphere, and helium is second. These two elements also predominate in the sun's interior, in the stars, and in the universe generally. Hydrogen contributes about 70 percent of the mass of the cosmic material, helium 28 percent, and the heavier elements the remainder. Exceptions to these proportions occur in the earth and other smaller bodies from which most of the lighter gases have escaped.

10.4 The chromosphere

The chromosphere appears as a red fringe around the dark disk of the moon when the moon completely conceals the photosphere at the time of total solar eclipse. On these occasions the bright line spectrum of the chromosphere can be observed; it has been known as the **flash spectrum,** because it flashes into view in the

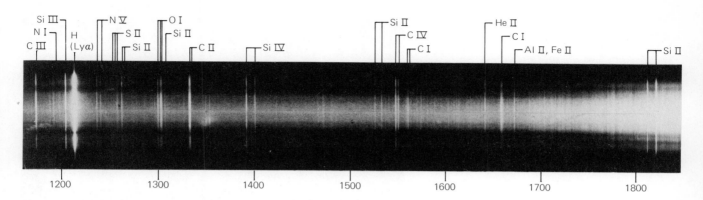

spectroscope near the beginning of the total eclipse, replacing the dark lines of the ordinary solar spectrum. The red Fraunhofer C line caused by hydrogen gives its color to the chromosphere and the prominences above it.

As it is photographed with a slitless spectroscope near the beginning or end of total solar eclipse, the spectrum of the chromosphere (Fig. 10.12) appears as a succession of images of the narrow crescent of the chromosphere left uncovered by the moon. The different lengths of the crescents show that some of the chemical elements that produce them are effective to greater heights above the photosphere than are others. Hydrogen, calcium, and helium give the longest images.

The strong helium lines of the flash spectrum, which resulted in the discovery of this element on the sun before it was recognized on the earth, are almost entirely missing in the dark-line solar spectrum. Aside from this and some other explainable exceptions, the lines of the chromospheric and ordinary solar spectra are similar, illustrating the rule that a luminous gas emits the same patterns of wavelengths that it absorbs from light passing through it.

It is important to understand the origin of the chromospheric light. The gas in the chromosphere, being cooler than the photosphere, selectively absorbs light passing through it. It immediately reradiates the light in all directions, including back toward the photosphere. Some light is reradiated in the initial direction, but it is greatly reduced in intensity compared to the unabsorbed light beside it in the spectrum (Fig. 10.13). The contrast is so great that we can only see this reradiated light during eclipses or when using special observing techniques.

Much of our knowledge of the exterior features of the sun is obtained with special apparatus used when the sun is not eclipsed. As an example, the **spectroheliograph,** an adaptation of the spec-

FIGURE 10.12
The solar "flash" spectrum. Top, just before second contact, center with Bailey's beads showing, and bottom, just after third contact. The pair of very bright lines on the left are the H and K lines of calcium and the bright line on the right is the F line (Hβ) of hydrogen. The other lines are due primarily to iron. (Hale Observatories photograph.)

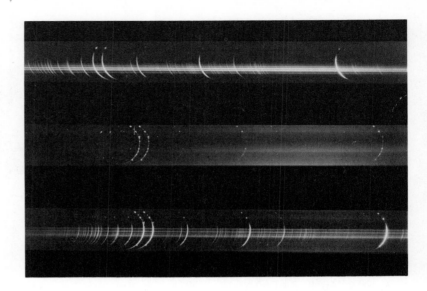

FIGURE 10.13
Formation of dark lines. The sum total of all of the light (a) falling on the gas is reemitted equally in all directions (b). Thus the reemitted light in the original direction is greatly reduced (c).

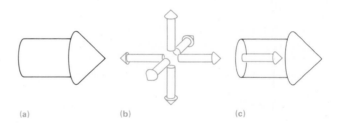

(a)　　　　　(b)　　　　　(c)

troscope, is an attachment to the telescope employed to record the chromosphere over the whole disk of the sun. The spectroheliograph is used to photograph the sun in the light of a single line of the spectrum, and therefore, of the chemical element in the sun's atmosphere that produces the chosen line. The "dark" lines are dark only by contrast with the brighter background of the solar spectrum; they contain the weaker light of the chromosphere that is to be recorded in the photograph, or **spectroheliogram.** The K line of calcium and the C line of hydrogen are generally used for the purpose with apparatus on the ground.

Photographs of the chromosphere (Fig. 10.14) show the bright **flocculi.** Flocculi are the light and dark markings seen on the photosphere when it is photographed in monochromatic light, as with a spectroheliograph. Where flocculi are bunched around sunspot groups or other active centers, they are known as **plages.** The plages are especially prominent in extreme ultraviolet and

X-ray photographs of the sun from spacecraft. On large-scale spectrograms showing a limb, the chromosphere appears uniformly above the surface, extending to a distance of about 6000 kilometers and exhibiting a ragged grass-like upper boundary. These spikes, or, more correctly, jets, are called **spicules.** Spicules appear and fade during a period of about 5 minutes. Spicule structure is probably important in the transport of radiation through the atmosphere of the sun into the corona. Each spicule reaches a height of about 10 seconds of arc above the chromosphere.

Monochromatic filters are also employed to obtain photographs in a narrow range of wavelength of the sunlight. They are effective for viewing the solar prominences and some features of the corona as well, which are normally seen only during a solar eclipse. Polarizers are often used with such filters to study the effects of the magnetic field on prominences.

10.5 The solar prominences and flares

The solar prominences are flame-like structures beyond the chromosphere seen during a total solar eclipse; their vivid red color is in striking contrast with the white glow of the corona. In the

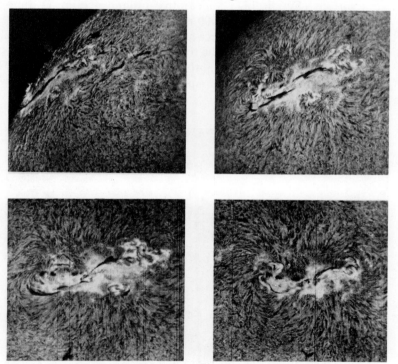

FIGURE 10.14
A section of the sun photographed in red hydrogen light (Hα) on 3, 5, 7 and 9 August 1915. The bright and dark stripes are referred to as flocculi. (Hale Observatories photograph.)

photographs in hydrogen light taken outside eclipse they often appear as long dark filaments against the brighter background of the chromosphere. Where they are carried beyond the disk by the sun's rotation, they appear bright against the sky. The prominences can be photographed in their hydrogen or calcium light, or can be viewed directly with the spectrohelioscope. Effective studies of their behavior have been made at the McMath-Hulbert Observatory and at high-altitude stations all over the world in successions of photographs taken on motion picture film. The projections of the films give dramatic portrayals of the prominence activities.

Most prominences are of the **active** type; they originate high above the chromosphere and pour their streamers down into it. **Quiescent** prominences are the least active and have the longest lives; their most common form is the "haystack." **Eruptive** prominences are among the rarer types. These rise from active material above the chromosphere, attaining high speeds and great altitudes before they vanish. A prominence of 1938 attained a speed of 720 kilometers a second. A prominence of 4 June 1946, rose to a distance of more than a million and a half kilometers above the sun's surface (Fig. 10.15). An eruptive prominence on 12 June 1970 attained a height of 300,000 kilometers in a very short period of time. A great arch prominence appeared on 11 August 1972 and was 15,000 kilometers high at the top of its loop.

Solar **flares** are remarkable outbursts generally in the vicinities of large and active sunspot groups of irregular polarity. Viewed

FIGURE 10.15
The magnificent solar·prominence of 4 June 1946. The size of the earth is indicated by the white dot for comparison. (High Altitude Observatory photograph.)

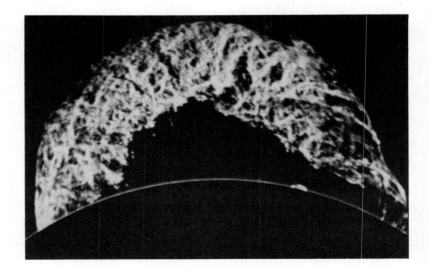

with the spectrohelioscope or a monochromatic filter they appear as brilliant areas of the chromosphere. They vary from smaller flares of a few minutes' duration to more intense and rarer ones that require up to 15 minutes to attain their greatest brightness and then disappear in an hour or more. They have been attributed to masses of glowing gas brought up from below the photosphere in the wakes of upheavals that cause the bursts observed with radio telescopes.

These hot gases emerge explosively and rise to considerable heights. The first report of their rapid ascent was given by Helen Dodson Prince at the McMath-Hulbert Observatory from her photographs with a motion picture camera. The flare of 8 May 1951, at the edge of the disk rose in the first minute of its life at the rate of 720 kilometers a second, comparable with the speeds of some eruptive prominences, and continued to the height of 48,000 kilometers. Great flares occurred on 2, 4, 7, and 11 August 1972, the one on 7 August being especially intense. It was the greatest flare since 1960, and very intense X-ray and radio activity were recorded. These events were followed by intense auroral activity in both the northern and southern hemispheres of the earth.

10.6 The radio sun

The reception of noise from the sun was first detected in 1942 by radar defense stations in Great Britain, where the source of the disturbance was traced to a large spot group near the central meridian of the sun. This accidental discovery provided a new means of studying the sun, which is being utilized with radio telescopes in various parts of the world. The radio emission is produced by the interactions of fast-moving charged particles from the photosphere with the atmospheric and coronal gases; it is strongest at wavelengths of 1 to 10 meters, where it originates at different levels of the corona, and it is weakest at centimeter ranges, which originate in the lower chromosphere.

With the shorter wavelengths the Australian radio astronomers have devised apparatus for rapidly scanning the whole disk. They record bright spots that are generally associated with visible sunspot groups but may last much longer than the visible spots. The radio emission has its least strength and is fairly constant from the **quiet sun,** that is, in years when sunspots are scarce.

From the **active sun,** when sunspots are numerous, **bursts** of irregular and much greater strength are superposed. The explanations of how they occur and of their relation to the visible flares are still tentative. J. P. Wild suggests that an upheaval in the sun causing a prominence or flare produces clouds of particles moving outward at a sixth the speed of light; when they reach the corona, they produce the radio flashes lasting a few seconds. The following shock wave caused by the upheaval may carry great numbers of trapped particles into the corona at the rate of 1600 kilometers a second, where they cause a radio **outburst** lasting from 10 to 20 minutes.

10.7 Terrestrial associations

The appearance of an intense solar flare is likely to be followed by a deterioration of our radio communications in the higher frequencies. Powerful ultraviolet radiations from the flare arrive with visible evidence of the flare itself, and the very swift particles that cause the radio flashes from the corona reach the earth less than an hour later. These disrupt ionized layers of the upper atmosphere, which normally keep reflecting our own radio beams back to the ground (see Section 1.9).

Less swift ionized particles (see Section 12.12) from the upheaval in the sun arrive here about a day after the solar flare was observed, and after the particles have produced the outburst of radio emission in their passage through the corona. They then interact with the earth's magnetic field, referred to as the geomagnetic field, excite the gases of our upper atmosphere, and set them glowing in an auroral display (see Section 1.14). The appearance of the aurora is generally accompanied by unusual gyrations of the magnetic compass, all of which informs us that a geomagnetic storm is in progress.

The passage of the intense streams of radiation and particles through the solar system are of great concern to those responsible for manned space flights. During severe solar activity the radiation dosages, even at the distance of the earth, are extremely high and may require considerable shielding to keep them at tolerable levels. Fortunately, the sunspot maximum during the Apollo program was not nearly as high as expected, and consequently flare activity was relatively low.

ECLIPSES OF THE SUN

An eclipse of the sun occurs when the moon passes directly between the sun and the earth, screening part or all of the sun's disk. The earth is then partly darkened by the moon's shadow.

The average length of the umbra, or conical shadow, of the moon is 373,000 kilometers, which is 4800 kilometers less than the

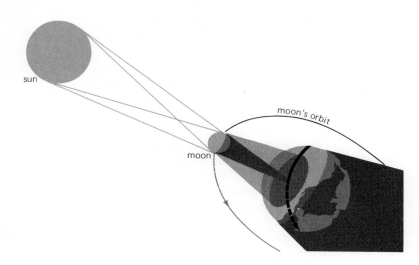

FIGURE 10.16
Path of total solar eclipse. The moon's shadow moves in an easterly direction over the earth's surface. The eclipse is total for an observer in the umbra, and is partial in the larger area of the penumbra.

average distance of the moon from the earth's surface. Generally, the umbra does not reach the earth's surface; however, it often does reach the earth because of the eccentricity of the moon's orbit around the earth and the earth's orbit around the sun. When the moon is nearest the earth and the earth is also farthest from the sun, the conical umbra of the moon's shadow falls on the earth 29,000 kilometers inside its apex.

10.8 Total and annular eclipses

A **total eclipse** of the sun occurs when the umbra of the moon's shadow extends to the earth's surface (Fig. 10.16). The area encompassed by the umbra rarely exceeds 240 kilometers in diameter when the sun is overhead. The observer within the area sees the dark circle of the moon completely hiding the sun's disk.

An **annular** (from Latin *annulus*, meaning ring) **eclipse** occurs when the umbra is directed toward the earth but is too short to reach it (Fig. 10.17). Within a small area of the earth's surface the moon is seen almost centrally projected upon the disk of the sun, but the moon appears slightly the smaller of the two, so that a ring of the sun's disk remains uncovered. Annular eclipses are 20 percent more frequent than total eclipses.

Around the small area of the earth from which the eclipse appears total or annular is the larger area, some 3200 to 4800 kilometers in radius, which is covered by the penumbra of the shadow. Here a **partial eclipse** is visible, and the fraction of the sun that is hidden decreases as the observer's distance from the center of the eclipse increases. Eclipses are entirely partial where

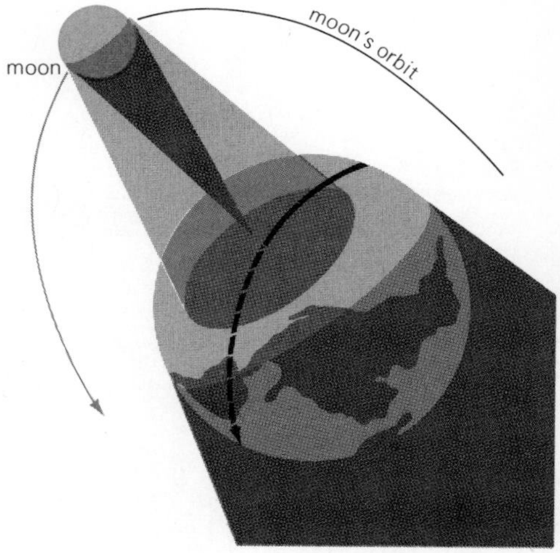

FIGURE 10.17
Annular eclipse of the sun. The umbra of the moon's shadow does not reach the earth's surface.

the axis of the shadow is directed slightly to one side of the earth. All total and annular eclipses are preceded and followed by partial phases.

The revolution of the moon causes its shadow to sweep eastward at the average rate of 3380 kilometers an hour. Because the earth is rotating toward the east at the rate of 1670 kilometers an hour at the equator, the speed of the shadow over the surface is 1700 kilometers an hour at the equator when the sun is overhead. The effective speed becomes greater with increasing distance from the equator, where the rotation is slower, and may become as fast as 8000 kilometers an hour when the sun is near the horizon.

Considering its high speed and small area, we see that the umbra can darken any part of its path for only a short time. The greatest possible duration of total solar eclipse at a particular place can scarcely exceed 7.5 minutes, and that of an annular eclipse can be only a little greater. The partial phase accompanying either type of eclipse may have a duration of more than 4 hours from beginning to end, but it is usually much less.

The path of a total eclipse, or of an annular eclipse, is the narrow track of the darkest part of the shadow over the earth's surface, from the time it first touches the earth at sunrise until it departs at sunset. Meanwhile the penumbra moves over the larger surrounding region in which the eclipse is only partial. Occasionally the umbra is long enough to reach the earth at the middle of its path but at the beginning and end fails to extend to the surface. In this event the eclipse is total around the middle of the day and is otherwise annular.

10.9 Eclipse seasons and predictions of eclipses

In order to eclipse the sun, the moon must be almost directly between the sun and the earth. This condition is not fulfilled every time the moon arrives at its new phase, because the moon's path around the heavens is inclined 5° to the ecliptic. Thus the new moon is more likely to pass north or south of the sun.

Eclipses occur during two **eclipse seasons**, about the times the sun is passing the two opposite nodes of the moon's path (Fig. 10.18). Owing to the rapid westward shifting of the nodes along the ecliptic (see Section 6.3), these seasons come about half a month earlier in the calendar from year to year; in 1965 they were around June and December. The length of the **eclipse year** is 346.62 days, which is the interval between two successive returns of the sun to the same node. Thus if the first season is early in the year, another season around the same node may begin before the end of the year.

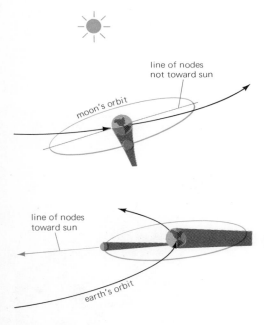

FIGURE 10.18
Eclipse seasons. Because the moon's orbit is inclined about 5° to the plane of the earth's orbit, eclipses can occur only at two opposite seasons when the sun is near the line of the nodes of the moon's path (b). At other times in the year the moon does not pass between the earth and the sun or into the earth's shadow (a).

Each solar eclipse season lasts a little more than a month, or somewhat more than the month of the moon's phases. During each interval the moon becomes new at least once, and may do so twice. Two eclipses of the sun are accordingly inevitable each year, one near each node. Five may occur, two near each node and an additional eclipse if the sun comes around again to the first node before the year ends. Similarly, it can be shown that three umbral eclipses of the moon are possible in the course of the year, although a whole year may pass without a single one.

Accurate predictions of solar eclipses, when they will occur, and where they will be visible are published in various astronomical almanacs a year or two in advance. Tracks of total solar eclipses for several years in advance are published in the U.S. Naval Observatory *Circulars*. The approximate times and places of eclipses from 1208 B.C. to A.D. 2163 can be found in Oppolzer's *Canon der Finsternisse*, which also contains maps showing the tracks of total and annular solar eclipses. A more current compilation covering the interval 1898 through 2510 is given in *Canon of Solar Eclipses* by J. Meeus, C. Grosjean, and W. Vanderleen.

The predictions of eclipses are made possible by knowledge of

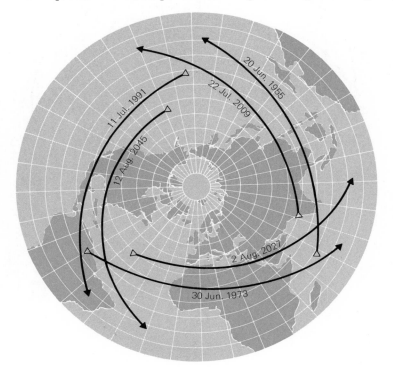

FIGURE 10.19
Paths of a family of solar eclipses through the year 2045. This family is slowly moving north.

the motions of the earth and moon. They are facilitated by a relation between the occurrences of eclipses, which has been known from very early times—at least since the ancient Babylonians. The **saros** is the interval of 18 years 11.3 days (or a day less or more, depending on the number of leap years included) in which eclipses of the same series are repeated. It is equal to 223 synodic months containing 6585.32 days, and is nearly the same as 19 eclipse years having 6585.78 days. After this interval the relative positions of the sun, moon, and node are nearly the same as before. The sun is about one diameter west of its former position relative to the node; accordingly the paths of the eclipses of a series are displaced progressively in latitude, being shifted gradually from pole to pole until the shadow fails to touch the earth and the particular series is completed.

FIGURE 10.20
Families of eclipses. A family is just beginning at the south pole (right). A family moving south is becoming annular (next page).

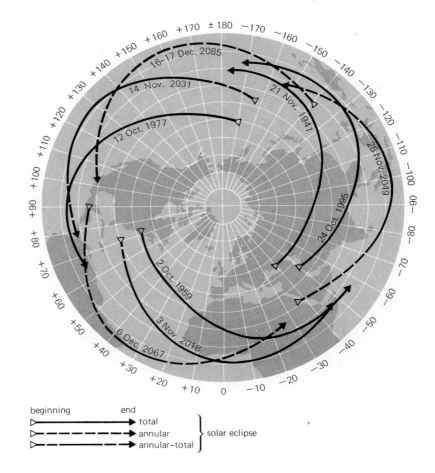

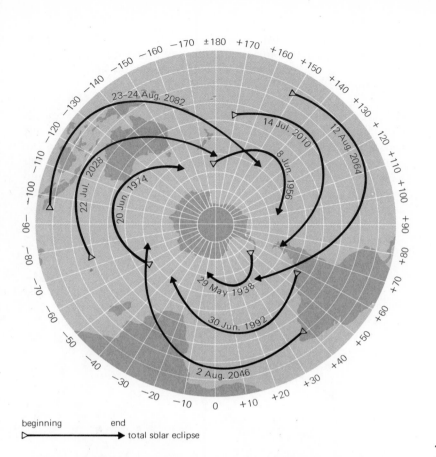

beginning end
▷   ⟶ total solar eclipse

The third of a day in the saros period causes the path of each eclipse to be displaced in longitude a third of the way around the earth with respect to its predecessor. After 3 intervals, about 54 years and 1 month, the path returns to about the same region as before (Fig. 10.19). Occasionally a series that starts as total eclipses becomes annular eclipses as the series progresses (Fig. 10.20). Many series are occurring at the same time as can be deduced from Figs. 10.19 and 10.20.

The dates, durations at noon, and land areas of the principal current total and annular eclipses are given in Table 10.2. The tracks are shown in Fig. 10.21.

10.10 Total solar eclipse

The sun in total eclipse is an impressive sight to be remembered always. The beginning of eclipse, called first contact, is manifested by the appearance of a dark notch at the sun's western edge.

TABLE 10.2
Total and annular solar eclipses[a]

Date	Duration	Region
TOTAL SOLAR ECLIPSES		
20 Jun. 1974	5	Australia
23 Oct. 1976	5	Africa, Australia
12 Oct. 1977	3	Central America, Northern S. America, Hawaiian Islands
26 Feb. 1979	3	Canada, Greenland
16 Feb. 1980	4	Africa, S. E. Asia
ANNULAR SOLAR ECLIPSES		
29 Apr. 1976		Ends in China
18 Apr. 1977		Ends in Indian Ocean
10 Aug. 1980		Ends in S. America

[a] Courtesy of R. Duncombe, U.S. Naval Observatory.

Gradually the sun's disk is hidden by the moon. When only a thin crescent remains uncovered, the sky and landscape have assumed a pale and unfamiliar aspect, because the light from the sun's rim is redder than ordinary sunlight.

The light fades rapidly as total eclipse approaches. There is a chill in the air, birds seem bewildered, and some flowers begin to close. Just before the last sliver of the sun breaks into brilliant beads and disappears the filmy corona bursts into view. When the sun disappears, called second contact, red prominences are often seen close to the edge of the eclipsing moon; some planets and bright stars may appear. Totality ends at third contact as abruptly as it began; the sunlight returns and the corona vanishes. Finally the moon moves off of the disk of the sun at fourth contact.

Recently a folklore has been established about watching eclipses

FIGURE 10.21
Eclipse tracks for the total solar eclipses through 1980.

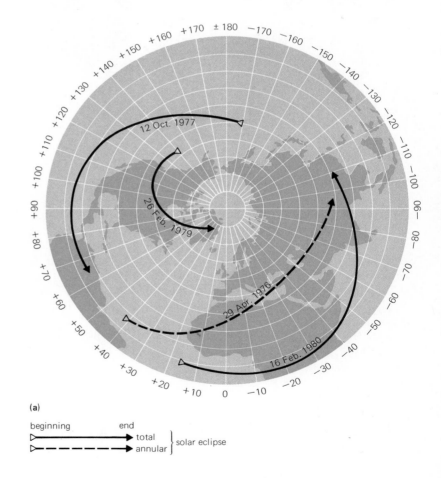

(a)

beginning   end

▷ ⎯⎯⎯⎯⎯⎯⎯▶ total ⎱ solar eclipse
▷ ⎯ ⎯ ⎯ ⎯ ⎯▶ annular ⎰

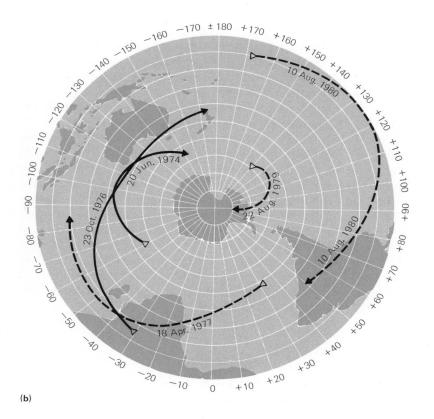

(b)

that is without basis in fact or reason. During the 1970 eclipse in the United States well-meaning parents, after an intensive advertising campaign, brought their children inside prior to totality denying themselves and their children one of the most awesome of natural phenomenon thinking that unusual radiation released during the eclipse would damage their children's eyes. The radiation present is the same during an eclipse as that present during every day of the year, and the most elementary precautions prior to second contact and after third contact serve to prevent damage to the eyes. Often athletes, such as tennis, soccer, and football players, glance directly at the sun without harm, so occasional glimpses of the sun are not harmful in themselves; however, prolonged staring at the uneclipsed sun clearly will damage the human eye and must be avoided.

The best way to view an eclipse is with the unaided eye; watching a stalking lion in its natural habitat or a ferocious electrical storm does not compare with the thrill and majesty of a total solar eclipse. The first total eclipse viewed by the author was viewed inadvertently through a pair of sunglasses. It was the source of great disappointment because 11 years elapsed before he had another chance to watch a total eclipse with enough composure to remove the pretotality phase protection devices. During the phases before second contact and after third contact pin-hole projection methods or small telescopic projection devices are indispensable. Occasional glimpses lasting a fraction of a second can be attempted, but using a totally exposed and completely developed piece of black-and-white photographic film (NOT COLOR FILM) is the safest procedure. The developed silver halide crystals are excellent neutral density filters and, in addition, reflectors of the long-wave (heat) radiation. With these precautions, the author has observed four total eclipses and two annular eclipses with no detectable, or even suspected, damage. S. A. Mitchell, with even more elementary precautions viewed the solar corona during totality probably more than any other person, with no resulting eye damage. He observed more than 40 solar eclipses. Information on how to photograph the various phases of an eclipse is available from the Eastman Kodak Company.

THE CORONA

The corona is the outer envelope of the sun and the chief contributor to the splendor of the total solar eclipse; its brightness is half that of the full moon. The optical corona is characterized by delicate streamers that vary in the sunspot cycle. Near sunspot maximum the form is roughly circular (Fig. 10.22); petal-like streamers all around give the appearance of a dahlia. Near sunspot minimum the corona is flattened in the polar regions (Fig. 10.23), where short, curved streamers are reminiscent of the lines of force around a bar magnet. Long streamers may reach out more than a million kilometers from the equatorial regions.

The light of the corona is composed of three components referred to as the F, K, and L corona. The F corona is reflected and diffracted sunlight, as the dark Fraunhofer lines in its spectrum show; it decreases slowly in intensity outward and mainly results from the dust ring that causes the zodiacal light (see Section 9.6). The K corona shows a continuous spectrum and is caused by scattering by electrons. The L corona, which gives a bright line spectrum, comes from the luminous gases of the corona itself, and is stronger near the sun's equator than at its poles. The bright lines in the spectrum of the corona were first identified by B. Edlén

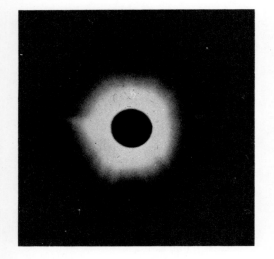

FIGURE 10.22
A picture of the 20 May 1947 total solar eclipse. The round corona is typical of sunspot maximum.

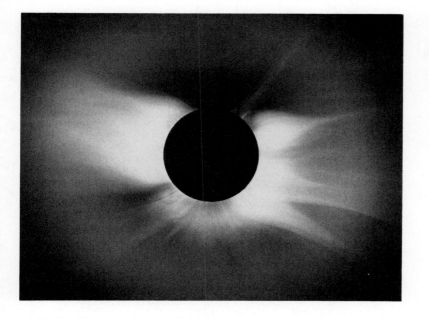

FIGURE 10.23
A beautiful photograph of the total solar eclipse of 30 June 1973. The long equatorial plumes and streamers are typical of sunspot minimum. Note the suggestion of the sun's magnetic field in the coronal lines at the sun's north and south pole. (Photography by High Altitude Observatory, a division of the National Center for Atmospheric Research, sponsored by the National Science Foundation.)

in 1942. They are emitted by shattered atoms of iron, nickel, calcium, and argon, which have been stripped of 9 to 15 electrons. Such disruption of the atoms would occur at a temperature of a million degrees Kelvin.

Features of the inner corona around the uneclipsed sun were first photographed in 1930 by B. Lyot with the **coronagraph** of the Pic du Midi Observatory. This type of telescope, having special precautions against bringing in direct sunlight to blot out the faint coronal light, is now in use at other high-altitude stations. With an appropriate filter, both still photographs and motion pictures of the corona are being made around the uneclipsed sun.

Wider extension of the corona than the photographs show has been recorded with radio telescopes. In Australia in 1960, O. B. Slee observed the dimming by the coronal material of radio emissions from the Crab nebula (see Section 13.14) and a dozen other radio sources (see Section 17.9) while the sun was passing in its annual circuit along the ecliptic. He reported that the surrounding material was dense enough to produce appreciable obstruction of the light over an elliptical area extending about 110 to 80 solar radii around the sun; the area was similar in shape to the optically observed corona at that stage in the sunspot cycle.

The coronal material streaming outward continually from the sun into space consists mainly of protons and electrons; it has been called the **solar wind** by E. N. Parker. According to Parker, the solar wind produces a loss of solar mass averaging about 10^{14} grams per second, it carries a magnetic field, causes the auroral displays in our atmosphere, and directs the tails of comets away from the sun's position (see Section 9.3).

QUESTIONS

1 How do we know that the sun is gaseous throughout?
2 Define the solar constant.
3 What is the photosphere of the sun?
4 Assume that the sun is a sphere. What is its equatorial velocity in kilometers per second?
5 Sunspots have temperatures in their centers of more than 4000°K. Why do they appear dark?
6 When should we expect the next sunspot maximum?
7 Can you propose a method to prove that the Fraunhofer B band is caused by terrestrial oxygen?
8 How high above the surface of the sun do spicules reach in kilometers?
9 What element is responsible for the Fraunhofer C line?
10 What are the conditions for a total eclipse of the sun? When is this eclipse an annular eclipse?

FURTHER READINGS

ABETTI, GIORGIO, *The Sun*, Macmillan, New York, 1957 (revised ed.).

BRANDT, JOHN C., *The Sun and Stars*, McGraw-Hill, New York, 1966.

BRANDT, JOHN C., AND PAUL A. HODGE, *Solar System Astrophysics*, McGraw-Hill, New York, 1964.

MENZEL, DONALD H., *Our Sun*, Harvard University Press, Cambridge, Mass., 1959 (revised ed.).

Cygnus the Swan as it appears in Bayer's Uranometria. The bright stars of Cygnus, α, β, γ, δ, form an easily recognized cross and are often referred to as the Northern Cross.

The Stars in Their Seasons

11

11

In the original sense the constellations are configurations of stars. The brighter stars form patterns of dippers, crosses, and the like. Some of the star figures we recognize today were well known to the people of Mesopotamia 5000 years ago, who had named them after animals and representatives of occupations, such as the herdsman and the hunter. The plan was later adopted by the Greeks who renamed some of the characters after the animals and heroes of their mythology.

Forty-eight constellations were known to the early Greeks and nearly all of these are described in the *Phenomena*, which the poet Aratus wrote about 270 B.C. (Several modern translations are available, see, for example, *Callimachies and Lycophron,* A. W. Mair, Transl., Putnam, New York, 1921, pp. 380–473.) The popularity of Aratus' poem did much to perpetuate the imagined starry creatures which he vividly described.

Interest in the celestial menagerie was also promoted by the publication of Ptolemy's *Almagest*, about A.D. 150, where the places of the stars are designated by their positions in the mythological figures. Then, too, famous artists of later times vied with one another to produce the liveliest pictures of the imagined creatures (Fig. 11.1).

These creatures do not appear on modern maps and globes of the heavens. They have nothing to do with the astronomy of today, except the connection with which they began—their names are still the names of constellations. Meanwhile, the constellations have increased in number and have taken on a new significance, although their oldest use as groups of stars survives as well.

The original constellations did not cover the skies of the Greeks completely; they did not include the duller areas where there were no striking configurations of stars to claim attention. At that time, only those stars within the imagined creatures belonged to constellations. In addition, the part of the heavens around the south celestial pole that did not rise above the horizon of the Greeks remained uncharted.

Celestial mapmakers of later times filled the vacant spaces with new constellations that they named after scientific instruments, birds, and other things having no connection with the creatures of the legends. Not all of them survived. At present we recognize 88 constellations, which completely cover the sphere of the stars from pole to pole. Seventy of them are visible, either wholly or in part, from the latitude of New York.

For the purposes of astronomy, the constellations are now definite divisions of the heavens marked off by boundary lines, just as states and nations are bounded. The boundaries first appeared in the star maps at the beginning of the nineteenth century. They were frequently irregular, making detours to avoid cutting across outstretched arms and paws of the legendary creatures. These devious dividing lines were straightened for most of the southern constellations by B. Gould in 1877, and finally for all the constellations by decision of the International Astronomical Union in

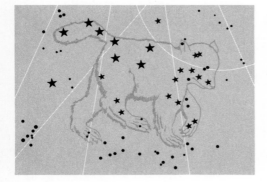

FIGURE 11.1
The Great Bear. The Big Dipper makes up part of this famous constellation.

1928. The boundaries now run only from east to west and from north to south, although they zigzag considerably so as not to expatriate bright stars and variable stars from constellations with which they have long been associated (Fig. 11.2).

11.1 Names of the stars

The brightest stars and other especially famous ones are known to us by personal names that have been handed down through the ages. Some of these names are of Greek and Latin origin. Some are derived from the Arabic; names such as Algol, Altair (*al* is the Arabic definite article), and many others are survivors from earlier times when astronomy was a favorite study of Mohammedan scholars. The stars were also named by shepherds, sailors, and nomads of the desert.

Procyon, meaning "before the dog," precedes Sirius the Dog Star in its rising. Aldebaran means "the follower"; it rises after the Pleiades. Antares is the "rival of Mars," because it is red. These are examples of how some of the stars were designated. Other star names have come to us in a different way.

In the earliest catalogs, such as Ptolemy's, the stars were distinguished by their positions in the imagined figures of heroes and animals. See Table 11.1. One star was the "mouth of the Fish"; another was the "tail of the Bird." Transcribed later into the Arabic, some of these expressions finally degenerated into single words. Betelgeuse, the name of the bright red star in Orion, was originally three words perhaps meaning the "armpit of the Central One."

11.2 Letter designations of the stars

The plan of designating the brighter stars by letters was introduced by Bayer, a Bavarian attorney, in 1603. In a general way, the stars of each constellation are denoted by small letters of the Greek alphabet (see Appendix) in order of their brightness, and the Roman alphabet is used for further letters. Where there are several stars of nearly the same brightness in the constellation, they are likely to be lettered in order from head to foot of the legendary creature. The full name of a star in the Bayer system is the letter followed by the possessive of the Latin name of the constellation. Thus Capella, the brightest star in Auriga, is alpha (α) Aurigae. The letters for some of the brighter stars are shown in the maps that follow.

Some of the fainter stars are known by their numbers according to a plan of numbering the stars in each constellation in order of right ascension. The star 61 Cygni is an example. Most faint stars,

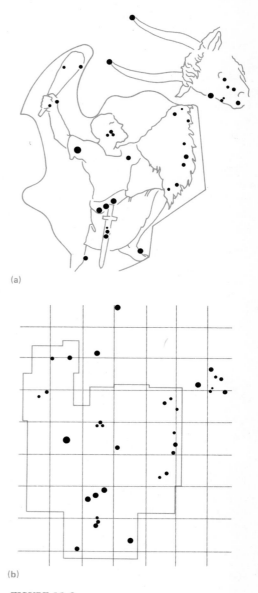

(a)

(b)

FIGURE 11.2
Old and new boundaries of Orion. Note the change to straight boundaries.

TABLE 11.1

Names of the constellations

Latin name	Possessive	English equivalent	Map
Androm'eda*	Androm'edae	Andromeda	4,5
Ant'lia	Ant'liae	air pump	
A'pus	A'podis	bird of paradise	
Aqua'rius*	Aqua'rii	water carrier	4
Aq'uila*	Aq'uilae	eagle	3,4
A'ra*	A'rae	altar	6
A'ries*	Ari'etis	ram	4,5
Auri'ga*	Auri'gae	charioteer	5
Boö'tes*	Boö'tis	herdsman	2,3
Cae'lum	Cae'li	graving tool	2,3
Camelopar'dalis	Camelopar'dalis	giraffe	
Can'cer*	Can'cri	crab	2,5
Ca'nes Vena'tici	Ca'num Venatico'rum	hunting dogs	2
Ca'nis Ma'jor*	Ca'nis Majo'ris	larger dog	5
Ca'nis Mi'nor*	Ca'nis Mino'ris	smaller dog	5
Capricor'nus*	Capricor'ni	sea-goat	4
Cari'na[a]	Cari'nae	keel	6
Cassiope'ia*	Cassiope'iae	Cassiopeia	1,4
Centau'rus*	Centau'ri	centaur	2,6
Ce'pheua*	Ce'phei	cepheus	1,4
Ce'tus*	Ce'ti	whale	4,5
Chamae'leon	Chamaeleon'tis	chameleon	
Cir'cinus	Cir'cini	compasses	
Colum'ba	Colum'bae	dove	5
Co'ma Bereni'ces	Co'mae Bereni'ces	Berenice's hair	2
Coro'na Austra'lis*	Coro'nae Austra'lis	southern crown	
Coro'na Borea'lis*	Coro'nae Borea'lis	northern crown	3
Cor'vus*	Cor'vi	crow	2
Cra'ter*	Crater'is	cup	2
Crux	Cru'cis	cross	6
Cyg'nus*	Cyg'ni	swan	3,4
Delphi'nus*	Delphi'ni	dolphin	4
Dora'do	Dora'dus	dorado	
Dra'co*	Draco'nis	dragon	1,3
Equu'leus*	Equu'lei	little horse	
Erid'anus*	Erid'ani	river	5,6
For'nax	Forna'cis	furnace	
Gem'ini*	Gemino'rum	twins	5
Grus	Gru'is	crane	4
Her'cules*	Her'culis	Hercules	3
Horolo'gium	Horolo'gii	clock	
Hy'dra*	Hy'drae	sea serpent	2
Hy'drus	Hy'dri	water snake	6
In'dus	In'di	Indian	
Lacer'ta	Lacer'tae	lizard	

TABLE 11.1 (CONTINUED)

Le′o*	Leo′nis	lion	2
Le′o Mi′nor	Leo′nis Mino′ris	smaller lion	
Le′pus*	Le′poris	hare	5
Li′bra*	Li′brae	scales	3
Lu′pus*	Lu′pi	wolf	3
Lynx	Lyn′cis	lynx	
Ly′ra*	Ly′rae	lyre	3,4
Men′sa	Men′sae	table mountain	
Microsco′pium	Microsco′pii	microscope	
Monoc′eros	Monocero′tis	unicorn	
Mus′ca	Mus′cae	fly	6
Nor′ma	Nor′mae	level	
Oc′tans	Octan′tis	octant	
Ophiu′chus*	Ophiu′chi	serpent holder	3
Ori′on*	Orio′nis	Orion	5
Pa′vo	Pavo′nis	peacock	6
Peg′asus*	Peg′asi	Pegasus	4
Per′seus*	Per′sei	Perseus	4,5
Phoe′nix	Phoeni′cis	phoenix	4
Pic′tor	Picto′ris	easel	
Pis′ces*	Pis′cium	fishes	4
Pis′cis Austri′nus*	Pis′cis Austri′ni	southern fish	4
Pup′pis[a]	Pup′pis	stern	5
Pyx′is[a]	Pyx′idis	mariner's compass	
Retic′ulum	Retic′uli	net	
Sagit′ta*	Sagit′tae	arrow	3,4
Sagitta′rius*	Sagitta′rii	archer	3
Scor′pius*	Scor′pii	scorpion	3
Sculp′tor	Sculpto′ris	sculptor's apparatus	4
Scu′tum	Scu′ti	shield	
Ser′pens*	Serpen′tis	serpent	3
Sex′tans	Sextan′tis	sextant	
Tau′rus*	Tau′ri	bull	5
Telesco′pium	Telesco′pii	telescope	
Trian′gulum*	Trian′guli	triangle	4,5
Trian′gulum Austr′le	Trian′guli Austra′lis	southern triangle	6
Tuca′na	Tuca′nae	toucan	6
Ur′sa Ma′jor*	Ur′sae Majo′ris	larger bear	1,2
Ur′sa Mi′nor*	Ur′sae Mino′ris	smaller bear	1,3
Ve′la[a]	Velo′rum	sails	2,6
Vir′go*	Vir′ginis	virgin	2
Vo′lans	Volan′tis	flying fish	
Vulpec′ula	Vulpec′ulae	fox	

* One of the 48 constellations recognized by Ptolemy.
[a] Carina, Puppis, Pyxis, and Vela once formed the single Ptolemaic constellation Argo Navis.

however, are designated only by numbers in one of the many catalogs to which the astronomer can turn for information about their positions, brightness, and other features. Two of these catalogs are most frequently used. One is the Bonner Durchmusterung. In it the sky is divided into strips of 1° in declination. Stars north of the celestial equator are designated by a plus; those south of the celestial equator are designated by a minus. The stars are then numbered in order of increasing right ascension in each strip starting with the number 1; this is closest to $0^h0^m0^s$ in right ascension. Thus the star BD +50° 3410 is the 3410th star in the +50° strip. The second of these catalogs is the Henry Draper catalog, which numbers all stars included in it in order of increasing right ascension. In this catalog the stars are indicated by the prefix HD and a running number. The star HD 206672 is the same as BD +50° 3410, which is also the star 80 Cygni.

As astronomy progressed a series of catalogs has been developed to serve special purposes. For example, double stars (see Section 13.1) are to be found in numerous catalogs as are variable stars (see Section 13.8). There are several catalogs of celestial radio sources, the most famous being the Cambridge series (2C, 3C, 4C) and the Parkes (PKS) catalog. Thus the object 3C273 is the 273rd object in the third Cambridge catalog of radio sources. X-Ray sources are cataloged by constellation in order of their discovery. Thus, Cygnus X-1 is the first X-ray source found in the constellation of Cygnus. We will refer to various special catalogs as we go along.

MAPS OF THE SKY

The six maps in this chapter show all parts of the celestial sphere. They are designed particularly for an observer in latitude 40°N, but are useful anywhere in middle northern latitudes. In order to avoid confusion, the many stars only faintly visible to the naked eye are not included, the names of inconspicuous constellations are generally omitted, and the boundaries between the constellations do not appear in the maps.

11.3 Directions in the sky

North in the sky is in the direction of the north celestial pole; south is near the opposite pole. The stars seem to circle from east to west daily. If these rules are remembered, there can be no confusion about directions in the sky.

The position of a star is clearly described when its directions in relation to other stars are given in this way. Its position is also

MAP 1
The northern constellations.

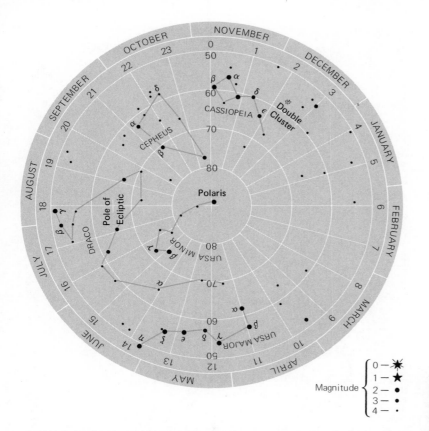

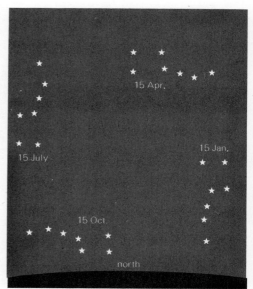

FIGURE 11.3
The Big Dipper at 9 o'clock in the evening at different seasons.

perfectly definite if we determine how it is related to two or more other stars. For example, the star may complete an equilateral triangle with two other stars already identified or it may be in line with them. We have noticed that a line through the pointer stars of the Big Dipper leads to the pole star. Such directions remain unaltered through the day and year.

Directions relative to the horizon, however, change as the stars go around the pole. From a star above the pole, north is down and west is to the left; from a star below the pole, north is up and west is to the right. Notice how the Big Dipper changes its position relative to the horizon during the night or from night to night as the seasons progress (Fig. 11.3). In the winter this figure of seven stars stands on its handle in the northeast at 9 o'clock, in spring it appears inverted above the pole, in summer the Dipper is seen descending bowl-down in the northwest, and in autumn it is right side up under the pole, where it may be lost from view for a time.

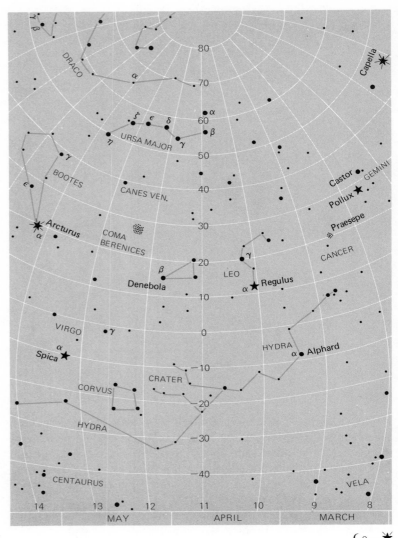

MAP 2
The stars of spring.

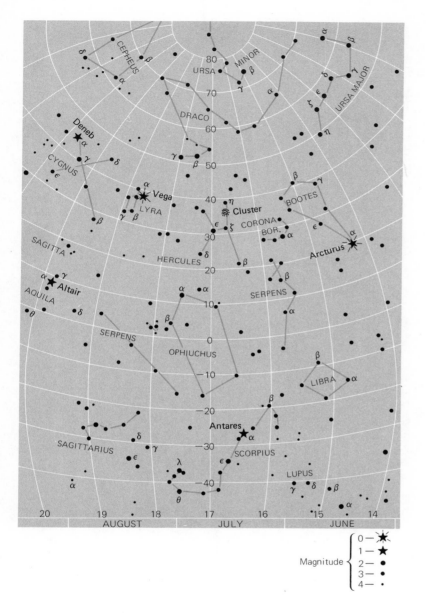

Magnitude $\begin{cases} 0 - \bigstar \\ 1 - \bigstar \\ 2 - \bullet \\ 3 - \bullet \\ 4 - \cdot \end{cases}$

MAP 3
The stars of summer.

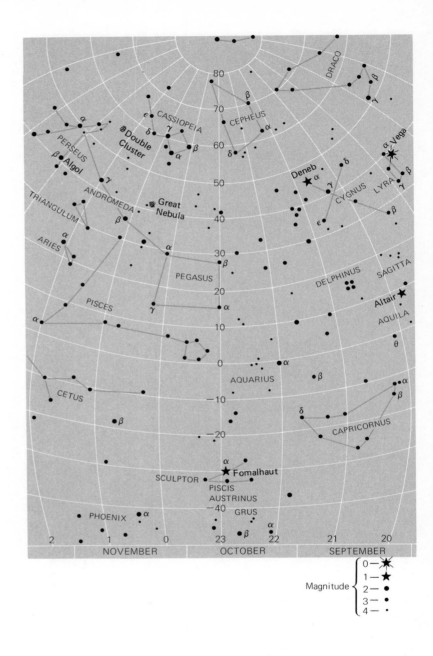

MAP 4
The stars of autumn.

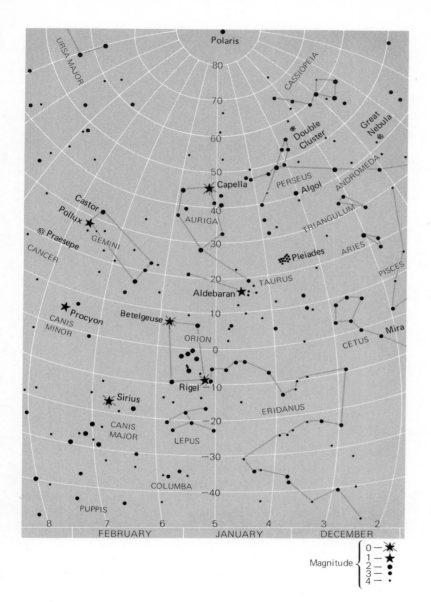

Magnitude $\begin{cases} 0 \\ 1 \\ 2 \\ 3 \\ 4 \end{cases}$

MAP 5
The stars of winter.

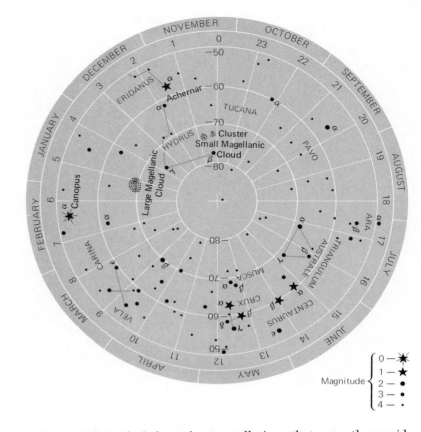

MAP 6
The southern constellations.

Maps 2 through 5 show the constellations that cross the meridian at 9 o'clock in the evening during each of the four seasons, all the way from the north celestial pole to the south horizon of latitude 40°N. The pole is near the top of each map. Hour circles radiating from the pole are marked in hours of right ascension near the bottom of the map, and circles of equal declination go around the pole.

To observe the constellations, select the map for the present season and hold it toward the south. The hour circle above the date of observation coincides with the celestial meridian at 9 o'clock in the evening on that date. Accordingly, the central vertical line in each of the four maps represents successively the positions of the meridian at 9 o'clock on 21 April, 21 July, 21 October, and 21 January. Remembering that the daily motions of the stars are from left to right for these maps and that a star comes to the meridian 2 hours earlier from month to month, we can locate the meridian in the maps for other times and dates.

Thus the maps are arranged to show what stars are crossing the celestial meridian at a particular time of the day and year. Any one of these stars can be identified in the sky if its distance from the zenith is also known at that time. The rule for finding the zenith distance is derived from Fig. 2.10. The zenith distance of a star at upper transit equals the observer's latitude minus the star's declination. If the distance is positive, the star is south of the zenith; if it is negative, the star is north of the zenith.

The zenith at the place of observation is on the circle having the same declination as the latitude of the place. As we are facing south, it would be necessary to lean backward to view the constellations in the upper parts of the maps. The northern constellations, however, are shown more conveniently in Map 1. They are repeated in the seasonal maps to display more clearly their relation to the constellations farther south.

The following examples illustrate the use of the maps:

1 Read from Map 2 the right ascension and declination of the star Regulus.

ANSWER Right ascension 10^h6^m, declination 12°N.

2 On 20 June 1960, the planet Jupiter was in right ascension 17^h55^m, declination 24°S. What was its position among the stars (Map 3)?

ANSWER It was in Sagittarius, west of the handle of the Milk Dipper.

3 On what date does the star Antares (Map 3) cross the meridian at 9 o'clock in the evening, standard time? What is then its distance from the zenith as observed in latitude 40°N?

ANSWER 12 July. The zenith distance is 66°S.

4 On what date does Orion (Map 5) rise at 9 P.M.?

ANSWER 1 November. Because Orion is near the celestial equator, it rises 6 hours earlier than the time of its meridian crossing, which is 9 P.M. on 1 February. Thus Orion rises at 3 P.M. on 1 February, and at 9 P.M. on 1 November.

11.4 The map of the northern sky

Map 1 shows the region of the heavens within 40° from the north celestial pole. The pole is at the center, closely marked by Polaris at the end of the Little Dipper's handle. Hour circles appear in this projection as straight lines radiating from the center; they are marked around the circumference of the map in hours of right ascension. The concentric circles are circles of equal declination at intervals of 10°; their declinations are shown on the vertical line.

This map is to be held toward the north. When it is turned so that the present month is at the top, the map represents the positions of the northern constellations at 9 P.M. standard time or, more exactly, local mean time if the correct part of the month is at the top. The hour circle that is then vertically under the date coincides with the celestial meridian at this time of night on this date. For a different time of night the map is to be turned from the 9 o'clock position through the proper number of hours—clockwise for an earlier time and counterclockwise for a later time.

If we wish, for example, to identify the northern constellations at 9 o'clock in the evening on 7 August, the map should be turned so that this date is at the top (at about the 18-hour circle). The Big Dipper is now bowl-down in the northwest, Cassiopeia is opposite it in the northeast, and so on. These northern constellations appear again in the maps for the different seasons. On all the maps the brightness of the stars is indicated by symbols representing their approximate magnitudes (see Section 12.17).

11.5 Stars of spring

The star maps are useful for identifying the prominent constellations in the sky and also for reference during the reading of this book. Our brief inspection continues with the stars of spring (Map 2), which are near the celestial meridian in the early evenings of this season.

In the early evenings of spring the Big Dipper appears inverted above the pole. This figure of seven bright stars is part of the large constellation Ursa Major, the Larger Bear, which extends for some distance to the west and south of the bowl of the Dipper; pairs of stars of nearly equal separations (Figs. 11.1 to 11.4) mark three paws of the ancient creature. Mizar, at the bend in the handle, has a fainter companion easily visible to the naked eye. With even a small telescope Mizar itself is revealed as a pair of stars.

Following the line of the Pointers northward we come to Polaris, the north star or pole star, less than 1° from the celestial pole. This useful star marks the end of the handle of the Little Dipper, the characteristic figure of Ursa Minor.

Leo is peculiarly associated with the spring season. Its familiar sickle appears in the east in the early evening as spring approaches, and it becomes the dominant figure in the south as this season advances. The bright star Regulus marks the end of the handle of the sickle figure, which with the right triangle to the east is the distinguishing feature of this constellation of the zodiac.

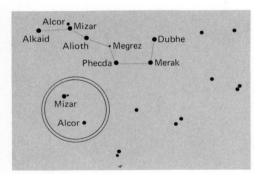

FIGURE 11.4
The constellation of Ursa Major giving the names of the stars of the Big Dipper. Compare this with Fig. 11.1. The insert shows Mizar and Alcor as viewed in a normal inverting telescope.

To the west of Leo is the dim zodiacal constellation Cancer with its Praesepe star cluster. To the east is the larger zodiacal constellation Virgo containing the bright star Spica. The course of the ecliptic here is very nearly the line from the Praesepe cluster through Regulus and Spica. Three-fifths of the way from Regulus to Spica it crosses the celestial equator at the autumnal equinox (see Fig. 11.5). The four-sided figure of Corvus in this region is likely to attract attention. When Corvus is at upper transit, at 9 o'clock in the evening about the middle of May, the Southern Cross is on the meridian 40° to the south, where it is invisible from all except viewers in the extreme southern part of the United States.

11.6 Stars of summer

A procession of familiar constellations, from Boötes to Cygnus, marches through the zenith in the evenings of summer in middle northern latitudes (Map 3). A fine region of the Milky Way, which we examine in a later chapter, extends from Cygnus past Aquila down to Scorpius and Sagittarius in the south (Fig. 11.6).

Boötes is overhead when the stars come out at the beginning of summer. Its stars outline the figure of a large kite with the brilliant Arcturus at the point where the tail is attached. This somewhat reddish star is pointed out by following the curve of the Big Dipper's handle around past its end. Arcturus' only peers in the whole northern celestial hemisphere are blue Vega and yellow Capella (see Chapter 12 for a discussion of star color and brightness).

Eastward from the kite figure and beyond the semicircle of Corona Borealis, the Northern Crown, we find Hercules, which is passing overhead at nightfall in midsummer. Some of its brighter stars may seem to suggest the figure of a rather large butterfly that is flying toward the west. The great cluster in Hercules, scarcely visible to the naked eye, is situated two-thirds of the way from the imagined butterfly's head along the west edge of the northern wing. Farther east on the line from the Crown through Hercules we come to Lyra. The figure here is a small parallelogram with a triangle attached at its northernmost point (Fig. 11.6). Vega marks a vertex of the triangle. The star at the north point of the triangle is the "double-double" epsilon Lyrae. This star is visible as two stars to the unaided eye but is more easily seen with binoculars; each star is again divided into two if observed with a telescope. The Ring nebula in Lyra (Fig. 11.7), visible only with the telescope, is about midway between the two stars in the southern side of the parallelogram.

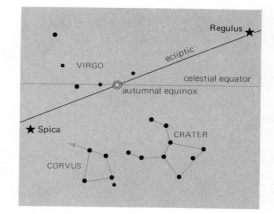

FIGURE 11.5
The position of the autumnal equinox. The autumnal equinox is roughly halfway between Spica and Regulus.

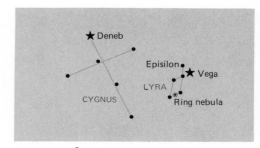

FIGURE 11.6
Cygnus and Lyra. This chart helps to locate the double-double star epsilon Lyrae and the Ring nebula.

FIGURE 11.7
The Ring nebula in Lyra. This beautiful planetary nebula (Chapter 14) is visible with a good small telescope. (Lick Observatory photograph.)

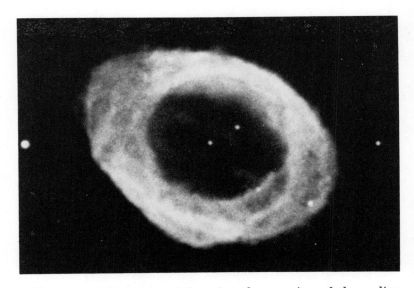

The formidable figure of Scorpius, the scorpion of the zodiac, dominates the southern sky at nightfall in early summer. Its bright Antares is a red supergiant star enormously larger than the sun.

Directly east of Antares, six stars of the zodiacal constellation Sagittarius outline the Milk Dipper, so named because it is in the Milky Way. In this direction we look toward the center of our Galaxy. The great star cloud of Sagittarius is an outlying portion of the galactic central region that is mainly obscured by intervening heavy clouds or cosmic dust. The Trifid nebula and the globular star cluster M 22 are some of the other features of this spectacular tract of the heavens. The place of the winter solstice is also near the handle of the Milk Dipper (Fig. 11.8).

11.7 Stars of autumn

Cassiopeia rises high in the north in autumn skies (see Map 4). The Northern Cross is overhead early in the season and, as it moves along, the Square of Pegasus becomes the dominant figure. Dim watery constellations spread across in the south, having only the star Fomalhaut to attract attention unless bright planets appear there as well.

The Northern Cross is overhead at nightfall in the early autumn in middle northern latitudes. This attractive figure, set in a fine region of the Milky Way (Fig. 11.9), is the characteristic feature of Cygnus, the Swan. Its brightest star, Deneb, marks the northern end of the longer axis of the Cross. Albireo (beta Cygni) decorates the southern end of this axis; it appears double with even a small

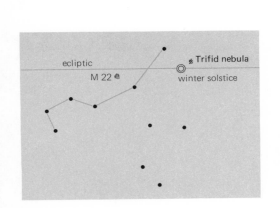

FIGURE 11.8
The Milk Dipper in Sagittarius. The winter solstice is located just to the right of the handle near the Trifid nebula.

telescope, a reddish star with a fainter blue companion. The Northern Cross shows the direction toward which the sun with its planetary system is moving in the rotation of our Galaxy.

The Square of Pegasus is peculiarly associated with our autumn skies. As this season approaches, we see it in the east in the early evening balanced on one corner. It crosses the meridian at 9 o'clock about 1 November. Four rather bright stars mark the corners of the large square which is the characteristic figure of Pegasus; they are alpha, beta, and gamma Pegasi, and in the northeast corner the alpha star of Andromeda. If we imagine that the Square of Pegasus is the bowl of a very large dipper figure, we find its handle extending toward the northeast (Fig. 11.10); its first three stars are the brightest members of Andromeda. A feature of this constellation is M 31, which is only faintly visible to the naked eye but is actually a spiral galaxy larger than our own. It is denoted

FIGURE 11.9
The Northern Cross. Beta Cygni at the foot of the Cross is out of the picture to the lower right. Note the North American nebula to the upper left. Note also the wreath-like nebula in the lower left. (Yerkes Observatory photograph.)

FIGURE 11.10
Region of the square of Pegasus. The spiral galaxies M31 and M32 are nearby (see Chapter 18).

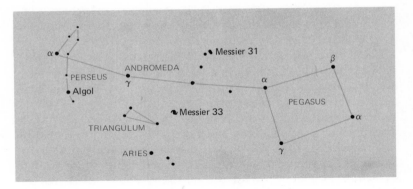

in Map 4 by its older name, the "Great Nebula" in Andromeda. Another galaxy is marked in the figure in the neighboring constellation Triangulum.

Southeast of the Square two streams of faint stars represent the ribbons with which the Fishes are tied. This dim constellation Pisces of the zodiac contains the zodiacal sign Aries, which has moved westward from its own constellation. Here we find the "first of Aries," or vernal equinox (Fig. 11.11). The line of the eastern side of the Square prolonged as far again to the south leads to this important point of the celestial sphere. The red variable star Mira shown in this figure is in the adjoining constellation Cetus.

11.8 Stars of winter

The bright winter scene is grouped around Orion, its most conspicuous figure. The zodiac is farthest north here; Taurus and Gemini have between them the position of the summer solstice.

Taurus contains two bright star clusters. The Pleiades cluster looks something like a short-handled dipper (Fig. 11.12). Seven of its stars are clearly visible to the naked eye, and two or three others twinkle into view. The conspicuous feature of the larger, Hyades cluster is in the form of the letter V which, including the bright star Aldebaran, represents the head of Taurus. At the tips of the horns are two fairly bright stars, the northern one of which is needed to complete the muffin figure of Auriga; its brilliant star Capella is near the zenith at 9 o'clock toward the end of January.

Gemini has an oblong figure. The heads of these Twins of the zodiac are marked by the bright stars Castor and Pollux. The position of the summer solstice (Fig. 11.13) is between their feet and the horns of Taurus. The Milky Way sweeps down from Auriga and Taurus past the feet of the Twins and on to the south horizon.

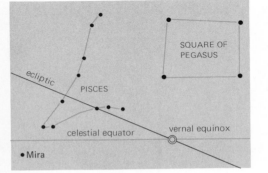

FIGURE 11.11
The position of the vernal equinox.

Orion, the brightest constellation, is peculiarly associated with the winter. Its oblong figure rises in the early evening as winter approaches, appears in the south at 9 o'clock about 1 February, and is setting in the twilight as spring advances.

This bright region of the heavens inspired a lively scene in the old celestial picture book. Orion, a mighty hunter accompanied by his dogs, stands with uplifted club awaiting the charging Taurus. Red Betelgeuse glows below his shoulder. Blue Rigel diagonally across the figure is somewhat the brighter of the two. Three stars

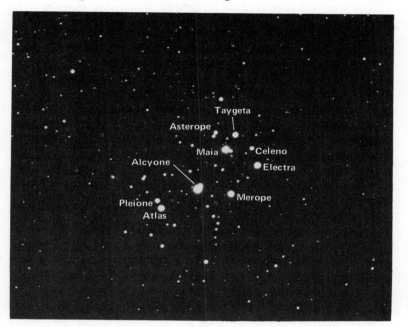

FIGURE 11.12
The fine Pleiades cluster with the names of the principal stars. This group of stars is easily visible to the naked eye. (Yerkes Observatory photograph.)

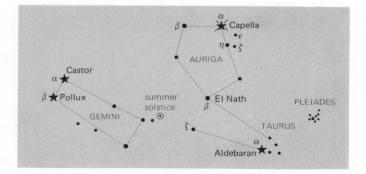

FIGURE 11.13
The region of Gemini, Auriga, and Taurus. The summer solstice is approximately half-way between Castor and Aldebaran.

near the center of the rectangle mark Orion's belt, and three fainter ones in line to the south (Fig. 11.14) represent his sword. The middle star of the three appears through the telescope as a trapezium of stars surrounded by the foggy glow of the great nebula in Orion.

The stars of Orion's belt are useful pointers. The line joining them directs the eye northwestward to the Hyades and southeastward to Sirius, the "Dog Star," the brightest star in the heavens and one of the nearest. Sirius, Betelgeuse, and a third bright star across the Milky Way form a nearly equal-sided triangle. The third one is Procyon, in Canis Minor.

11.9 The southern constellations

Map 6 shows the part of the heavens that does not come up into view in latitude 40°N. It contains the Southern Cross, the two

FIGURE 11.14
The constellation of Orion tangled in nebulae. A blue-sensitive emulsion was used, hence the extreme brightness of Rigel and Bellatrix compared to Betelgeuse, which is midway between these stars in visual brightness. (Hale Observatories photograph.)

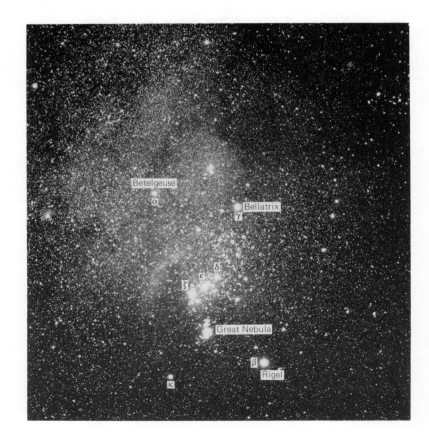

Magellanic Clouds, a number of brilliant stars, and a fine region of the Milky Way.

Crux, the Southern Cross, is a small figure of four stars, which resembles a kite as much as a cross. It becomes entirely visible south of latitude 28°N. Its brightest star, alpha Crucis, is about as bright as Aldebaran. The Magellanic Clouds, companions of our spiral galaxy, do not rise anywhere in the United States. From more southern latitudes they are clearly visible to the unaided eye. Canopus, almost directly south of Sirius, is second to it in brightness. Alpha Centauri is third in order of brightness among all the stars, and is nearest of all to the sun.

The south celestial pole is situated in the dim constellation Octans. There is not a star as bright as Polaris within 20° from this point. Sigma Octantis, a star barely visible to the naked eye, is 50′ from the pole.

PLANETARIUMS

The word **planetarium** may refer either to the complex instrument that projects a changing picture of the heavens on the interior of a hemisphere, or else to the building or chamber that houses the apparatus. The planetarium is playing an ever increasing role in public education in astronomy.

Among the larger planetariums in the United States having domes 18 meters in diameter and greater are the Adler Planetarium in Chicago, the Fels Planetarium in Philadelphia, the Griffith Observatory in Los Angeles, the American Museum–Hayden Planetarium in New York, the Buhl Planetarium in Pittsburgh, the Morehead Planetarium in Chapel Hill, the Morrison and Charles Hayden Planetarium in Boston, and the Air Force Planetarium near Colorado Springs. These and many others are open to the public on a regular schedule basis and their demonstrations change four to six times per year. In addition all of them feature special science exhibits of equipment and demonstrations of basic principles.

Planetariums in increasing numbers are offering instructive replicas of the heavens and of the motion of the celestial bodies. They are often used in connection with introductory astronomy courses having laboratory exercises. Laboratory classes are at the mercy of the weather and are difficult at best in large cities, so the planetarium offers a chance to make certain observations under known conditions. The students can plot the motions of the planets, receive training in meteor counting and plotting, etc. The planetarium is also proving to be a very useful tool in the teaching of mathematics and celestial navigation to pilots and astronauts.

QUESTIONS

1 What are the two principal methods for designating the brighter stars?
2 If you hear the designation Vela X-1, what does it mean?
3 When is the star Capella directly above the pole star at 9 P.M.? When is it due east of the pole star at 9 P.M.?
4 When is the Large Magellanic cloud directly above the south pole at 9 P.M.? When is it directly below the south pole at 9 P.M.?
5 If you continue the arc of the handle of the Big Dipper, to what bright star does it lead (approximately)?
6 Star Map 5 seems to have more bright stars than the others. Can you explain this by observing this region of the night sky?
7 A total lunar eclipse occurs just before midnight on 21 June. What constellation is the moon likely to be in?
8 How far south must you be to see the Southern Cross?
9 Mark the approximate south pole of the ecliptic on Map 6. What object is nearby?
10 Note the large wreath in the lower left corner of Fig. 11.9) Place it in the proper star map.

FURTHER READINGS

ALLEN, RICHARD H., *Star Names*, Dover, New York, 1963.
BAKER, ROBERT H., *Introducing the Constellations*, Viking, New York, 1957, revised edition.
BURNHAM, ROBERT, *Celestial Handbook*, Northland Press, Flagstaff, Ariz., 1966.
MAYALL, NEWTON, MARGARET MAYALL, AND JEROME WYCKOFF, *The Sky Observer's Guide*, Golden Press, New York, 1959.
NORTON, ARTHUR P., *A Star Atlas*, 14th ed., Sky Publishing Co., Harvard Observatory, Cambridge, Mass., 1959.
See also the simpler *Popular Star Atlas*.
WEBB, THOMAS W., *Celestial Objects for Common Telescopes*, Dover, New York, 1963, revised by Margaret W. Mayall.

A nineteenth-century star map from Ball's Atlas of Astronomy.

The Stars Around Us

12

12

The distances of the nearer stars are shown by the amounts of their slightly altered directions as the earth revolves around the sun. The motions of the stars are revealed by their progress against the background of more remote stars and by displacements of the lines in their spectra. Stars in the sun's neighborhood are often bright enough to permit the study and classification of their spectra. Using such observable data and their relative brightness, stars may be compared with one another and with the sun.

The distance of a celestial object may be found by observing its parallax, which has been defined as the difference between the directions of an object as viewed from two different places a known distance apart. The moon's parallax (see Section 6.1) is large enough to be measured from widely separated stations on the earth. The difference between the directions of even the nearest star, however, observed from opposite points on the earth is not greater than the width of a period on this page if the period were viewed from a distance of about 2000 kilometers. To detect the parallaxes of stars we require the far greater separation of points of observation afforded by the earth's revolution around the sun.

12.1 Parallaxes of the stars

The earth's revolution causes apparent slight oscillations of the nearer stars (Fig. 12.1). The extent of the displacement for a particular star is found by comparing its positions in a series of photographs taken 6 months apart; the positions are determined with reference to other stars that appear nearby but are likely to be too far from us to be considerably displaced. Because part of the observed displacement may be caused by the movement of the star itself, it is necessary to repeat the photographs at half-year intervals until the annual parallax oscillation can be disentangled from the straight-line motion of the star.

The parallax of a star, in the usual meaning of the term, is its heliocentric relative parallax. It is the greatest difference between the star's directions from the earth and sun during the year, with slight correction to keep the earth at its average distance from the sun, and is about half the greatest parallax displacement of the star. When the parallax has been measured, the star's distance can be calculated.

In this book we shall use the symbol π for parallax in keeping with the practice of those who measure stellar trigonometric parallaxes. The student should have no difficulty telling when the symbol has its mathematical meaning—that of the ratio of the circumference of a circle to its diameter. When the symbol stands alone it is understood to be the **relative parallax** of the star, that is, the star's parallax relative to the reference stars used. When a correction is applied to allow for the finite distance of the reference stars, a very small correction obtained from large statistical studies, we say that we have obtained the **absolute parallax** (π_{abs}) of the star.

The distances to the stars are so great in ordinary units that new terms must be devised that make the numbers more manageable. A convenient unit is the distance (d) at which a star will have a parallax of 1 second of arc and is called the **parsec** (pc). By definition, 1 parsec equals 206,265 astronomical units. Since 1 astronomical unit is about 8.3 light minutes, the time it takes light to travel 1 astronomical unit, we can compute that 1 parsec equals 3.26 **light years** (ly).

$$d_{pc} = \frac{1}{\pi}$$

$$d_{ly} = \frac{3.26}{\pi}$$

1 ly = 9.46×10^{12} km

The moon is 1.3 light seconds away. The sun is 8.3 light minutes from the earth. Pluto is almost 5.5 light hours from the sun. The nearest stars, as we will soon see, are in the alpha Centauri system with a parallax of 0.761 second of arc. According to the above its distance in parsecs is 1/0.761 or 1.31 parsecs. To get this in light years we merely multiply by 3.26 and find that the nearest star is 4.28 light years away. Space is essentially devoid of stars and the Solar System is miniscule indeed.

Later in this book, numbers involving distances in parsecs become large and we resort to distances given in thousand parsec units, or kiloparsecs (kpc). Finally, even kiloparsecs get out of hand and we resort to a million parsecs, or the megaparsec (Mpc), as our unit.

12.2 The nearest stars

When Copernicus proposed his heliocentric theory it was realized that if it were true, the stars should reflect the earth's annual motion around the sun. Parallaxes were searched for but could not be detected, therefore the theory was thought to be wrong. Man simply did not realize how far away even the nearest stars were. The first measured parallax was not announced until 1838 by F. W. Bessel, long after the heliocentric theory had been proved by other unanticipated methods. The technique used by Bessel and other early measurers of parallaxes was to measure very accurately the transit times of several faint stars around the suspected parallax star. Over the course of 18 months he noted the minute annual change in position (a total of 0.04 second of time) of 61 Cygni. Current methods replace this technique with a long-focus telescope and the photographic plate, but the principle remains the same.

The sun's nearest neighbor among the stars is the bright double

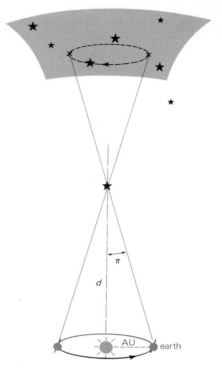

FIGURE 12.1
The parallax of a star. As the earth revolves around the sun the nearer star appears to oscillate relative to the more remote stars. Half of the angle is called the parallactic angle.

star alpha Centauri. The two stars have a faint companion situated about 2° from them; it is known as "Proxima," because it seems to be slightly nearer us than to them. The parallax of these stars is 0".761, so that their distance is 4.3 light years, or about 41×10^{12} kilometers. If the size of the sun is represented by a period on this page, alpha Centauri would be shown on this scale by two similar dots 8 kilometers away. This is a fair sample of the wide separation of the stars around us.

At least 26 stars are known to be within 12 light years from the sun. These stars are listed in Table 12.1. The column marked spectral type refers to the color of the stars and is explained in Section 12.20. Very briefly, most stars seem to arrange themselves

TABLE 12.1
The nearest stars

	Visual magnitude	Absolute magnitude	Spectral type	Parallax	Proper motion
α Centauri C	10.7 *	15.1 *	M5eV	0".763	3".68
α Centauri A	0.00	4.38	G2V	.761	—
α Centauri B	1.38	5.76	K4V	.761	—
Barnard's star[c]	9.53	13.21	M5V	.552	10.30
Wolf 359	13.66	16.80	M6eV	.431	4.84
Lalande 21185[c]	7.47	10.42	M2V	.402	4.78
Sirius A	−1.46	1.41	A1V	.377	1.32
Sirius B	8.67	11.54	w.d.A5	.377	—
L726-8A	12.41	15.28	M6eV	.365	3.35
L726-8B	12.95	15.87	M6eV	.365	
Ross 154	10.6 *	13.3 *	M5V	.345	0.74
Ross 248	12.25	14.75	M6eV	.317	1.82
ε Eridani	3.74	6.14	K2V	.305	0.97
L 789-6	12.58	14.93	M6eV	.302	3.27
Ross 128	11.13	13.50	M5V	.301	1.40
61 Cygni A[c]	5.20	7.53	K5V	.292	5.22
61 Cygni B	6.03	8.36	K7V	.292	—
ε Indi	4.73	7.01	K5V	.291	4.67
Procyon A	0.35	2.64	F5V	.287	1.25
Procyon B	10.8 *	13.1 *	w.d.F	.287	—
Σ 2398A	8.92	11.19	M3V	.284	2.29
Σ 2398B	9.68	12.23	M4V	.284	
Groombridge 34A	8.10	10.35	M1V	.282	2.91
Groombridge 34B	11.07*	13.32*	M6V	.282	—
Lacaille 9352	7.36	9.59	M2V	.279	6.87
τ Ceti	3.54	5.72	G8V	.273	1.92

SOURCE After P. van de Kamp.
[c] astrometric companion
* uncertain values

quite nicely on a graph from bright, hot, blue stars to faint, cool, red stars, and this smooth arrangement is called the **main sequence.** Although the bright stars alpha Centauri, Sirius, and Procyon are included in this list, the majority of the nearest stars are too faint to be seen without a telescope. We conclude that the stars differ greatly in intrinsic brightness, so that a bright star in our skies may not be nearer than a faint one.

The direct method of determining the distances of stars by observing their parallaxes is limited to the nearer ones. At the distance of 300 light years, which is only a step into space, the parallaxes become too small to be measured reliably by this technique. Ways of finding the distances of more remote stars will be described later.

THE STARS IN MOTION

The motions of the stars in different directions relative to the sun are shown by the very gradual changes in their places in the heavens and by Doppler displacements of their spectral lines. The sun is also moving among the stars around it. Like the sun, the stars are rotating on their axes.

12.3 Proper motions of the stars

In 1718 Edmund Halley of Halley's comet was the first to explain that the stars are not stationary. He observed that Sirius and some other bright stars had drifted as much as the apparent width of the moon from the places assigned them in Ptolemy's early catalog. Meanwhile, the proper motions of all lucid stars and of many fainter ones have become known by comparison of the records of their positions at noted different times.

The present technique makes use of an astrograph (see Section 5.5); plates are taken at widely separated times (ten years or more). The plates often covering fields up to 10×10 degrees contain the images of many thousands of stars, and when the two plates are alternately superposed optically, the stars that have moved appear to jump back and forth and attract the astronomer's attention. W. Luyten, H. Giclas, and many others have used this highly successful method for compiling catalogs containing thousands of stars, most of them quite faint, having significant proper motions.

The **proper motion** of a star is the angular rate of its change of place in the plane of the sky per year. This change becomes so slow at great distances from us that the remote stars can serve as landmarks to show the progress of the nearer stars. Barnard's star (Fig. 12.2) has the largest proper motion. In Fig. 12.2 three plates, September 1948, March 1949, and September 1949, are superposed so that the two background stars are fixed. This reveals the east–west parallactic motion of the star and its large south–north proper motion. Named after the astronomer who first observed its swift flight, this faint star in Ophiuchus moves among its neighbors in the sky at the rate of 10″.3 a year, or as far as the apparent width of the moon in 175 years. If all the stars were moving as fast as this and at random, the forms of the constellations would be altered appreciably during a lifetime. Luyten asserts that the known motions of only about 330 stars exceed 1″ a year and that the average for all stars visible with the naked eye is not greater than 0″.1 a year. In the course of a century the progress of very few stars is enough to be detected without a telescope.

12.4 Radial velocities of the stars

The proper motion of a star tells us nothing of its movement toward or away from us. The **radial velocity,** or velocity in the line of sight, is revealed by the Doppler effect (see Section 5.12)

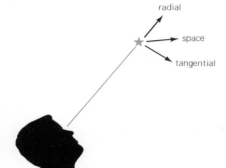

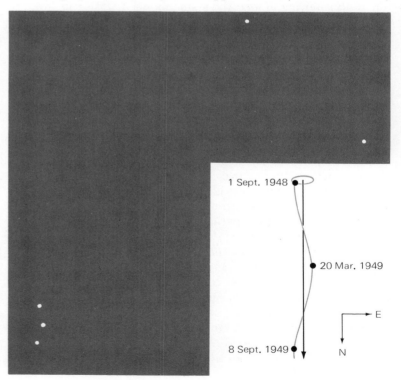

FIGURE 12.2
Proper motion and parallax of Barnard's star. This is a composite of three different plates taken at six-month intervals. The reference stars have been carefully superposed and the picture demonstrates proper motion and parallax. (Sproul Observatory photograph.)

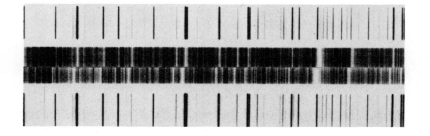

FIGURE 12.3
Doppler displacements in the spectra of the stars HD 161096 (top) and HD 66141 (bottom). The shift in the spectra are obvious. The upper spectrum and the comparison lines, top and bottom, have been carefully aligned.

in its spectrum. If the lines are displaced toward the violet end of the spectrum, the star is approaching us; if they are displaced toward the red end, the star is receding from us. The amount of the displacement is proportional to the speed of approach or recession.

The spectra of two stars are shown in Fig. 12.3 with the bright lines of the iron arc above and below them. Dark lines in the upper spectrum are displaced to the violet (left); the star is approaching us at the rate of 108 kilometers a second. The same dark lines in the lower spectrum are displaced to the red (right); the star is receding at the rate of 116 kilometers a second. These are unusually swift motions. The radial velocities of most stars in the sun's neighborhood do not exceed 60 kilometers per second.

12.5 The sun's motion

The sun's motion among the stars in our neighborhood is indicated by the common drift of these stars in the opposite direction in addition to their individual motions. The stars are spreading from the standard **apex of the sun's way** (Fig. 12.4), the point in the heavens toward which the sun's motion is directed, and are closing in toward the **antapex**, the opposite point. This apparent drift of the celestial scenery may be observed by examining the proper motions of many stars in various parts of the heavens. It is also observed in the radial velocities, which have their greatest values of approach in the region of the apex and of recession around the antapex; these show the speed of the solar motion as well.

Relative to the stars around it, the sun with its family of planets is moving at the rate of 19.4 kilometers a second toward a point in the constellation Hercules, in right ascension 18^h0^m and declination $+30°$, about $10°$ southwest of the bright star Vega. When the drift of fainter stars is included, the apex shifts toward Cygnus in which the sun's motion is directed in the rotation of the Galaxy

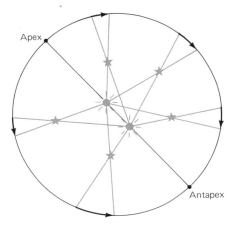

Apex

Antapex

FIGURE 12.4
Apparent drift of stars away from the apex of the sun's way. Note that stars near the apex or antapex show the least drift.

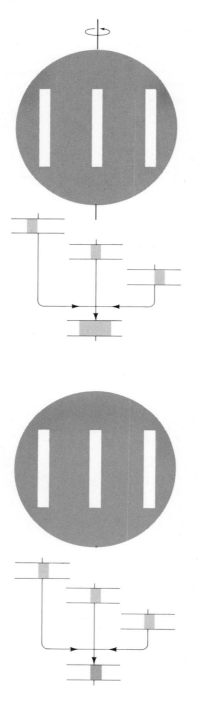

(Section 17.7) at about 230 kilometers per second. When the drift of special types of stars and objects are considered, the apex of the sun's way shifts accordingly. This tells us that these special types of stars and objects form significant groups differing from the average motion of all stars together. For example, the motion of the sun with respect to the short-period variable stars (see Section 13.9) is about 130 kilometers per second whereas its motion with respect to the globular clusters (see Section 16.8) is 175 kilometers per second.

12.6 Rotations of the stars

Since the pioneer work of O. Struve and C. T. Elvey at Yerkes Observatory, about 1930, the rotations of stars have been studied extensively by means of the widening of the lines of their spectra. Starlight comes from a source that is partly approaching and partly receding from the earth except when the axis of rotation is directed toward the earth. Thus the lines are widened by the Doppler effect by an amount that depends on the speed of the rotation and the direction of the axis. We can picture the line widening as if our spectroscope could see first the approaching limb of the star, then the center, and finally the receding limb as shown in the diagram in the margin. Actually, since the stars are seen as point sources our spectrograph records the line from every point of the star. The integrated effect is a broadened, shallow line as shown. If the star is not rotating or if its axis rotation is pointed toward us the integrated effect is a narrow deep line as shown in the lower part of the diagram.

Blue stars of the main sequence (see Section 12.20) are likely to have high speeds of rotation, some exceeding 300 kilometers a second at their equators. As an example, note the wide lines in the spectrum of Altair (Fig. 12.5); the period of rotation of this star is 6 hours, as compared with about a month in the case of the sun. The narrower lines in the spectrum of Vega, also a blue star, would suggest that its axis is directed more nearly toward us if its rotational period is also 6 hours.

Yellow and red stars of the main sequence rotate more slowly, except when they belong to close binary systems; perhaps much of the original spins of the single stars now appear in the revolutions of their planets. Giant stars generally have slower rotations than corresponding main-sequence stars. Conservation of their angular momentum requires them to rotate more slowly as they expand in their evolution from the main sequence to the giant stage. An excellent demonstration of this physical law is seen by watching a good figure skater when she increases or slows down

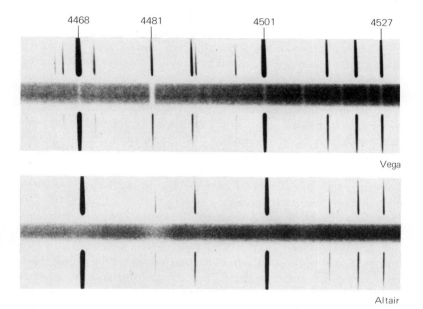

Vega

Altair

FIGURE 12.5
Negative prints of spectra of Vega and Altair. The broadened lines of Altair show the rapid rotation of this star. (Yerkes Observatory photograph.)

her rate of spin by simply retracting or extending her arms during the spin.

12.7 Magnetism of stars

While the detailed physics of the polarization of radiation passing through a magnetic field or material in a magnetic field is beyond the scope of this text, the basic concept is fairly easy to understand. Think of starlight as being composed of radiation vibrating along axes having a completely random distribution and let the stellar or interstellar particles align themselves with the magnetic field. Clearly the radiation vibrating in the same plane will pass through unobstructed whereas that vibrating in the perpendicular plane will be obstructed and not pass through. Measuring the amount of polarization tells us something about the strength of the magnetic field causing the polarization, the particle size, and the particle density of the material itself. This form of polarization is known as linear polarization and is revealed by the Zeeman effect. In addition radiation can be elliptically polarized or circularly polarized depending upon the magnetic field and the angle at which the radiation passes through the field.

The Zeeman effect (see Section 10.1) in stellar spectra was discovered by H. W. Babcock at the Mount Wilson Observatory in 1946. His catalog of 1958 lists 89 stars definitely showing this evidence of magnetism, and many other stars that probably show

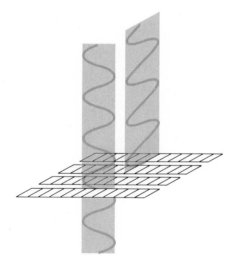

this effect. The majority are stars of type A (see Section 12.15), which have in their spectra narrow and unusually intense lines of metals such as chromium and strontium, and of rare earths such as europium. They are presumed to be rapidly rotating stars viewed pole-on.

The magnetic fields vary in strength, and some show periodic reversals of polarity. As an example, the polar magnetic field of the star α^2 Canum Venaticorum ranges in cycles of 5.5 days between extremes of $+5000$ and -4000 gauss, values that are comparable with the strongest magnetic fields of sunspots. At the first extreme the lines in the spectra of the metals have their greatest intensity, and at the second extreme the lines of the rare earth elements are the most intense. The explanation is proposed that these are effects of oscillations in the outer layers of the stars analogous to those in the magnetic cycle of sunspots.

J. Kemp reasoned that certain types of stars known as white dwarfs should show circular polarization because they should have very strong magnetic fields. A percentage of the white dwarfs ought to have their magnetic axes oriented in such a way that we can observe the effect. After studying several stars Kemp and his colleagues detected one, Grw $+70°$ 8247 (Grw is an abbreviation meaning the Greenwich Catalog), indicating that it had a magnetic field on the order of 10^7 gauss or larger! This discovery was followed by the detection of very large field strengths in several more white dwarfs.

STELLAR SPECTRA

The spectra of stars are characterized by patterns of dark lines, and in some cases of bright lines as well, on otherwise continuous bright backgrounds. The patterns are different for stars having different surface temperatures. As noted earlier, different molecules and atoms produce characteristic lines on any particular spectrum. In order to understand this process more thoroughly, we will first examine the chemical makeup of these substances and then look at some spectra in detail.

12.8 Constituents of atoms

The smallest particle of any of the 92 chemical elements (103 including man-made elements) that retain the properties of that element is known as an **atom**. Atoms are composed essentially of electrons, protons, and neutrons. The **electron** is the lightest of these constituents and it carries unit negative charge of electricity.

The proton is 1836 times as heavy as the electron and carries unit positive charge. The **neutron** has about the same mass as the proton and is electrically neutral.

The atom consists of a nucleus surrounded by electrons. The nucleus ranges from a single proton in the ordinary hydrogen atom to an increasingly complex group of protons and neutrons in the heavier chemical elements. Each added proton contributes one unit to the positive charge on the nucleus. In the normal atom the nucleus is surrounded by negatively charged electrons equal in number to the protons, so that the atom as a whole is electrically neutral. (See Fig. 12.7.)

Atoms can combine to form compounds that are also electrically neutral. Compounds will have different properties than those of the atoms that comprise them. The smallest part of a compound that still retains the properties of the compound is known as a **molecule.** An example is carbon dioxide, which is composed of an atom of carbon and two atoms of oxygen.

12.9 The chemical elements

Table 12.2 lists the names, symbols, atomic numbers, and atomic weights of the 26 lighter elements, from hydrogen to iron. All elements heavier than helium are referred to as heavy elements by astronomers.

The **atomic number** of an element is the number of protons in the nucleus and also the number of electrons around the nucleus of the normal atom. All atoms having the same atomic number belong to the same chemical element. Of the 103 recognized elements, those beyond uranium, number 92, were first identified in various radiation laboratories.

The **atomic weight** is the relative mass of the atom; the unit employed in the table is one sixteenth the mass of the average oxygen atom, taken as 16.0000. The weights given here are generally the averages for two or more different kinds of atoms, or **isotopes,** of the same element having different numbers of neutrons in their nuclei.

Hydrogen comprises about 90 percent of all the atoms and 70 percent of all the mass of material in the stars. Helium is second with about 9 percent of the atoms and 28 percent of the mass. All the other elements together contribute only a little more than 1 percent to the mass of the stars.

12.10 Model of the hydrogen atom and its spectrum

The ordinary hydrogen atom consists of one proton attended by a single electron. A pictorial model proposed by Niels Bohr in 1913

TABLE 12.2
The lighter chemical elements

Element	Symbol	Atomic number	Atomic weight
Hydrogen	H	1	1.0080
Helium	He	2	4.0028
Lithium	Li	3	6.939
Beryllium	Be	4	9.013
Boron	B	5	10.812
Carbon	C	6	12.0116
Nitrogen	N	7	14.0073
Oxygen	O	8	16.0000
Fluorine	F	9	18.9992
Neon	Ne	10	20.184
Sodium	Na	11	22.991
Magnesium	Mg	12	24.313
Aluminum	Al	13	26.982
Silicon	Si	14	28.09
Phosphorus	P	15	30.975
Sulfur	S	16	32.066
Chlorine	Cl	17	35.454
Argon	A	18	39.949
Potassium	K	19	39.103
Calcium	Ca	20	40.08
Scandium	Sc	21	44.958
Titanium	Ti	22	47.90
Vanadium	V	23	50.944
Chromium	Cr	24	52.00
Manganese	Mn	25	54.940
Iron	Fe	26	55.849

NOTE To place on C^{12} scale multiply atomic weight by 0.99996; to place on physical scale multiply by 1.00028.

helps in an understanding of the atomic spectrum of hydrogen. In this model the electron revolves around the proton in one of many permitted orbits or energy levels. The atom can absorb energy from a radiation field, for example from a continuum of light that strikes it, only in the precise amount required to raise the electron from a lower orbit or energy level to a higher one. This **absorption** produces a **dark line** at the appropriate wavelength in the spectrum of the light. If the electron then falls back to the original lower level the energy is reemitted at the same wavelength.

At first one might think that this process should cancel out and leave the spectrum as it was. However, the light abstracted in the first process is removed from the line of sight to the observer, but the reemitted light is sent out in all directions, including the direction back to the source. Thus, only a small percentage of the light is reemitted toward the observer and the dark line is observed.

Although one hydrogen atom with its single electron absorbs only one dark line at a time, the many atoms in the atmosphere of a star together can produce the series of all permitted lines. The particular series where the electrons are raised from the second orbit of the Bohr model (Fig. 12.6) is known as the **Balmer series**. Its first line, Hα, is in the red region of the spectrum and its second line, Hβ, is in the blue-green region. From there the lines continue with diminishing spaces between them to a head in the ultraviolet. More than 30 lines of the Balmer series are recognized in the spectra of blue stars.

The **Lyman series**, based on the innermost Bohr orbit, is a similar array of lines in the far ultraviolet; it is recorded as a series

FIGURE 12.6
Conventional representation of the hydrogen atom. Possible orbits of the electron around the nucleus are shown as circles and represent the various energy levels the electron occupies. Changes to and from the levels give rise to emission and absorption series of hydrogen.

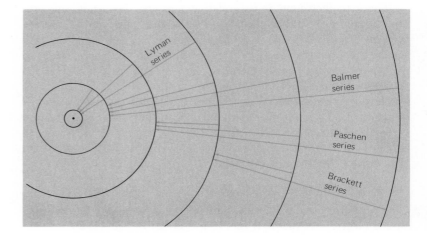

of bright lines in the spectrum of the sun photographed from rockets and satellites. The **Paschen series** and others are in the infrared.

It is apparent that transitions from the third energy level to the second (Hα) are less energetic than those from the second to the first (Lα). Similarly, those from the fourth to the third (Pα) are less energetic than Hα, and so on. Transitions from the very high levels to levels nearby are consequently much less energetic and may fall in the radio region of the spectrum. The transition from level 110 to level 109 has been observed at λ5.99 centimeters from the hydrogen atom in the low density interstellar space.

Another energy transition of hydrogen that has had enormous consequences for astronomy occurs in the lowest level or ground state. In our model we may picture the nucleus and electron as spinning, or rotating, as the electron revolves about the nucleus. Quantum theory tells us that the electron must have constant spin but the spin may be in the same direction as that of the nucleus (parallel) or opposite to the direction of the nucleus (antiparallel). E. Fermi first pointed out that the energy in these two conditions is different. H. van de Hulst noted that when the atom flips from parallel to antiparallel spin the radiated energy has a wavelength of 21 centimeters and this should be detectable if the amount of neutral hydrogen is sufficient. This single transition has been the most powerful tool in determining the structure of the Galaxy over the past decade.

12.11 Atomic models of other elements

Each step in the succession of elements heavier than hydrogen adds one proton and usually one neutron to the nucleus of the atom and one electron in the outer structure. Thus the normal helium atom has two protons, two neutrons, and two electrons, and lithium has three of each. The electrons of the growing population take their places systematically in **shells,** which are the same as the Bohr orbits. The shells are filled in order of distance from the nucleus by $2n^2$ electrons in each, where n is the number of the Bohr orbit. Thus the first shell is filled by two electrons, the second by eight electrons, and so on. A filled shell will not receive more electrons and is reluctant to release any that it possesses.

The electron structures of a few normal atoms are shown in Fig. 12.7. The helium atom has its two electrons locked in its filled innermost shell, which is not easily broken. The neon atom has two filled shells and is also relatively inactive. The lithium and sodium atoms, which are very active, each have in an outer shell a single electron that can be readily removed. These few examples

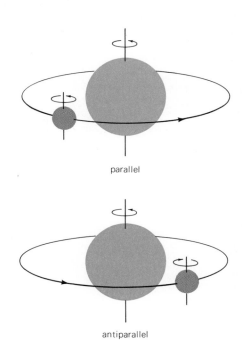

parallel

antiparallel

FIGURE 12.7
Shell model of some of the lighter atoms.
In this model the orbits are great circles on
the various shells. The more complete the
shell the tighter, that is, the smaller is the
shell.

helium (2)   neon (10)

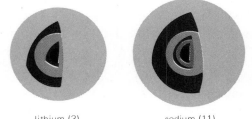

lithium (3)   sodium (11)

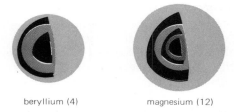

beryllium (4)   magnesium (12)

will suggest that atoms of the different chemical elements in the atmospheres of stars in the same conditions may not be equally effective in producing lines in their spectra. We see presently that the different patterns of lines in stellar spectra are determined mainly by the different temperatures of the stars.

12.12 Neutral and ionized atoms

The **neutral atom** has its full quota of electrons. The **normal atom** has its electrons at the lowest possible levels, as in Fig. 12.7, whereas the **excited atom** has absorbed energy to raise one or more electrons to higher levels. We designate the neutral atom by its symbol plus the Roman numeral I.

The **ionized atom** has lost one or more electrons. It has absorbed enough energy to expel these electrons beyond its outermost orbit. The process of expelling electrons is called ionization. They have become **free electrons**, free to move independently until they are captured by other ionized atoms. The singly ionized atom has lost a single electron and thereby carries a single unit positive charge of electricity. The doubly ionized atom has an excess of two positive charges. Each successive ionization makes the atom more difficult to excite or to ionize further. We designate a singly ionized atom by its symbol and the Roman numeral II, a doubly ionized atom by its symbol and the Roman numeral III, etc.

We have seen that the sun is a glowing gaseous globe with a surface temperature of nearly 6000°K. The energy distribution approximates very nearly that derived by M. Planck for black bodies, that is bodies that absorb and reemit all energy falling upon them and the energy distribution depends only upon temperature. Thus we refer to the sun and stars as thermal sources. The total energy emitted (F) per square centimeter is proportional to the fourth power of the temperature. This is known as the Stephan-Boltzmann relation.

$$F = \sigma T^4$$
$$\sigma = 5.67 \times 10^{-5}$$

The wavelength (λ_{max}) of the maximum of the distribution in centimeters is inversely proportional to the temperature. This is known as the Wien displacement law.

$$\lambda_{max} T = 0.2898$$

Note that as T becomes large λ_{max} shifts toward shorter and shorter wavelengths. Hence, for $T = 3°$, λ_{max} is close to 1 mm; for $T = 3000°$, λ_{max} is close to 9600 Å; and for $T = 6000°$, λ_{max} is close to 4800 Å.

As the temperature rises the thermal agitation becomes increasingly severe. This agitation excites greater numbers of atoms and eventually ionization begins and increases. The effect of thermal agitation is greater (1) as the temperature of the gas increases and thus the violence of the atomic collisions is greater; and (2) as the electrons are more loosely held by the atom for a particular temperature.

The energy required to excite or to ionize an atom is usually given in electron volts (eV). For hydrogen an energy of 10.2 eV is required to excite the atom from the ground state to the first excited level, and further excitation to the next level requires only an additional 1.9 eV. The energy required to ionize the hydrogen atom is 13.6 eV, hence the level to and from which all of the Balmer lines arise is only 3.4 eV from ionization. The first level of excitation for sodium, calcium, and iron requires only 3 eV or less, whereas that of helium is about 20 eV. These are a few of the elements prominent in the visible spectrum. The values, in electron volts, for the first stage of ionization are: sodium, 5.1 eV; calcium, 6.1; iron, 7.9; and helium 24.6. For all elements the corresponding numbers are smaller for higher degrees of excitation than the first and larger for higher stages of ionization than the first. For example, to remove the first electron from helium requires 24.6 eV as we have noted. To remove the second electron from helium requires 54.5 eV.

12.13 Molecular spectra

In a somewhat analogous manner to the hydrogen atom we may think of a simple molecule as being made up of two atoms, for example, carbon and nitrogen (CN), and revolving about each other. Such a molecule may vibrate as it rotates. The atoms may revolve about each other in certain orbits, much as the electron about an atomic nucleus, and the vibration may occur only at certain frequencies. We say these motions are quantized, and they give rise to discrete lines in the spectrum. The energy radiated or absorbed when the molecule changes levels is referred to as the **rotation spectrum** if no vibration occurs or as the **vibration–rotation spectrum** of the molecule if both are occurring. Also, changes in the combined electron configurations may occur and these are referred to as **electronic transitions** of the molecule. The first two spectra have the transition energies so close together that they give the appearance of bands in the star's continuum. The last transitions may occur close together to form a band. Just as often, however, they will give rise to widely spaced lines and be mistaken for some unidentified atomic transition.

Many molecules have been identified in stellar spectra as we have already noted in the sun. The most interesting as far as stars are concerned are the two-atom, or **diatomic**, molecules of CN (cyanogen), TiO (titanium oxide), VO (vanadium oxide) and H_2 (molecular hydrogen). More complex molecules have been detected in interstellar clouds in the low pressure conditions that exist there; we will note these later (Chapter 14) and call attention to their impact on our changing ideas on the origin and evolution of stars and "life" itself.

After this introduction to atomic and molecular spectroscopy, it is in order to describe and interpret the classification of stellar spectra.

12.14 Photographs of stellar spectra

Photographs of stellar spectra are taken generally in two different ways. One method employs the **slit spectroscope** at the focus of the telescope. It permits wider separation of the spectrum lines and the recording of a laboratory spectrum adjacent to the star's

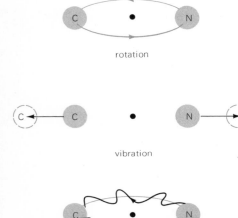

rotation

vibration

rotation plus vibration

spectrum (Fig. 12.9). This method is accordingly useful for measuring wavelengths and Doppler shifts of the lines, but it is time-consuming because it gives the spectrum of only one star at a time.

In the second method, a large prism of small angle is placed in front of the telescope objective, so that the whole apparatus becomes a spectroscope without slit or collimator. A single photograph shows the spectra of all stars of suitable brightness in the field of view of the telescope (Fig. 12.8). This **objective prism** type is useful in qualitative studies of many stars, as in the classification of stellar spectra.

Studies of stellar spectra in objective prism photographs were in progress at Harvard Observatory as early as 1885. A result of these studies is the Henry Draper Catalogue, mainly the work of A. J. Cannon, which with its extensions lists the positions, magnitudes, and spectral types of 400,000 stars. The Harvard types, together with the more recent luminosity classes (see Section 12.21), remain the criteria in the present classification of stellar spectra.

12.15 Types of stellar spectra

The Harvard classification arrays the majority of the stars in a single continuous sequence with respect to the patterns of lines in their spectra. Seven divisions of the sequence are the principal spectral types and are designated by the letters O, B, A, F, G, K, and M. (A simple mnemonic for this sequence is **Oh Be A Fine**

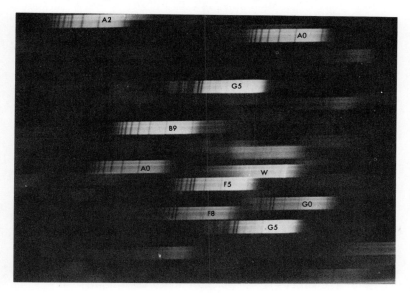

FIGURE 12.8
Objective prism spectra of stars in the region of Cephus. Note the emission line W type star. (Warner and Swasey Observatory photograph.)

Girl, Kiss Me.) The various divisions are further subdivided on the decimal system. Thus an A5 star is halfway between A0 and F0. Some features of the principal types, as they appear in the violet and blue regions of the spectrum, are the following:

Type O. Lines of ionized helium, oxygen, and nitrogen are prominent in the spectra of these very hot stars along with lines of hydrogen.

Type B. Lines of neutral helium are most intense at B2 and then fade, until at B9 they have practically vanished. Hydrogen lines increase in strength through the subdivisions. Examples are Spica and Rigel.

Type A. Hydrogen lines attain their greatest strength at A2 and then decline through the remainder of the sequence. Examples are Sirius and Vega. Thus far the stars are blue.

Type F. Lines of metals are increasing in strength, notably the Fraunhofer H and K of ionized calcium. These are yellowish stars. Examples are Canopus and Procyon.

Type G. Metallic lines are numerous and conspicuous in the spectra of these yellow stars. The sun and Capella belong to this type.

Type K. Lines of metals surpass the hydrogen lines in strength. Bands of cyanogen and other molecules are becoming conspicuous. These cooler stars are reddish. Examples are Arcturus and Aldebaran.

Type M. Bands of titanium oxide become stronger up to their maximum at M7. Vanadium oxide bands strengthen in the still cooler divisions of these red stars. Examples are Betelgeuse and Antares.

Four additional and less populous types branch off near the ends of the sequence. **Type W** near the blue end comprises the Wolf-Rayet stars, having broad bright lines in their spectra. Near the red end, **types R** and **N** show molecular bands of carbon and carbon compounds, and **type S** has conspicuous bands of zirconium oxide.

The reader might suspect that such an apparently haphazard arrangement of letters for spectral types is not the way the system began. Originally the letters were arranged alphabetically and the principal criterion was the strength of the hydrogen lines. Thus the strongest hydrogen lines are seen in A stars. When it was realized that the general alignment was one of temperature, it became apparent that the O stars, which were very blue and hence very hot, had to be moved to the beginning of the classification. Similarly, the B stars had to be placed between the O and A stars. Many of the other classes were found to consist of so few stars that they were either small numbers of peculiar stars or minor mistakes that were corrected upon reexamination.

12.16 Sequence of stellar spectra

The sequence of stellar spectra (Fig. 12.9) is in order of increasing redness, and therefore of diminishing unit surface brightness of the stars. In general the progression in the line patterns is not caused by differences in chemical composition; helium is not more abundant in B stars, nor is hydrogen in A stars. The lines of the different chemical elements become strongest at temperatures where their atoms in the stellar atmospheres are most active in absorbing the starlight.

The compact neutral helium atoms imprint their dark lines at the high temperatures of B stars. Hydrogen is most active at the more moderate temperatures of A stars, although the great abundance of this element makes its lines visible throughout the sequence. Neutral atoms of the metals are more easily excited, so that their lines are strongest in the spectra of the cooler yellow and red stars. At higher temperatures the atoms of metals are ionized and have their stronger lines in ultraviolet and X-ray regions of the spectrum. Atoms can assemble into molecules in

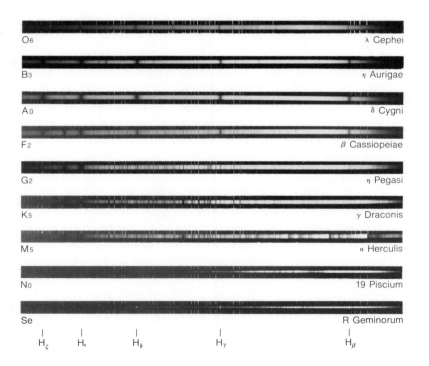

FIGURE 12.9
The principal types of stellar spectra in the region λ3900 Å to λ4800 Å. Note how the hydrogen lines strengthen from O6 to A0 and then weaken. Molecular bands are evident at K5 and very strong at M5. (Photograph from the Hale Observatories.)

the relatively cool atmospheres of red stars; bands of titanium oxide are conspicuous in the spectra of M stars as the accompanying photographs show.

MAGNITUDES OF THE STARS

In his star catalog compiled eighteen centuries ago, Ptolemy gave estimates of brightness of the stars to aid in their identification. For these estimates the stars were divided into six classes, or **magnitudes,** in order of brightness. The brightest stars were assigned to the 1st magnitude. Somewhat less bright stars, such as the pole star, were of the 2nd magnitude. Each succeeding class contained fainter stars than the one before, until the 6th magnitude remained for stars barely visible to the unaided eye.

The original magnitudes of the lucid stars continued in use until recent times, and the plan was extended somewhat arbitrarily to include fainter stars brought into view with telescopes. The present plan of magnitudes was proposed in 1856, when the brightness of stars had begun to be determined more accurately and was becoming an important factor in astronomical inquiries.

12.17 Apparent magnitudes and colors

Two stars differ by exactly 1 magnitude if their ratio of brightness is the number whose logarithm is 0.4; this ratio is the number 2.512. Thus a star of magnitude 2.0 is about 2.5 times as bright as a star of magnitude 3.0, more than 6 times as bright as one of magnitude 4.0, and so on. The brightness ratio can be expressed as:

$$\frac{B_1}{B_2} = 2.512^{m_2 - m_1}$$

Stars down to the 19th magnitude are visible with the 200-inch Palomar telescope, and stars almost as faint as magnitude 24 can be photographed with this telescope.

Because the very brightest of the original 1st-magnitude stars are visually more than 2.5 times as bright as a standard star of magnitude 1.0, such as Spica, they have now been promoted to brighter magnitudes. The magnitude of Vega is 0.0 and that of Sirius is minus 1.4.

It should be noted that the magnitude system is the consequence of the logarithmatic response of the human senses to the appropriate stimulus. It is a curious fact that our sensors and the interpretation in our minds say that any stimulus 100 times more intense is only 5 times more effective. Thus a star emitting 100 photons per unit area is sensed to be only 5 magnitudes brighter than one emitting 1 photon per unit area. This is known as Fechner's law. The astronomical scale runs contrary to what one would expect for historical reasons, thus the star with the larger positive number is the fainter to our senses. In radio astronomy we give "brightnesses" in terms of flux units or amount of energy received, thus the smaller the number the fainter the object.

These **apparent magnitudes** refer to the observed brightness of stars. They may be determined to tenths of a magnitude by the trained eye, to hundredths of a magnitude in photographs, and to thousandths by more precise photoelectric methods.

The **visual magnitude** of a star refers to its brightness as observed with the eye, which is most sensitive to yellow light. The **photographic magnitude** refers to the star's brightness as recorded in a blue-sensitive photograph. The magnitudes differ in the two cases by an amount that depends on the color of the star. Note in Fig. 12.10 how much fainter, as indicated by its smaller image, a red star appears as photographed in blue than in yellow light.

Present studies often require the magnitudes of a star in several colors. The magnitudes may be measured in photographs with suitably stained plates exposed behind filters that transmit the light of different regions of the spectrum, or more precisely with the photomultiplier tube and filters. The magnitudes are generally measured around specified wavelengths in the ultraviolet, blue, yellow or visual, red, and infrared, and are denoted by the letters U, B, V, R, and I in a system introduced by H. L. Johnson. Another system introduced by Bengt Strömgren is the u, v, b, y, etc.,

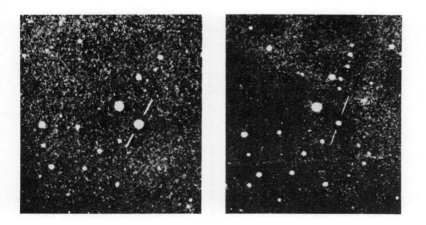

FIGURE 12.10
A red star appears fainter on a blue-sensitive plate (left) than it does on a yellow-sensitive plate (right). (Yerkes Observatory photograph.)

which is somewhat more precisely defined. The magnitudes of standard stars whatever the system used are being carefully determined in the different colors, so that all investigators may keep to the same color systems.

The **color index** (C.I.) of a star is the difference of its photographic magnitude (m_{pg}) and its visual magnitude (m_{pv}). Its value is often given as the difference, $B - V$, between the blue and visual magnitudes, but strictly speaking this is a color not the color index. The color index is then a numerical expression of the star's color. Taken as zero for stars of spectral type A0, the color is accordingly negative for the bluest stars and attains a positive value of 2 magnitudes or more for the reddest ones. The observed color index of a star may exceed the normal index for the particular spectral type because of reddening of the starlight by intervening cosmic dust.

12.18 The brightest stars

The twenty-two stars listed in Table 12.3 are brighter than apparent visual magnitude +1.5 and are sometimes called "stars of the first magnitude," although they range through nearly 3 magnitudes. The visual (V) and blue (B) magnitudes were measured photoelectrically with the appropriate filters by H. L. Johnson. The colors are shown by the differences $B - V$, which can be read from the table. The bluest stars, having the largest negative colors are alpha and beta Crucis and Spica. The reddest are Betelgeuse and Antares, which have the largest positive colors. The spectral types and luminosity classes are on the Morgan-Keenan system (see Section 12.24); where the star is double, they refer to the brighter star of the pair.

The brightest stars have a considerable range in distance from us, as the parallaxes in the table show. Alpha Centauri, Sirius, and Procyon are among the nearest stars. Rigel, Canopus, and Deneb are the most remote in the list; they are evidently very luminous to shine so brightly in our skies. The significance of the last column of the table is explained in the next section. *The Yale Catalogue of Bright Stars* lists all of the stars brighter than 6.5 apparent magnitude and selected stars down to 7.5 magnitude. It lists a total of 9110 stars.

Sixteen of the brightest stars are visible in their seasons throughout the United States. The remaining six become visible south of the following northern latitudes; Canopus, 38°; Achernar, 33°; alpha and beta Centauri and beta Crucis, 30°; alpha Crucis, 28°, or about the latitude of Tampa, Florida.

TABLE 12.3
The brightest stars

Name	Spectrum	Apparent visual magnitude (V)	Apparent photographic magnitude (B)	Parallax	Visual magnitude
Sirius[vb]	A1 V	−1.44	−1.45	0″.376	+1.4
Canopus	F0 II	−0.72	0.88	.018*	−4.4
α Centauri[vb]	G2 + K1	−0.27	0.44	.761	+4.2
Arcturus	K1 III	−0.05	1.19	.091	−0.2
Vega	A0 V	0.03	0.03	.123	+0.5
Capella[sb]	G8 + G0	0.09	0.90	.071	−0.6
Rigel[sb]	B8 Ia	0.11	0.06	.004*	−7.0
Procyon[vb]	F5 IV-V	0.36	0.77	.287	+2.7
Achernar	B5 IV	0.49	−0.66	.028	−2.2
β Centauri	B1 II	0.63	0.39	.008*	−5.0
Betelgeuse[v]	M2 I	0.4	1.89	.006*	−5.9
Altair	A7 IV-V	0.77	0.99	.204	+2.3
Aldebaran	K5 III	0.80	2.35	.048	−0.8
Acrux[vb]	B1 + B3	0.83	0.37	.012	−3.7
Antares[vb,v]	M1 Ib + B	0.94	2.77	.008*	−4.7
Spica[sb]	B1 V	0.97	0.74	.015	−3.1
Fomalhaut	A3 V	1.16	1.25	.143	+1.9
Pollux	K0 III	1.15	2.16	.093	+1.0
Deneb	A2 Ia	1.25	1.33	.002	−7.2
Mimosa	B0 III	1.29	1.04	.008*	−4.3
Regulus	B7 V	1.34	1.23	.038	−0.8
ε Canis Majoris[vb]	B2 II	1.48	1.31	.005*	−5.0

vb = visual binary
sb = spectroscopic binary
v = variable star
* uncertain values

12.19 Absolute magnitudes

The apparent magnitude of a star relates to its observed brightness. This depends on the star's intrinsic brightness, or **luminosity,** and its distance from us. One star may appear brighter than another only because it is the nearer of the two; thus Sirius appears brighter than Betelgeuse, although the latter is actually many times the more luminous. In order to rank the stars fairly with respect to luminosity, it is necessary to calculate how bright they would appear if they were all at the same distance. By agreement the standard distance is 10 parsecs, at which the parallax is 0″.1.

The **absolute magnitude** is the magnitude a star would have at the distance of 10 parsecs, or 32.6 light years. The relation is:

$$M = m + 5 - 5 \log r$$

where M is the absolute magnitude and m is the apparent magnitude at the distance r in parsecs. The absolute magnitude has the same character as the apparent magnitude from which it is derived, whether visual, photographic, or some other kind. Rearranging the relation slightly we can write:

$$m - M = 5 \log r - 5$$

where $m - M$ is known as the **distance modulus**. This term and its significance will be discussed later.

The sun's absolute visual magnitude is +4.8. At the moderate distance of 32.6 light years the sun would appear as a star only faintly visible to the unaided eye. Note in the last column of Table 12.3 that the apparently brightest stars are all intrinsically brighter than the sun.

12.20 Stars of the main sequence

When the spectral types of stars around us are plotted with respect to the absolute magnitudes of these stars, as in Fig. 12.11, the majority of the points are arrayed in a band running diagonally across the diagram. This band is known as the **main sequence** and the graph is known as the **H-R diagram**, after E. Hertzsprung and H. N. Russell who first introduced it. The middle line of the band drops from about absolute magnitude −3 for blue stars to fainter than +10 for red stars.

The sun, a yellow star of type G2 and absolute visual magnitude +4.8, is a main-sequence star, as we see. Blue stars of the sequence are more luminous than the sun because they are hotter and larger. Red stars of the sequence are less luminous than the sun because they are cooler and smaller. Remembering that the sun at the standard distance of 10 parsecs (32.6 light years) would be only faintly visible to the unaided eye, we understand why the red stars of the sequence are generally invisible without a telescope.

Since its introduction, the spectrum–absolute magnitude diagram or the similar color–magnitude diagram has played a leading part in directing the studies of the stars. E. Hertzsprung had previously drawn attention to a sharp distinction between red stars of high and low luminosity, and had named them giant and dwarf stars, respectively. The term **dwarf star** is commonly used

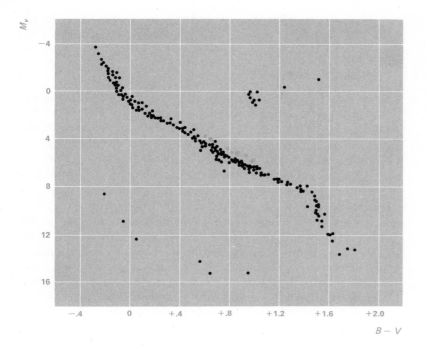

FIGURE 12.11
The Hertzsprung-Russell diagram for relatively nearby stars. Most stars are arranged along the main sequence or an almost horizontal giant branch to the right. The white dwarfs occupy the lower part of the diagram. Historically the H-R diagram is a plot of spectral type against apparent or absolute magnitude. Spectral type is related to color or temperature. In most modern diagrams color is preferred.

today to denote main-sequence stars fainter than about absolute magnitude $+1$, but this distinction is not generally applied to the white dwarf stars.

Giant stars, such as Capella and Arcturus, are considerably more luminous than main-sequence stars of the same spectral type. They are giants in size as well as in brightness. All stars of the same type have the same order of surface temperature, and therefore of surface brightness per square meter. If one type M star, for example, greatly surpasses another type M star in luminosity at the same distance, its surface must contain many more square kilometers and its diameter must be much the greater. Giants in the sun's neighborhood are mainly red stars.

The dots near the top of Fig. 12.11 represent the very large and luminous **supergiants.** Those near the upper right corner denote the largest stars of all. The red star Betelgeuse is one of these; its diameter is several hundred times that of the sun. **Subgiants** fall between the giants and the main sequence. **Subdwarfs** appear somewhat below this sequence although, technically they lie to the left of the main sequence. Dots near the bottom of the diagram represent a few of the white dwarf stars (see Section 15.10), some of which are yellow and even reddish. These very faint stars

TABLE 12.4

Effective temperatures and color indexes of stars

Spectrum	Temperature (°K)	Color (B − V)
MAIN SEQUENCE		
O5	35,000	−0.45
B0	21,000	−0.32
B5	13,600	−0.16
A0	9,700	0.00
A5	8,100	+0.15
F0	7,200	+0.30
F5	6,500	+0.44
G0	6,000	+0.57
G5	5,400	+0.70
K0	4,700	+0.84
K5	4,000	+1.11
M0	3,300	+1.39
M5	2,600	+1.61
GIANTS		
F5	6,470	—
G0	5,400	+0.65
G5	4,700	+0.84
K0	4,100	+1.06
K5	3,500	+1.40
M0	2,900	+1.65
M5	2,710	+1.85

are smaller than the larger planets. For example, Sirius B, the most famous of the white dwarfs, has a diameter of only 11,000 kilometers but a surface temperature of 32,000°K.

12.21 Luminosity classes of stars

The surface temperatures and normal colors of stars at intervals along the spectral sequence are shown in Table 12.4. The temperatures, as assigned by P. C. Keenan and W. W. Morgan, are based on an adopted temperature of 5730°K for the sun, type G2. A few adjustments have been made in the table to bring it into agreement with later work by N. G. Roman. Diminishing temperature is mainly responsible for the succession of spectral patterns, as has been said. Note that the temperatures of yellow and red giants are lower than those of main-sequence stars of the same spectral type. The reason is that the giant stars are less dense, so that their atoms attain a particular degree of ionization at a lower temperature. Although the spectra of giants and main-sequence stars of type K0, for example, are generally similar, particular features may be considerably different. Some lines become stronger and others weaker as the stars are more luminous. Thus a complete description of a star's spectrum is obtained by adding its **luminosity class** to its spectral type.

The luminosity classes of Morgan and Keenan are numbered in order beginning with the most luminous stars, which have the least dense atmospheres. The numeral I refers to supergiant stars, Ia for the more luminous and Ib for the less luminous supergiants; II refers to bright giants; III to normal giants; IV to subgiants; V to main-sequence stars; and VI to subdwarfs. Thus the two-dimensional designation in Table 12.3 for Deneb is A2 Ia; Antares, M1 Ib; epsilon Canis Majoris, B2 II; Aldebaran, K5 III; Procyon, F5 IV; and Vega, A0 V.

12.22 Unusual stars

Many stars do not fit neatly into the spectral sequence and luminosity classes that we have just described. Most conspicuous are the different types of variable stars that we discuss in Chapter 13. The remaining nonconforming stars usually have circumstellar shells, envelopes of one sort or another, conspicuous radiation of an unusual nature, or peculiar abundances of elements. Among these stars are the planetary nebulae, T Tauri stars, X-ray stars, and infrared stars.

One of the most conspicuous of unusual stars is a group called **shell stars** whose prototype is P Cygni. These stars appear to be early B type stars of absolute magnitude about −4, with a shell giving rise to very bright emission lines. Apparently related are the **Wolf-Rayet stars,** which are late O-type stars or early B-type stars, with absolute magnitudes around -5. The Wolf-Rayet stars divide into two classes, Wn and Wc depending upon whether nitrogen or carbon is the more obvious in their emission line spectra. D. N. Limber explains the Wolf-Rayet stars as rapidly rotating stars that suddenly contract leaving an extensive shell that expands outward at large velocities, 500 kilometers per second or larger. The nuclei of planetary nebulae (see Section 14.4) resemble Wolf-Rayet stars, but must be in a later stage of evolution because their absolute magnitudes are around $+1$. The shells of planetary nebulae expand much more slowly also.

Another group of unusual stars are peculiar stars of type A, designated Ap stars and referred to as **A peculiar stars.** About 10 percent of all A-type stars show any of the following peculiarities: abnormal abundances of heavier elements, strong magnetic fields, variability in their light and magnetic fields, and spectral variability. It is believed that these stars have nuclear reactions taking place on their surfaces—a process referred to as **spalation.** This would explain the abnormal abundances of heavy elements because they are being formed near the surface. Normally the heavy elements are formed in the core and cannot come to the surface.

In the giant and supergiant region of the H-R diagram lie the carbon stars that, as we have seen, divide into classes N and S. The carbon stars of class N, which are irregular variables, always seem to show an abnormal abundance of technetium and hence have been called **technetium stars.**

We have only touched upon a few of the more obvious unusual stars. P. van de Kamp reminds his students that no matter how normal a particular star may be, as soon as you begin studying it it reveals differences and peculiarities that cause it to stand out from all other stars and make it more interesting. Stars are like people in that each one is different; it is only their very gross properties that allow us to classify them.

QUESTIONS

1 What is meant by the term heliocentric parallax?
2 Why do you think Bessel selected 61 Cygni to attempt his first measurement of a stellar parallax?
3 What do you need to know to convert a star's proper motion into kilometers per second?

4 Explain spectral line broadening due to stellar rotation.

5 What are the basic particles from which an atom is composed?

6 What does the symbol Fe XII mean?

7 What is the maximum wavelength, as defined here, for an A0 V star with a temperature of 10,000°K?

8 What are the advantages of slip spectrograms and objective prism spectra?

9 Discuss the Harvard spectral classification.

10 Make your own H-R diagram from Tables 12.1 and 12.3. Discuss your diagram.

FURTHER READINGS

DOIG, PETER, *Stellar Astronomy*, Draughtsman Publ. Co., London, 1927.

DUFAY, JEAN, *Introduction to Astrophysics: The Stars* (O. Gingerich, Transl.), Dover, New York, 1964.

SWIHART, T. L., *Astrophysics and Stellar Astronomy*, Wiley, New York, 1968, Chap. I.

VAN DE KAMP, PETER, *Basic Astronomy*, Macmillan, New York, 1952.

A zodiacal construction by Kepler to show that Aries, Sagittarius, and Leo are dry constellations, thus refuting the great Arabic astrologer Albumasar. The appearance of Kepler's supernova in Sagittarius reinforces Kepler's position.

13

Binary stars are physically associated pairs of stars, and their study yields a direct approach to determining the masses of stars. In some pairs the components are far enough apart to appear separated through a telescope; in other pairs the presence of the companions can be detected only with a photographic plate through minute periodic changes in position; in still others the members can be detected only with a spectroscope. Some spectroscopic binaries mutually eclipse as they revolve so that they vary periodically in brightness. Many single stars are intrinsically variable in light because they are pulsating. The variations of large red stars are ascribed at least in part to their variable veiling by clouds rising high above them. Spectacular variations are exhibited by eruptive stars.

13.1 Visual double stars

Visual double stars appear as single stars to the unaided eye but separate into pairs when observed with the telescope. Mizar, at the bend in the Big Dipper's handle (Fig. 11.4), was the first of these, in 1650, to be reported casually by several observers; the only connection between the two stars of such pairs was believed to be that they chanced to have nearly the same direction from us. Castor, beta Cygni, and alpha Centauri were other early known examples. By 1803 Herschel had determined from repeated observations that the two stars of Castor were revolving around a common center between them. He then made the distinction between "optical doubles" (stars that seem to be doubles because of their proximity to each other) and "real doubles" (stars that revolve around a common center). The latter, now called **visual binary stars,** are in the great majority.

A total of 65,000 visual binaries are recognized, mainly by the common proper motions of the component stars. They are tabulated in detail in the *Double Star Index and Observation Catalogue* maintained by the U.S. Naval Observatory. About 2500 binaries have had time since their discoveries to show evidence of orbital motion, and 20 percent of these have progressed far enough in their revolutions to permit reliable determinations of their orbits. Pairs having the smallest separations, which are likely to go around most rapidly, can be observed satisfactorily only with long-focus telescopes.

The presence of additional stars in binary systems, forming **multiple systems,** is not exceptional. Alpha Centauri is representative of a common type of triple systems. The double-double epsilon Lyrae is a familiar example of quadruple systems.

13.2 The apparent and the true orbit

The position of the fainter star of a visual binary relative to the brighter star is denoted by the position angle and distance. The **position angle** (θ) is the angle at the brighter star between the directions of the fainter star and the north celestial pole; it is reckoned in degrees from the north around through the east. The **distance** (ρ) is the angular separation of the two stars. The position is measured micrometrically either directly at the telescope or in a photograph where the separation is wide enough.

When the recorded positions have extended through a considerable part of the period of revolution they are plotted after

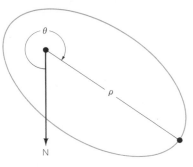

making corrections for precession. The relative **apparent orbit** of the fainter star is an ellipse that best represents the observed positions and in which Kepler's law of areas (see Section 7.3) is maintained in the motion around the brighter star (Fig. 13.1). The apparent ellipse is projected on the plane tangent to the sky, and from it the true ellipse having the brighter star at one focus is readily calculated.

The **elements** of the relative **true orbit** are the specifications of the true ellipse, such as its eccentricity, period, time of periastron passage, longitude of periastron, position angle of the node, inclination to the plane of the sky, and angular size, or the linear size, if the distance from us is also known. When the revolutions of both stars have been observed with reference to stars nearby in the photographs, the separate orbits can be determined; these differ from the relative orbit only in size.

Some characteristics of a few visual binaries are shown in Table 13.1. The first two binaries in the list have unusually short periods of revolution for visual binaries. The angular semimajor axis of an orbit may be converted to astronomical units by dividing the tabular value by the corresponding parallax. Thus the mean distance between the stars of the binary BD −8° 4352 is slightly more than the earth's distance from the sun, and that of the components of Castor is twice Pluto's distance from the sun. The separate masses of the stars are given in terms of the sun's mass, except that the value in parentheses is the sum of the two masses.

A study of the table leads to some interesting general conclusions. Visual binaries do not differ too greatly in brightness; this

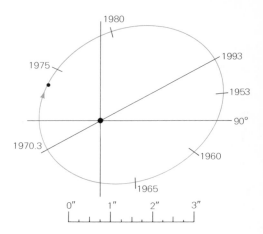

FIGURE 13.1
The apparent orbit of the faint binary Kruger 60. The apparent orbit can be evaluated to reveal the true orbit which, if the distance to the pair is known, gives the masses of the stars.

TABLE 13.1
Parameters for selected visual binaries

Name	Visual magnitudes		Period (year, P)	Semimajor axes	Eccentricity	Parallax	Solar masses	
	m_1	m_2		(α)	(e)	(p)	m_1	m_2
BD −8° 4352	9.7	9.8	1.7	0″.22	0.05	0″.157	0.45	0.45
δ Equulei	5.2	5.3	5.7	0 .26	0.42	.056	1.57	1.51
42 Comae	5.0	5.1	25.8	0 .67	0.49	.054	1.40	1.46
Procyon	0.4	10.7	40.6	4 .55	0.40	.287	1.76	0.65
Kruger 60	9.8	11.4	44.5	2 .39	0.42	.254	0.26	0.16
Sirius	−1.4	8.5	50.1	7 .50	0.59	.376	2.12	1.04
α Centauri	0.0	1.2	79.9	17 .58	0.52	.761	1.05	0.89
ξ Boötes	4.7	6.8	150 *	4 .88	0.51	.149	0.84	0.72
Castor	2.0	2.8	420 *	6 .30	0.33	.074	(3.50)	

* uncertain values

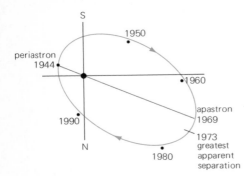

FIGURE 13.2
The apparent orbit of the famous binary Sirius. Note that the greatest separation in the apparent orbit does not correspond to the time of periastron. This is due to the fact that the orbit is not perpendicular to the line of sight.

FIGURE 13.3
Sirius A and B. Sirius B is almost nine magnitudes fainter than Sirius A, hence it is hard to see and very difficult to photograph. In this photograph a hexagonal diaphragm was used as an apodizer allowing the faint star to be photographed. (Sproul Observatory photograph by S. L. Lippincott and J. K. Wooley.)

is an observational selection effect because we cannot easily see stars of great brightness difference close together. Procyon and Sirius are notable exceptions. The orbital eccentricities, except for very short-period stars, tend to be larger than 0.3. The masses of the various stars taken singly do not differ by a factor of 10 from that of the sun.

One of the nearest stars, the brilliant Dog Star, Sirius, drifts in the heavens three-fourths the apparent width of the moon in 1000 years. As early as 1844, F. W. Bessel announced that it is pursuing a wavy course instead of having the uniform motion of a single star. He concluded that Sirius is mutually revolving with a traveling companion; its orbit with the still unseen companion was derived in 1851. This was the first **astrometric binary**.

The companion was first seen by the optician A. Clark in 1862, who was testing a new lens, now the objective of the 18½-inch telescope at the Dearborn Observatory. The companion revolves in a period of 50 years and in an orbit of rather high eccentricity (Fig. 13.2). Despite the glare of Sirius, the companion is clearly visible with large telescopes except when the two stars are least separated. The latest periastron passage occurred in 1944 and the widest separation will come next in 1974, although apastron occurred in 1969. This happens because the widest separation in the apparent orbit does not occur at the same time as the widest separation in the true orbit. The companion of Sirius (Fig. 13.3) was among the first white dwarf stars to be known (see Sections 12.20 and 15.10).

The companion of Procyon was likewise discovered by Bessel from its gravitational effect on the proper motion of the bright star; it was first observed with the telescope in 1896. A similar case is the companion of the faint red star Ross 614, which was detected by D. Reuyl in 1936 at McCormick Observatory. A definitive study by S. L. Lippincott at Sproul Observatory led to the photographing of its companion by W. Baade with the 200-inch telescope in 1955.

Unseen companions of other stars have been studied at Sproul Observatory by their gravitational effects (perturbations) upon the motion of the visible star. An especially remarkable example is the companion of Barnard's star (see Section 12.3), detected by van de Kamp. The companion's estimated mass is only 1.5 times the mass of Jupiter. It revolves with its primary in a period of 24 years. This companion is tentatively called a planet because, according to our present knowledge, it does not shine by its own light. Van de Kamp even advances the idea that the perturbation may result from two smaller objects.

13.3 Masses of binary stars

The combined mass of a visual binary can be evaluated when the orbit of the binary has been determined. The method is another application of the general statement of Kepler's harmonic law, which we have already used to find (see Section 7.5) the combined mass of a planet and its satellite. Applying the same law and choosing our units carefully, we can write the law such that the combined mass of the pair of stars equals the cube of their linear mean distance apart (a) in astronomical units divided by the square of the period (P) of revolution in years.

$$m_1 + m_2 = \left(\frac{\alpha}{\pi}\right)^3 \times \frac{1}{P^2}$$

We have converted the angular separation (α) in the true orbit into astronomical units by dividing by the parallax (π) of the system. By selecting the earth–sun system to establish our units (we can neglect the mass of the earth because it is almost a factor of 10^6 smaller than the sun) we have established one solar mass as our unit of mass, which is much more convenient than grams, kilograms, or metric tons in this case.

Note that we have just contrived to write our relations in an especially convenient way. While the absolute value of the mass of the sun may change according to variations in the value of the year or astronomical unit, the solar mass does not change—hence our various tables written in terms of solar masses remain valid.

As an example of the use of the above relation, the sum of the masses of the binary alpha Centauri is calculated from the data of Table 13.1 as follows:

$$m_1 + m_2 = \frac{17.58^3}{(79.9)^2(0.761)^3} = 1.94$$

Thus the combined mass of alpha Centauri is almost twice the sun's mass.

The sum of the masses is all that can be determined from the relative orbit. When, however, the revolutions of both stars have been observed, the individual masses become known, because the ratio of the masses is inversely proportional to the ratio of the distances of the two stars at any time from the common center around which they revolve.

$$\frac{m_1}{m_2} = \frac{d_2}{d_1} = \frac{\alpha_2}{\alpha_1}$$

Solving this relation together with the harmonic law will yield the individual masses.

In Table 13.1 a few interesting anomalies occur. We recall from Table 12.1 that the companions of Sirius and Procyon are very faint. We also recall that the companions have colors (surface temperatures) very similar to the primaries. The only way that we can explain these facts is if the companions are very small, much smaller than the sun yet with masses similar to that of the sun. The fact that two of these white dwarfs are in our small sample indicates that white dwarfs must be fairly common.

13.4 Spectroscopic binaries

Spectroscopic binaries are mutually revolving pairs that appear as single stars with the telescope. They are recognized by the periodic oscillations of the lines in their spectra caused by the Doppler effect as the two stars alternately approach and recede from us in their revolutions. When the two stars of a binary differ in brightness by as much as 1 magnitude, only the spectrum of the brighter star is likely to be observed. Capella and Spica are examples of spectroscopic binaries among the brightest stars.

The brighter star of Mizar's visual pair was the first spectroscopic binary to be detected. In the early studies of stellar spectra at Harvard Observatory in 1889, it was noticed that the dark lines in the spectrum of this star were double in some photographs and single in others (Fig. 13.4).

When the displacements of the lines in the spectrum of a binary have been measured at intervals throughout the period of revolution, the **velocity curve** can be drawn to show the variation of the radial velocity during the period. From this curve it is possible to determine the orbit projected on a plane through the line of sight. The inclination of the true orbit to this plane is not determined from the spectrum; it may be derived from the light curve when the pair is also an eclipsing binary.

Generally, the orbital velocities of spectroscopic binaries are quite large even when we observe only the projected component. This is an instrumental selection effect because we determine the velocity component by measuring the Doppler shifts of the lines. For example, using the moderately high dispersion of 30 Å/mm and an ability to resolve 0.5 angstrom (5×10^{-9} cm) on the plate at $\lambda 5000$ Å in the spectrum we find, using

$$\frac{\Delta\lambda}{\lambda} = \frac{v}{c}$$

that the projected orbital velocity, v, must be 30 km/sec where c is the velocity of light in kilometers per second ($c \simeq 300,000$ km/sec).

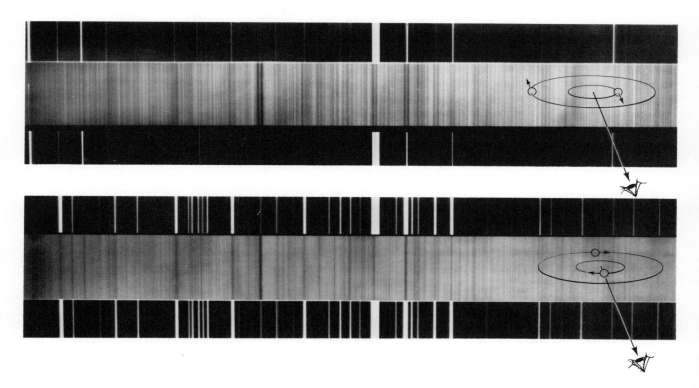

13.5 The mass–luminosity relation

Studies of binary stars by Eddington in 1924 led to the discovery of a relation between the masses and luminosities of stars in general. The more massive the star, the greater is its absolute brightness. The relation

$$\log m = \frac{4.6 - M_b}{10}$$

is shown by the blue line in Fig. 13.5, where the logarithm of the mass in terms of the sun's mass is plotted with respect to the star's absolute bolometric magnitude. **Bolometric magnitude** refers to the star's radiation in all wavelengths with allowance for its absorption by the earth's atmosphere.

The mass of a single star may be read from the curve if the bolometric magnitude is known and if the star is a main-sequence star for which the relation holds. Among special kinds of stars that do not conform are the white dwarf stars, which are much fainter than main-sequence stars of equal mass.

FIGURE 13.4
Doppler shifts in the spectra of the binary Mizar. When the components are moving toward and away from us the lines shift to the blue and red, and hence appear as double. When the components are moving across the line of sight the lines appear single.

FIGURE 13.5
The mass–luminosity diagram. Visual binaries are plotted as yellow circles and spectroscopic binaries as blue circles. This nice empirical relation does not hold for non main-sequence stars nor does it hold for white dwarfs—three of which are plotted below the curve. (Diagram adapted from one by K. A. Strand.)

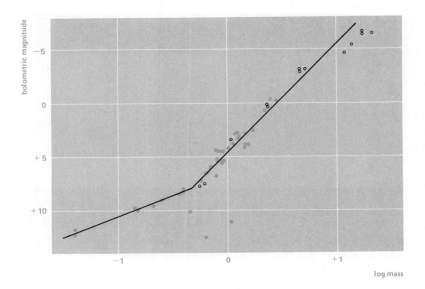

The mass–luminosity relation, the harmonic law, and the relation $M = m + 5 + 5 \log \pi$ (see Section 12.19), allows us to obtain the parallax (π) for binary stars where the separation, period, and apparent magnitudes are known. Such a parallax is called a **dynamical parallax**. Estimating the sum of the masses and using the harmonic law we find a provisional parallax. Using this value of the parallax and the apparent magnitudes we determine the absolute magnitudes, which can now be entered in the mass–luminosity relation to give us a better estimate of the masses. Using this better estimate of the masses we go through the procedure again and again until the value of the parallax does not change.

VARIABLE STARS

Variable stars are stars that vary in brightness. Their fluctuations in light are studied by comparing their magnitudes repeatedly with those of stars having constant brightness. The comparisons are often made in photographs taken at different times, or with the photoelectric tube when precision is required.

13.6 Light curve

The **light curve** shows the variation in magnitude of a star with respect to time. If the same variation is repeated periodically, the

elements of the light variation can be derived; these are the **epoch**, the time of a well-defined place on the curve such as minimum or maximum brightness, and the **period** of variation. The curve for a single cycle may then be defined more precisely by plotting all the observed magnitudes with respect to **phase**, the interval of time expressed either in days or in fractions of the period since the epoch preceding the time of each observation.

As an example, the elements of the variable star eta Aquilae are recorded as:

$$\text{maximum brightness} = 2,414,827.15 + 7^{\text{d}}.1767 \times E$$

where the first number is the epoch expressed in Julian days and the second is the period in days. In order to predict the times of subsequent maxima we have simply to multiply the period by $E = 1, 2, \ldots$ and to add the products to the original epoch. The **Julian date** is the number of days that have elapsed since Greenwich mean noon of an arbitrary zero day. It is a convenient device often used to avoid the complexity of the calendar system. The Julian dates are tabulated for each year in the *American Ephemeris and Nautical Almanac*. For example, J.D. 2,436,934.5 corresponds to 1 January, 1960, at Greenwich mean midnight. The Julian date for mean midnight 1 January 1975, will be J.D. 2,442,413.5.

13.7 Designation and classes of variable stars

The designation of variable stars follows a plan that started simply enough but became complicated when the discoveries of these stars ran into thousands. Unless the star already has a number in the Bayer system (see Section 11.3), it is assigned a capital letter, or two, in the order in which its variability is recognized. For each constellation the letters are used in the order: R, S, . . . , Z; RR, RS, . . . , RZ; SS, . . . , SZ; and so on until ZZ is reached. Subsequent variables are AA, AB, . . . , AZ; BB, . . . , BZ; etc. By the time QZ is reached (the letter J is not employed), 334 variable stars have been named in the constellation. Examples are R Leonis, SZ Herculis, and AC Cygni. Following QZ the designations are V 335, V 336, and so on; an example is V 335 Sagittarii. The *General Catalogue of Variable Stars*, prepared by B. V. Kukarkin, P. P. Parenago, and associates at the Sternberg Astronomical Institute, Moscow, gives the names, elements of the light variations, and other information about recognized variable stars in our galaxy. The second edition (1958) lists 14,708 variable stars. The 1961 additions included approximately 800 more stars. The *Catalogue* also lists more than 10,000 suspected variables. All variable stars, according to the *Catalogue*, may be divided into three main classes: eclipsing, pulsating, and eruptive variables. Each of these is further subdivided into several types.

ECLIPSING VARIABLES

Eclipsing variable stars are binaries having the axis of their orbits so nearly perpendicular to the center of the earth that the two stars mutually eclipse twice in the course of each revolution. They appear with the telescope as single stars that become fainter while the eclipses are in progress. The periods in which they revolve and fluctuate in brightness range from 82 minutes for WZ Sagittae to 27 years in the very exceptional case of epsilon Aurigae. AM Canum Venaticorum is believed to be an eclipsing binary with a period of only 17.5 minutes. The average period lies between 2 to 3 days for this class of stars. This is the result mainly of observational selection because we tend to notice rapidly blinking stars more easily than those that blink only at long intervals. The very short period eclipsing binaries show a flickering in their light curves identical to that seen in the remnant stars of novae supporting the idea that the nova event is somehow related to double stars. HZ Herculis is such a binary, one component of which is a pulsar (see Section 13.14).

Algol (beta Persei) is representative of eclipsing pairs in which the stars are nearly spherical. In pairs such as beta Lyrae the stars are elongated one toward the other by mutual tidal action, so that additional variations in brightness are produced by the different presentations of the ellipsoidal stars to us, from end-on at the eclipses to broadside between eclipses. In short-period pairs of the W Ursae Majoris type the components are so nearly in contact that the tidal effect is extreme. The spectra of some eclipsing pairs reveal gas streams that issue from the two stars and swirl in the directions of their revolutions.

Combined photometric and spectroscopic observations of eclipsing stars lead to evaluations of their linear dimensions and masses—important data in studies of the constitution of stars.

In some systems the time of minimum (i.e., the period) changes in a periodic manner. This is often the result of the system rotating around a common center with a third unseen star. In some cases the variation results from a rotation of the lines of the apsides due to an eccentric orbit. When this is the case, it is possible to determine the distribution of the density in the stars involved. In some other systems the periods may change abruptly and erratically. Such changes are surely associated with mass loss, as F. B.

Wood has shown, from one or both components and/or mass being exchanged between the components. The study of these changes is still in its early stages and may hold some insight to the problems of stellar evolution.

Finally, from photometric observations of eclipsing binaries we can observe the limb darkening of stars. Only eclipsing binary observations yield this important information excepting, of course, the case of the sun where limb darkening is directly observable. The limb darkening for both components of Algol (0.76 in each case) is well determined.

Algol is a familiar example of eclipsing variable stars and was the first of these to be recognized, in 1783. In the severed head of Medusa, which Perseus carried in the old picture book of the skies, this "Demon Star" winks in a way that might have seemed mysterious until the reason for its winking came to be understood. The brighter star of Algol revolves in a period of 2^d21^h (Fig. 13.6) with a companion that is 20 percent greater in diameter but 3 magnitudes fainter. Once in each revolution the companion passes in front of the brighter star, partially eclipsing it for nearly 10 hours. At the middle of this eclipse the light of the system is reduced to a third of its normal brightness.

The small drop in the light midway between the primary eclipses occurs when the companion is eclipsed by the brighter star. The light of the system rises slightly toward the secondary eclipse, because the hemisphere of the companion that is then turned toward us is made the more luminous by the light of the brighter star. This is called the **reflection effect.**

The diameter of the brighter star of Algol is 3 times the sun's diameter. The centers of the two stars are 7.5×10^6 kilometers apart, or somewhat more than a tenth of Mercury's mean distance

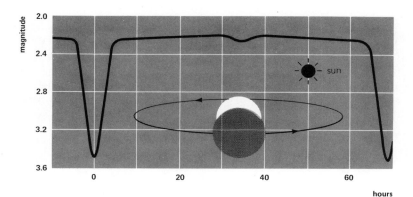

FIGURE 13.6
The observed light curve and schematic drawing of the binary Algol. The size of the sun on the same scale is shown. (Light curve and orbit by J. Stebbins.)

FIGURE 13.7
*The light curve of the binary HR 6611.
Note the difference in the relative depth of
secondary eclipse as compared with Algol.
This indicates that these two stars are very
nearly equal in size, shape, and brightness.
(Diagram by R. Zissell.)*

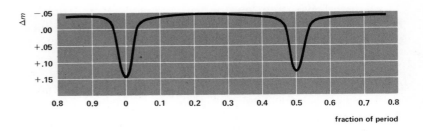

from the sun. The orbit is inclined 8° from the line of sight. The orbital velocity is about 200 kilometers per second.

A binary similar to Algol, but where both stars are about equal in size, mass, etc., to each other is HR 6611. This system has been thoroughly studied by R. Zissell. The larger and brighter star is 3 times the size of the sun. They revolve in a period of 3.89498 days at a distance of 0.08 AU. The light curve (Fig. 13.7) reveals no flat or total eclipse takes place. The relative system is shown to scale in Fig. 13.8.

PULSATING VARIABLES

Intrinsic variable stars fluctuate in brightness from causes inherent in the stars themselves and not because of eclipses. Some of these stars are variable because they are alternately contracting and expanding, becoming hotter and cooler in turn. Pulsating stars include prominently the cepheid and RR Lyrae variables.

13.8 Cepheid and RR Lyrae variable stars

Cepheid variable stars take their name from one of their earliest known examples, delta Cephei. This star fluctuates regularly in cycles of 5^d9^h, brightening more rapidly than it fades (Fig. 13.9). The velocity curve is nearly the mirror image of the light curve. Near maximum brightness the spectrum lines are displaced farthest to the violet, showing that the gases in front of the star are approaching us in the pulsation at the greatest speed. Near minimum light the lines are displaced farthest to the red, showing that these gases are receding from us at the greatest speed; the star is then redder than at greatest brightness and its spectrum has changed to the pattern of a cooler star.

Cepheid variables are yellow supergiants; the more numerous **classical cepheids** resemble the prototype. Their periods range from a day to many weeks and are most frequent around 5 days.

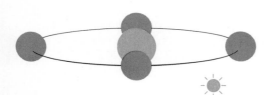

FIGURE 13.8
*The apparent orbit of the binary HR 6611.
The sun is shown to scale for comparison.
Mercury would be 40 solar diameters away
from the sun. (Diagram by R. Zissell.)*

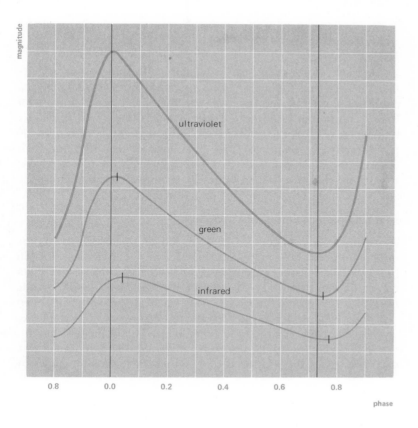

magnitude

ultraviolet

green

infrared

0.8   0.0   0.2   0.4   0.6   0.8

phase

FIGURE 13.9
*The light curve of delta Cephei. At longer
wavelengths (infrared) the light curve lags
behind that of shorter wavelengths (ultraviolet). The maximum is not as high in the
infrared as in the ultraviolet. (Curve generated from observations by J. Stebbins.)*

The visual range of the light variation is often around 1 magnitude. Those in our own Galaxy congregate toward the Milky Way.
Other examples are eta Aquilae, zeta Geminorum, and Polaris,
the latter having an exceptionally small range of variation.

The **type II cepheids** have been recognized more frequently in
the globular clusters and toward the center of the Galaxy. Their
periods are generally between 12 and 20 days. The light curves
have broader maxima and are more nearly symmetrical. An example is W Virginis. Both types of cepheids are observed in other
galaxies. RR Lyrae variables are named after one of their brightest
examples; they are also known as **cluster variables** because they
were first observed in large numbers in the globular clusters. They
are sometimes referred to as **short-period** variables because their
periods are less than one day. They are blue giants varying in
brightness in periods around half a day and with ranges up to
1.5 magnitudes or more. The light curves of these stars with the
shorter periods are likely to be almost symmetrical; at half a day

the curves become abruptly asymmetrical with very steep upslopes and extreme ranges (Fig. 13.10), effects that moderate as the periods are longer.

RR Lyrae variables are pulsating stars with some characteristics like those of the cepheids. Their spectrum lines oscillate in the period of the light variation and the stars are bluer at maximum than at minimum brightness. These stars are recognized in greater numbers than are the cepheids even though they are much the less luminous. Because of this, the distance required for their observation is shorter than that for cepheids. Not one is visible to the unaided eye. Their brightest examples are the prototype and VZ Cancri, both of the 7th magnitude.

13.9 The period–luminosity relation

The periods of cepheid variables are longer because the stars are more luminous. Originally established by H. Leavitt for stars in the Small Magellanic cloud between 1908 and 1912, the relation was calibrated first by H. Shapley in 1913 and later by W. Baade in 1952. The former curve for the classical cepheids was then raised 1.5 magnitudes in the diagram and was replaced at the lower level by the curve for the type II cepheids. The horizontal line for the RR Lyrae variables was left unaltered. This stage in the development of the relation is represented with certain changes by

FIGURE 13.10
The light curve of RR Lyra as observed by T. Walraven. The rapid rise to maximum light is typical of this type of variable star. (Reproduced by permission of Astronomische Bulletin Nederlands.)

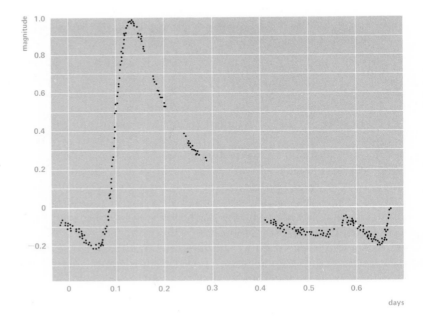

the center of the bands in Fig. 13.11. The bands for the cepheids show how the logarithm of the period was then believed to vary with the median absolute magnitude of the star. The **median magnitude** is the average of the brightest and faintest magnitudes. The spread around these lines in the figure illustrates a more recent version of the relation.

The way for this revision of the period–luminosity relation was prepared by H. C. Arp's measures of cepheids in the Small Magellanic cloud. Arp found that the relation must be represented by a band at least 1 magnitude wide rather than by a simple curve. Alan Sandage supplied the reason for the band. From a relation already known between the period (P) and mean density (ρ)

$$P \sqrt{\rho} = \text{constant}$$
$$= 0.05 \text{ day}$$

he deduced that the absolute magnitude of a cepheid depends not only on the period of its pulsation but also on the star's mean surface temperature, or color. This would apply to all pulsating stars. The next step began with J. B. Irwin's discovery of two cepheids in galactic clusters and the later finding of a few additional cases. Accurate distances and colors of cepheids that can be obtained in these clusters seem to sustain the revised relation.

When the period and mean color of a cepheid of either type is observed, the star's median absolute magnitude can be read from the appropriate diagram; its value may be taken from the central line of the band for cepheids having the largest variations in brightness. When the median apparent magnitude is also observed, the star's distance, r, in parsecs may be calculated by the formula (see Section 12.19)

$$\log r = \frac{m - M + 5}{5}$$

For the RR Lyrae variables the median magnitudes are independent of the period, and the value of M, formerly taken as zero, has now become about $+0.7$ on the average. All such distances require correction where cosmic dust intervenes (see Section 14.2).

Because of their high luminosity, cepheid variables are useful as distance indicators for the nearer galaxies, whereas the less luminous RR Lyrae variables are employed generally as indicators for distances of objects in our own galaxy.

The period–luminosity relation is central to many studies and hence the stars forming it are under constant study. Any change in the relation changes the distances accordingly. For example, a shift of 0.1 magnitude changes the distances by almost 5 percent.

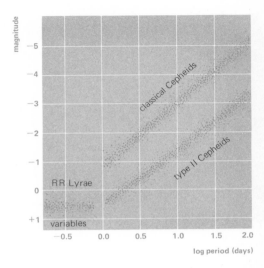

FIGURE 13.11
The period–luminosity relation for pulsating variable stars. Note the spread around the average at any given period and that the classical cepheids are about 1.5 magnitudes brighter than the type II cepheids. The median magnitude of the RR Lyrae variables is near $+0.6$. (Drawn from data by H. C. Arp and M. K. Hemenway.)

13.10 Mira-type variables

Many red supergiant and giant stars are variable in brightness in a roughly periodic manner. The periods range from a few months to more than 2 years. The visual variation averages 5 magnitudes and may exceed 10 magnitudes in the extreme case of chi Cygni (Fig. 13.12); yet the total radiation, as measured by the heating of a thermocouple at the focus of a telescope, varies only about 1 magnitude. These **Mira-type**, or **long-period variables**, are often regarded as pulsating stars and are so classified in the *General Catalogue*.

In addition to the dark lines and bands that characterize the spectra of red stars, these spectra show bright lines at certain phases, particularly lines of hydrogen. They are likely to be somewhat displaced to the violet at the maximum brightness of the stars, but a one-way shifting of the lines and not an oscillation is suggested.

Mira (omicron Ceti) is the best known and at times the brightest of these variables. It was in fact the first variable star to be recognized, aside from a few novae (see page 296), and was accordingly called **stella mira** (wonderful star). This red supergiant is at least 10 times as massive as the sun, and its diameter is 300 times the sun's diameter. The average period of the light variation is 330 days. The greatest brightness in the different cycles ranges generally from the 3rd to the 5th visual magnitude and the least brightness from the 8th to the 10th magnitude, where the star is accordingly invisible to the unaided eye.

There is considerable doubt that the fluctuation of Mira is caused mainly or even at all by simple pulsation. P. W. Merrill expressed his preference for a main hypothesis of successive "hot fronts" that move outward from below the photosphere and disappear at the uppermost levels of the atmosphere. Here the dissipating waves may cause the gases to condense into droplets, which veil the photosphere until they finally evaporate.

The total energy change of Mira is not as great as the visual brightness range would indicate. A change of 5 magnitudes would suggest an energy change of a factor of 100. However, the bolometric magnitude of Mira changes by only slightly more than 1 magnitude or a total brightness change of a factor of about 3. This is because Mira has a low surface temperature—somewhat less than $2000°K$. At this temperature the wavelength (see Section 12.11) of the energy distribution maximum is at $14,500Å$, which is in the infrared region of the spectrum at a wavelength twice that observable by the human eye.

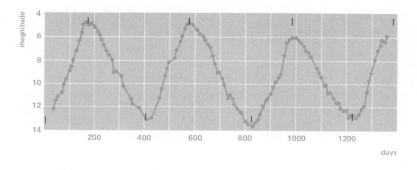

FIGURE 13.12
Light curve of the Mira-type variable chi Cygni. Note that the variation is roughly periodic and that the maxima and minima vary. (Determined by the American Association of Variable Star Observers.)

13.11 Irregular and semiregular variables

Many red supergiants and giants vary irregularly in narrower limits, often not exceeding half a magnitude. Betelgeuse is the brightest of the irregular variables.

The red supergiant alpha Herculis is another example. This star varies between visual magnitudes 3.0 and 4.0 in irregular cycles of several months' duration. Its distance is 500 light years, and its diameter is several hundred times the sun's diameter. This star is surrounded by an expanding shell of patchy clouds, according to A. J. Deutsch, which rise to heights above the surface of at least 700 times the earth's distance from the sun. At very high levels they become clouds of solid particles, which disappear by dilution as they move outward and are replaced by other clouds. This is an example of mass being lost from a star. Their partial veiling of the photosphere is believed to contribute to the variability of alpha Herculis and other large red stars.

RV Tauri stars are semiregular yellow and reddish supergiants that form a sort of connecting link between the cepheids and the Mira-type variables. Other types of semiregular variables having small memberships are recognized.

ERUPTIVE VARIABLES

Eruptive, or explosive, variable stars include novae, flare stars, and T Tauri stars (see Section 15.1). Although C. Payne-Gaposchkin prefers to refer to the latter two types of stars as spasmotic variables. **Novae** are stars that rise abruptly from relative obscurity and gradually decline to their former faintness. They are designated either by the word nova followed by the possessive of the constellation name and the year of the outburst, or more recently by letters along with other variable stars. Thus Nova Herculis 1934 is also known as DQ Herculis. In addition to typical novae, recurrent novae, and dwarf novae, there are the even more spectacular supernovae.

13.12 Typical novae

Prior to its one recorded outburst, a typical nova is smaller and denser than the sun. It is often pictured as a semidegenerate star that is collapsing to become a white dwarf. When more energy is liberated by the contraction than the small surface can radiate, the star blows off the excess energy together with a small fraction of its gaseous material in a succession of violent explosions. The total amount of material ejected is about 1/10,000 of the star's mass. Clouds of gas emerge with speeds that have exceeded 3000 kilometers per second, as shown by the Doppler shifts of the spectrum lines. With the growing volume of these hot gases the star may rise 12 magnitudes, or about 60,000 times in brightness.

The emerging gas is opaque at first, as D. B. McLaughlin has explained. The dark spectrum lines are displaced strongly to the violet, giving the illusion that the whole star is swelling enormously. Soon after maximum brightness of the star the expanding gas becomes more nearly transparent. Bright undisplaced lines then appear in the spectrum (Fig. 13.13), much broadened because the light comes from parts of the gas that are approaching and other parts that are receding from us in the expansion. The broad bright lines are bordered at the violet edges by narrow dark lines absorbed as before by the gas immediately in front of the star. When the envelope has become still more tenuous, the spectrum of the nova changes to the bright line pattern of an emission nebula (see page 306) except that the lines are wider. Meanwhile, the brightness fades as the envelope is dissipated by expansion, until after 20 to 40 years the star returns to about its original status.

The typical nova event may be divided into pre- and postmaximum phases. The premaximum phase consists of the **prenova star,** which many astronomers suspect shows a flickering light variability prior to the event, the **initial rise** which is very rapid probably lasting only hours to a day at most, a **premaximum halt** in the rise about 2 magnitudes below its greatest brightness, and then a rather slow **final rise** to maximum. These phases are not very well studied because novae seldom are noticed before they reach maximum light. The next stage is the **principal maximum.** After maximum light the nova begins an exponential decline, finally ending with the star's return to its original appearance prior to the nova event. The postmaximum **early decline** is characterized by the spectral changes discussed above. After declining about 3 or 4

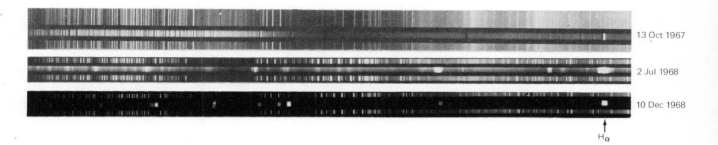

13 Oct 1967

2 Jul 1968

10 Dec 1968

$H\alpha$

magnitudes in brightness a **transition phase** sets in where the light curve may suddenly drop to the prenova brightness, or it may fluctuate violently, or it may show a smooth decline with only slight fluctuations. After this the nova makes its **final decline** to the **postnova** star. The postnova star, when studied intensively, proves to be a binary star in almost all cases, as was first pointed out by R. Kraft, and may show violent flickering. In any event, the typical nova phases scale with unusual precision. For example, the transition phase takes about 20 percent of the nova event duration whether it is 4 months, in which case the transition phase lasts about 24 days, or 4 years, in which case the transition phase lasts about 290 days. The similarities of nova light curves were first pointed out by D. B. McLaughlin.

Five typical novae in the present century became stars of the first magnitude or brighter. Nova Persei 1901 rose to apparent magnitude +0.1, as bright as Capella. Nova Aquilae 1918 reached magnitude −1.4, as bright as Sirius. Nova Pictoris 1925, Nova DQ Herculis (1934), and Nova CP Puppis (1942) became, respectively, as bright as Spica, Deneb, and Rigel. Although it is estimated that 25 novae burst out yearly in the Milky Way, the majority escape detection and most of the observed ones are invisible to the unaided eye.

The envelopes produced by eruptions of novae have sometimes become large enough to be observed directly with the telescope. Nova Aquilae 1918 had a spherical envelope; the envelope began to be visible 4 months after the outburst and increased in radius at the rate of 1″ a year. Spectroscopic observations showed a velocity for the material of 1600 kilometers per second, which yields a distance of 1100 light years to the nova.

$$d_{\text{pc}} = 0.21 \frac{v}{\mu}$$

$$d_{\text{ly}} = 0.68 \frac{v}{\mu}$$

FIGURE 13.13
Three spectra of Nova Delphini 1967. In October the spectrum resembled that of P Cygni, a well known shell star. By July 1968 the spectrum shows very broad, violet displaced emission lines. As the ejected material cools the lines narrow as seen in the spectrum of December 1968. The emission line on the extreme right is Hα. (Courtesy of J. Gregar, Ondrejov Observatory.)

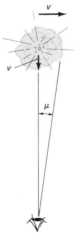

In 1940, the vanishing envelope had a radius exceeding 5000 times the earth's distance from the sun. The envelopes around typical novae have generally disappeared after a few years. Their short durations contrast with the longer lives of supernova envelopes.

13.13 Recurrent and dwarf novae

Recurrent novae, which have two or more recorded outbursts, also differ from typical novae in their more moderate rise in brightness (around 7 magnitudes). An example is RS Ophiuchi, normally around the 12th magnitude; it rose abruptly to the 4th magnitude in 1898 and 1933, and to the 5th magnitude in 1958.

Dwarf novae have some characteristics of typical novae. An example is SS Cygni (Fig. 13.14). Normally around apparent magnitude 12, this star brightens abruptly about 4 magnitudes at irregular intervals and declines in a few days. A. H. Joy's discovery at Mount Wilson Observatory in 1943 that SS Cygni is a spectroscopic binary star may have prepared the way for eventual understanding of all novae. Other typical and dwarf novae have since proved to be members of binary systems. R. P. Kraft has discussed the possibility that membership in a particular type of double system may be a necessary condition for a star to become a nova.

There is some indication that the amount of brightening of a nova is directly related to the average interval between the outbursts. If the relation holds for typical novae, the intervals between their outbursts must be many thousands of years. Thus the eruptions of stars such as Nova Aquilae 1918 might be expected to be repeated eventually.

FIGURE 13.14
Light curve of SS Cygni a recurrent nova during 1965. The numbers along the top are in Julian days beginning with JD 2435100. The numbers at the bottom are the tabulation of eruptions and the letter–number combination indicate the type of eruption. (Determined by the American Association of Variable Star Observers as plotted by M. W. Mayall.)

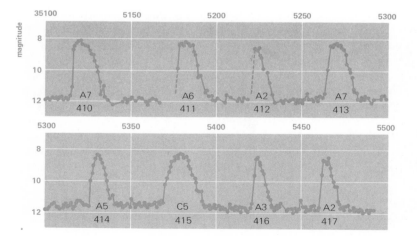

13.14 Supernovae in our Galaxy

Supernovae are stars considerably more massive than the sun, which explode once in the course of their lifetimes. During their explosions they become many million times more luminous than the sun and blow into space gaseous material amounting to at least one solar mass on each occasion. Allowing for the many that are too remote to be conspicuous, I. S. Shklovsky estimates that supernovae are exploding in the system of the Milky Way at an average rate of one in 30 to 60 years. Novae and supernovae in other galaxies are discussed in Chapter 18.

The apparently brightest supernova on record became as bright in the sky as the planet Venus. It flared out in Cassiopeia in 1572 and was observed by Tycho Brahe; after 18 months it disappeared to the unaided eye. "Kepler's star" in Ophiuchus in 1604 rivaled Jupiter in brightness. A supernova in Taurus in the year 1054 became as bright as Jupiter and remained visible about two years. Supernovae divide into two groups: Group I attains an absolute magnitude of −16 or −17, while Group II reach an absolute magnitude of −14.

Distances to supernovae can be independently obtained from their expanding shells as explained above for ordinary novae. Another method for determining distance makes use of the velocity of light if the event happens to illuminate interstellar material in the vicinity. In this case the radial velocity is not necessary so long as the expansion rate of the illuminated ring is measured in seconds of arc per year (μ).

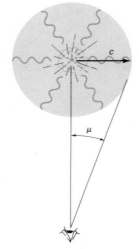

$$d_{\mathrm{pc}} = \frac{63000}{\mu}$$

Supernovae events are quite different from that of the ordinary novae discussed above. It is clear that some cataclysmic event has occurred, and the star involved is shedding a large portion of its mass. We will theorize later that any star possessing too much mass as it reaches the end point of its evolution is forced to release the excess mass through just such a violent process.

About 2 dozen remnants of supernova envelopes recognized or suspected in the Milky Way are listed by D. E. Harris. The first known of these is the Crab nebula (Messier 1), the counterpart of a source of strong radio radiation called Taurus A, which is shown in Fig. 13.15. This conspicuous nebula is still increasing in radius at the rate of 1100 kilometers per second, around the site of the supernova of 1054. At a distance of 3500 light years, its present linear diameter is 6 light years. The Crab nebula consists

FIGURE 13.15
The Crab nebula (= M1 = NGC 1952 = radio source Taurus A) in Taurus. This is the remnant of the supernova event observed by the Chinese in A.D. 1054. The immediate impression is one of a violent event. (Lick Observatory photograph.)

of a homogeneous central region surrounded by an intricate system of filaments. Fragments of the envelopes of the supernovae of 1572 and 1604 have been photographed with the Hale telescope. The fragments are still spreading.

The familiar loop of nebulosity in Cygnus (Fig. 13.16) is an example of the remnants of more ancient supernovae in the Galaxy. These appear in the photographs as complex patterns of filamentary nebulosity. The emissions from supernova remnants in the radio wavelengths are mainly nonthermal, like those produced by the synchrotron type of accelerator in radiation laboratories.

13.15 Pulsars

In late 1967, J. Bell and A. Hewish and their colleagues at Cambridge discovered a new type of variable star, which they announced six months later. They discovered these stars in their radio records because of their sharp pulses of energy with very regular but extremely short periods (3 seconds and less) and hence gave them the name **pulsar.** The identification of such objects is given as a letter, C for Cambridge, N for National Radio Astronomy Observatory, etc., followed by P for pulsar, followed by four digits to indicate the right ascension in hours and minutes. Thus CP 1919 is the pulsar discovered at Cambridge at 19^h19^m.

Many nova and supernova remnants are binary systems and, because two pulsars are coincident with supernova remains (the Crab Nebula and Vela, for example), it is natural for one to think that these may be binaries with very short regular periods. One pulsar is associated with HZ Herculis, a binary with a period of 1.7 days, and has a 1.24-second period. This binary is an X-ray source and a nova remnant as well. Neglecting effects that arise from high velocities, stars the size of the sun can have periods on the order of several hours. (Velocities of half the velocity of light or greater are referred to as **relativistic velocities.**) Degenerate stars, such as white dwarfs with diameters on the order of 15,000 kilometers could have periods on the order of a minute or less. Neutron stars (Chapter 15) with diameters on the order of 15 kilometers can have periods on the order of a few thousandths of a second (see Section 15.11). We really cannot ignore the relativistic effects so we prefer to think of pulsars as neutron stars spinning on an axis at very high rates and with extremely strong magnetic fields having an axis different from the rotation axis. The magnetic fields serve to accelerate the electrons injected into them

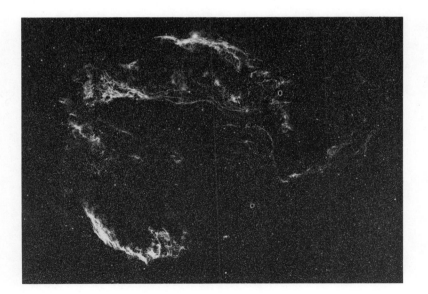

FIGURE 13.16
Filamentary nebula in Cygnus referred to as the Loop nebula. This is the remnant of a supernova event which occurred 70,000 years ago. For scale see Fig. 11.9. (Photograph from the Hale Observatories.)

and the electrons then emit the observed nonthermal synchrotron radiation.

Such stars are possible end products of stellar evolution and can have the intense magnetic fields required. Picture the sun as having a magnetic field on its surface of so many lines of force. Now let the sun shrink to white dwarf dimensions and the lines of force must move much closer together increasing the strength of the field. Since the areas of spheres scale as the square of their diameters we can write for the sun to white dwarf case $(15 \times 10^5 / 15 \times 10^3)^2$ or a resulting field 10^4 times greater than the solar field. We have seen (see Section 12.20) that fields of 10^6 times that of the sun and larger have been observed in some white dwarfs. Scaling farther to the size of a neutron star we can expect fields of at least $(15 \times 10^5 / 15)^2$, or 10^{10} greater than that of the sun. The student is cautioned not to take these simple calculations too seriously since the pulsars seem to be the result of cataclysmic events, and there is no reason to believe that the magnetic field remains intact during such an event. The point of this brief diversion using numerical examples is that the observed values are in keeping with the theoretical predictions.

A neutron star such as we have described will have its radiation emitted in a highly directional way and with a short period (Fig. 13.17). We must be very nearly in the plane of the emitted radiation to observe it at all. Only about 60 pulsars are known so even if the special orientation is required we can assert that the pulsar phase in the life of a star is rather rare and/or that it lasts for a very brief time astronomically speaking.

13.16 Flare stars

Some red main-sequence stars are subject to repeated intense outbursts of very short duration reminiscent of the solar flares (see Section 10.5). At least 20 stars of this type have been listed. An example was the sudden brightening by 1.5 magnitudes of the normally fainter component of the visual binary Krüger 60 (Fig. 13.18) observed by P. van de Kamp and S. L. Lippincott at Sproul Observatory. Another example is Proxima Centauri.

Flare stars are designated in the *General Catalogue* as UV Ceti-type variables after a typical representative first recorded by W. J. Luyten in 1948. This star is the fainter component of the binary Luyten 726-8. The main outbursts of UV Ceti occur at average intervals of 1.5 days, when the rise in brightness of the star is generally from 1 to 2 magnitudes. On one occasion in 1952, however, an increase of 6 magnitudes was observed, the greatest flare on record for any star. Only a small fraction of the star's

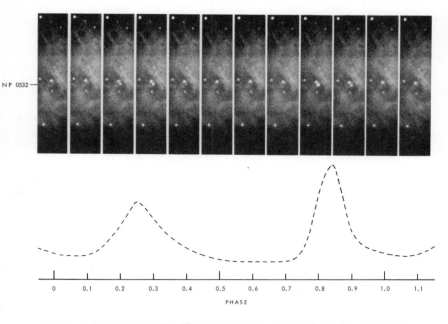

FIGURE 13.17
Pulsar NP0532 in the Crab nebula observed at high speed. The star's period is 0.33 second and the light curve is shown below for reference. (Kitt Peak National Observatory photograph by H-Y. Chin, R. Lynds, and S. P. Maran.)

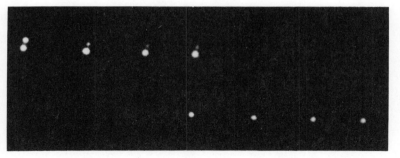

FIGURE 13.18
Kruger 60B flaring on 26 July 1939. The last of four successive exposures (left) shows that the faint companion became as bright as the primary in only a few minutes time. (Sproul Observatory photograph.)

surface is affected, as in the case of a solar flare. Between the main outbursts the light of the flare star varies continuously and irregularly in smaller amplitude.

Recently, observations by B. Lovell and F. Whipple show that the flare stars, particularly UV Ceti, flare in the radio region of the spectrum as well. In this way, the flare star's outbursts are different from even the most violent flare on the sun since the power emitted by UV Ceti in the radio wavelengths is enormous compared to a solar flare. Even more powerful flares have been detected from flare stars in the Orion nebula by O. B. Slee and C. S. Higgins. These stars are considerably further away and hence must radiate prodigious amounts of energy in order to be detected at all here in our Solar System.

QUESTIONS

1 What are the three major classes of binary stars?
2 What is the difference between the relative orbit and the true orbit?
3 Give three methods for detecting unseen companions.
4 What is meant by a dynamical parallax, and how is it determined?
5 Assume that AM Canum Venaticorum is a contact binary of equal size stars. What are the diameters of the stars?
6 Why are RR Lyrae stars often called cluster variables?
7 How is the period–luminosity relation used?
8 What are the nine stages, or phases, common to all novae?
9 Suppose that you take measurements of the expanding ring of light from a nova indicating that it is spreading at an annual rate of 30 seconds of arc. How far away is the nova?
10 How are pulsars designated?

FURTHER READINGS

BINNENDIJK, LEENERT, *Properties of Double Stars*, University of Pennsylvania Press, Philadelphia, 1960.

MERRILL, PAUL W., *The Nature of Variable Stars*, Macmillan, New York, 1938.

PAYNE-GAPOSCHKIN, CECELIA, *Variable Stars and Galactic Structure*, Athlone Press, London, 1954.

——— *The Galactic Novae*, Wiley-Interscience, New York, 1957.

SASLAW, W. C. AND K. C. JACOBS, EDS., *The Emerging Universe*, University Press of Virginia, Charlottesville, 1972, Chap. 3.

SWIHART, T. L., *Astrophysics and Stellar Astronomy*, Wiley, New York, 1968, Chap. II.

A sketch of M 51 by Lord Rosse drawn in the early nineteenth century. The attention to detail is not as good for this galaxy as it was for many others, but the idea of streams of nebulosity is conveyed.

Cosmic Gas and Dust

14

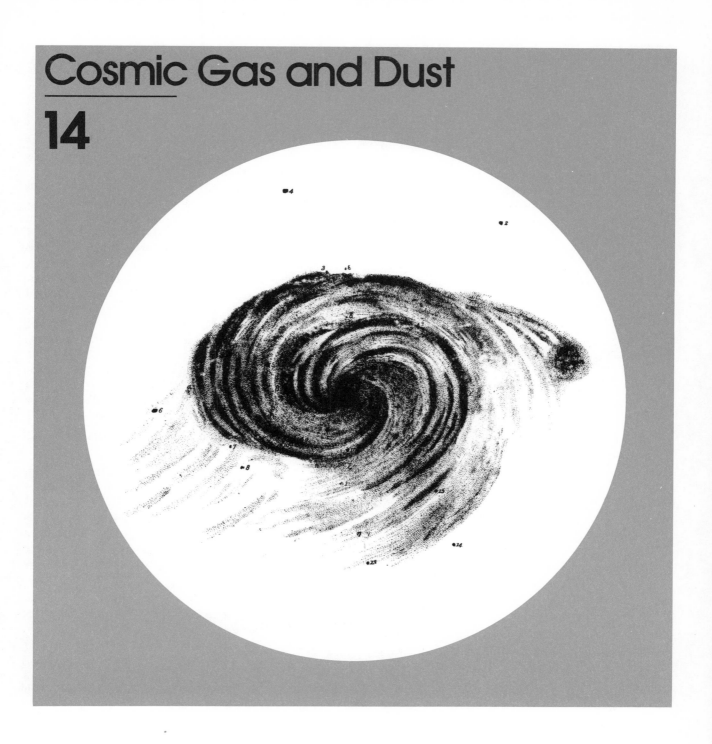

14

Nebulae in general are clouds of cosmic gas and dust. Diffuse nebulae are condensations of the interstellar material that is abundant in our Milky Way and in the arms of spiral galaxies. Such nebulae are believed to supply the material from which stars are born. Planetary nebulae constitute a second category. These are glowing gaseous envelopes expanding around certain very hot stars; they are examples of the return of gas by stars to the interstellar clouds.

DIFFUSE NEBULAE

Diffuse nebulae have irregular forms and often large angular dimensions. Some resemble the cumulus clouds of our atmosphere, whereas others have a filamentary structure that is reminiscent of our high cirrus clouds (these are shown in Fig. 10.1). Shocks and compressions of the colliding turbulent material and effects of magnetic fields can account for the complex structures and also for the light of nebulae in certain cases.

Some diffuse nebulae are made luminous by the radiations of stars in their vicinity. Generally in the absence of involved or neighboring stars the nebulae are practically dark. This relation was first explained by E. Hubble, who also showed that the quality of the nebular light depends on the temperature of the associated stars. Where the star is as blue as type B1, the nebular spectrum differs from that of the star, being mainly a pattern of bright lines. Where the star is cooler than B1, the light is mainly reflected starlight, so that the spectrum resembles that of the star. Thus there are two types of bright diffuse nebulae with respect to the quality of their light: emission nebulae and reflection nebulae. The dark nebulae, in the absence of a nearby star, reveal themselves by dimming and/or reddening the light of stars located beyond them, by imposing their absorption lines upon that of a more distant star, or by emitting radiation in the radio region of the spectrum.

14.1 Emission nebulae

The extreme ultraviolet radiations of very hot stars contain enough energy to remove electrons abundantly from atoms of certain elements in the gases of the surrounding nebulae. As the ionized atoms capture other electrons, the nebulae emit light differing from that of the stimulating stars. The spectrum of the emitted light, as seen in Fig. 14.1, is more conspicuous than that of the starlight reflected by the same nebula for two reasons: (1) The

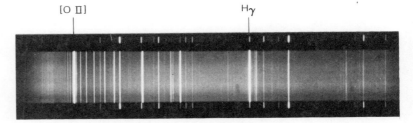

FIGURE 14.1
A low dispersion spectrogram of the Orion nebula by D. E. Osterbrock. The forbidden doublet of oxygen [O II] and the hydrogen gamma line are indicated. The comparison lines are hydrogen and helium.

emitted light is concentrated in a few bright lines of the spectrum, whereas the reflected light is dispersed over the entire spectrum; (2) much of the reflecting dust may have been blown away from the vicinity of the star by the star's radiations, hence reducing the amount of reflected light. Because diffuse emission nebulae and the associated blue stars in our Galaxy are features of the large scale pattern (spiral arms), they have been useful in the optical tracing of the arms (see Section 17.6).

Prominent in the spectra of emission nebulae are "forbidden lines," so called because they are not likely to be observed in ordinary laboratory conditions. These bright lines of oxygen, nitrogen, and some other elements remained unidentified until they were explained theoretically by I. S. Bowen in 1927. In our discussion of atoms and atomic spectra in Chapter 12, we pointed out that only a photon of the proper energy can excite an atom from a given level to a higher level. The forbidden oxygen and nitrogen lines observed in the nebulae are high-excitation lines, that is, they require high-energy photons to excite them. High-energy photons of the proper energy did not seem to exist. Bowen noted that an emission line from singly ionized helium (He II) in the far ultraviolet ($\lambda 304$ Å) agreed precisely with a photon from the unexcited state of doubly ionized oxygen (O III). Helium is quite abundant, hence a source for the photons was found. A little further work by Bowen showed that one of the O III emission lines coincided with photons from the unexcited state of doubly ionized nitrogen (N III), thus explaining a long standing mystery.

This line of reasoning has been applied by other astronomers to explain unusual lines, not only in interstellar clouds, but in planetary nebulae, circumstellar rings and shells, in novae and supernovae, etc. A similar line of reasoning may explain some lines observed from interstellar clouds in the radio region of the spectrum. Despite their stronger showing in the spectra, oxygen and nitrogen are less abundant in these nebulae than are hydrogen and helium; in collisions with other atoms, however, they are able to utilize greater quantities of energy provided by the exciting starlight.

A strong pair of oxygen lines at wavelengths 4950 Å and 5007 Å in the green region of the spectrum impart the characteristic greenish hue to emission nebulae. In another oxygen pair at $\lambda 3726$ Å and $\lambda 3729$ Å in the ultraviolet, the strength of one line relative to the other depends on the density of the gas. This relation has been employed by D. E. Osterbrock and others to determine the densities of these nebulae.

The Great Nebula in Orion, Fig. 14.2, is the brightest diffuse emission nebula in the direct view with the telescope. Scarcely visible to the unaided eye, it surrounds the middle star of the three in Orion's sword. With the telescope the nebula appears as a greenish cloud around the star, which itself is resolved into the familiar Trapezium of type O stars. In the photographs the Orion nebula is spread over an area of the sky having twice the apparent diameter of the moon. At its distance of 500 parsecs the corresponding linear diameter is 8 parsecs, or about the distance of Vega from the sun.

By the relation mentioned above, Osterbrock finds that the gas of the Orion nebula has an unusually high density, about 20,000 atoms per cubic centimeter, in one of the brightest central regions. The density is reduced to 300 atoms per cubic centimeter at a point halfway from the center to the edge. While these numbers sound large they really represent very low densities. If we assume that all of the material is hydrogen then we are considering densities of 3×10^{-20} grams per cubic centimeter and 5×10^{-22} grams per cubic centimeter, respectively. We should compare this with the atmosphere of the sun which is 10^{-7} grams per cubic centimeter and the atmosphere of the earth which is 10^{-3} grams per cubic centimeter.

Among other conspicuous diffuse emission nebulae are M8 in Sagittarius and the nebula surrounding the star eta Carinae, the latter being a most remarkable nebula. A feature of many emission nebulae is small almost spherical globules seen as **dark globules** (Fig. 14.3), against the glowing nebulae. B. J. Bok has called attention to these interstellar objects and points out that they must be very dense accumulations of gas and dust in order to be so opaque. It is thought that the dark globules are places where groups of stars are about to be formed.

14.2 Reflection nebulae

Where stars involved in interstellar material are cooler than type B1, the nebulae around them glow with the starlight reflected by their dust. The bright nebulae surrounding stars of the Pleiades are examples of reflection nebulae. These have the same spectra as the associated stars as was shown by V. M. Slipher long ago. We have already noted that we occasionally see nebular material by the reflection of light from novae and supernovae. The densities of reflection nebulae are on the order of 10^{-22} gram per cubic centimeter.

The similarity in color of reflection nebulae and of the stars

FIGURE 14.2 [OPPOSITE]
The Great Nebula in Orion (NGC 1976, top) and NGC 1977 (bottom). These nebulae are faintly visible to the unaided eye and are superb examples of emission nebulae.

FIGURE 14.3
*The Lagoon nebula (M 8 = NGC 6523)
in Sagittarius. Note irregular dark globules
between us and the emission nebula. (Lick
Observatory photograph.)*

responsible for their shining is shown in photographs taken with filters of a dusty region of Scorpius. Here the reflection nebulae surround stars like the glows around street lamps on a foggy night. The nebular light around the red star Antares is scarcely noticeable with blue filters but becomes conspicuous with red ones. The opposite is true for the nebulae around blue stars in the vicinity.

It is important that the reader not be misled by the previous paragraph. A reflection nebula always shows the characteristic spectrum of the star illuminating it. However, its actual color is somewhat bluer than that of the star rendering it visible. The reason for this is that some of the red light passes through and is not reflected, causing the reflected light to be a little bluer than the star itself.

14.3 Dark nebulae

Dark nebulae are clouds of gas and dust that are sufficiently dense to obscure the stars behind them and have no stars near enough to light them effectively. Their faint illumination by the general star fields can be detected only by measurements of high precision. Dark nebulae make their presence known optically by obscuring whatever lies behind them as we see in Fig. 14.4. The rifts they imprint on the bright background of the Milky Way are conspicuous in the photographs. Some rifts, such as the Coalsack near the Southern Cross, are easily visible to the naked eye and have accordingly been known for a long time; but their interpretation as dark clouds rather than as vacancies is fairly recent.

The darkest clouds in the Milky Way are relatively near us at distances of 300 to 1500 light years. At greater distances their contrast with the bright background is diluted by increasing numbers of stars in front of them. Few dark nebulae are recognized optically in our galaxy at distances exceeding 5000 light years. Even so, these dark nebulae are of great interest as we will see later in this chapter because they show evidence of complex molecules in interstellar space.

PLANETARY NEBULAE

These nebulae are envelopes around blue stars at their centers. They are called **planetary nebulae** simply because, through the telescope, some appear as greenish disks similar to the disks of Uranus and Neptune. The disks may be amorphous or irregular, and some have the appearance of rings; a familiar example is the Ring nebula in Lyra (Fig. 14.5).

14.4 Features of planetary nebulae

At least 500 planetary nebulae are recognized. They range in size from the ring-like NGC 7293 in Aquarius, having half the moon's apparent diameter, to objects so reduced in diameter by distance that they are distinguished from ordinary stars only by their peculiar bright line spectra. Their linear diameters range from 20,000 to more than 100,000 times the earth's distance from the sun.

The central stars of planetary nebulae are about as massive as the sun, but are much smaller and denser than the sun. Having surface temperatures of 50,000°K or more, they furnish a rich supply of ultraviolet and X-ray radiation causing the illumination of these emission nebulae. Because their radiations are mainly in the ultraviolet, the central stars are less easy to see than are the nebulae themselves; the stars come out clearly in the photographs taken in blue light, however.

Planetary nebulae having a rich supply of high-energy photons show the forbidden lines typical of emission nebulae, particularly the oxygen and nitrogen lines noted above. Again, although these lines are prominent, the abundances of these elements are quite normal. In fact the various elements appear in about the same amounts as we find in other nebulae, H II regions, etc. The distribution of planetary nebulae is interesting. They seem to concentrate in one general region of the sky that corresponds to the direction of the center of the Milky Way. Other types of stars and objects have similar distributions. These objects are very old and, therefore, we conclude that the central stars of planetary nebulae are old stars also. From their small numbers we can conclude that the planetary nebula phase in the lifetime of the star is quite short.

Planetary nebulae are expanding around their central stars, as Doppler effects in their spectra show. Although they resemble the envelopes around novae in this respect, the planetaries are expanding more slowly and their lifetimes are much longer. Their moderate speeds of expansion in radius range from 10 to 50 kilometers a second. After lifetimes as long as 20,000 years they begin to disintegrate by breaking into separate clouds of gas, whereas the envelopes of normal novae have disappeared only a few years after emerging from the stars.

There is also a pronounced difference in the amount of material involved. The mass of a planetary nebula is about 1/10 the sun's mass; but the mass of a normal nova envelope does not exceed 1/10,000 the sun's mass.

FIGURE 14.4 [OPPOSITE]
The Horsehead nebula in Orion. The illuminated nebula is quite thin as can be seen by counting stars in the light and dark portion. (Photograph from the Hale Observatories.)

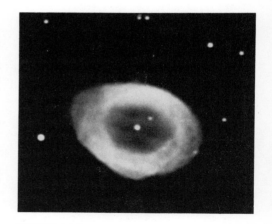

FIGURE 14.5
The Ring nebula in Lyra. This beautiful planetary nebula is an excellent example of a star returning material to the interstellar medium. See color plates. (Dominion Astrophysical Observatory photograph.)

THE INTERSTELLAR MATERIAL

In addition to the more obvious bright and dark nebulae, certain regions of our Galaxy contain an abundance of tenuous gas and dust. The interstellar dust dims and reddens the light of more distant stars. The gas imprints dark lines in the spectra of stars beyond it. The presence of optically invisible neutral hydrogen in the gas is recorded with radio telescopes. The average density of the interstellar gas is around 10^{-24} gram per cubic centimeter and interstellar dust is 10^{-26} gram per cubic centimeter.

14.5 Interstellar spectra

Interstellar lines in the spectra of stars, first noted and correctly interpreted by J. Hartmann in 1904, are abstracted from the starlight by the gas through which the light passes. These dark lines are generally much narrower than the lines in the spectra of the stars themselves and have different Doppler displacements than those of corresponding star lines. Among the chemical constituents of the interstellar gas indicated by these lines are atoms of sodium, potassium, calcium, iron, and titanium, and molecules of cyanogen and various hydrocarbons. Hydrogen atoms are doubtless very abundant, but in these conditions their lines would appear only in the generally unobservable extreme ultraviolet region of the spectrum and at $\lambda 21$ centimeters.

The division of interstellar lines into two or more components was recognized some time ago, particularly by W. S. Adams (Fig. 14.6). The division is given important interpretation by the more recent studies of G. Münch; his photographs of stellar spectra with the 200-inch telescope show that the principal components of the interstellar lines are absorbed by gas in two intervening arms of our Galaxy. Very precise measures of the interstellar hydrogen lines at $\lambda 21$ centimeters confirm and extend Münch's work (Fig. 14.7).

14.6 Interstellar hydrogen

Where a cloud of interstellar gas surrounds a hot star, the gas is set glowing within a radius around the star that depends on the temperature of the star and the density of the gas. This sphere is referred to as the Strömgren sphere. Outside this region the gas is normally dark. Because hydrogen is the most abundant chemical element, it has been the custom to speak of the two parts of such a cloud as the H II and H I regions, respectively.

In the part of the cloud that is nearer the star, hydrogen atoms are ionized by the star's ultraviolet radiation. Because the atoms

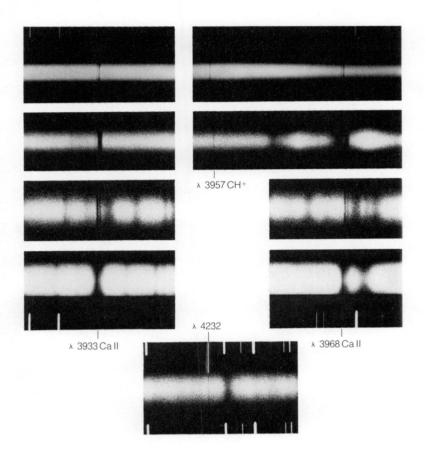

FIGURE 14.6
Interstellar lines in the spectra of various stars. The interstellar lines are extremely sharp and are split into several lines indicating many thin clouds between us and the star. (Photograph from the Hale Observatories.)

λ 3957 CH+

λ 3933 Ca II   λ 4232   λ 3968 Ca II

that have lost electrons soon capture others reemitting the energy, the gas is thereby made luminous. In addition to the conspicuous emission nebulae, many faint H II regions are detected in photographic surveys with wide-angle cameras and plates especially sensitive to the spectral region of the red line of hydrogen. Such surveys have been useful in tracing the spiral arms of our Galaxy and other nearby galaxies.

The bright material in the inner region of a cosmic cloud is likely to expand and to collide with the gas in the dark region outside. Such collisions may be vigorous enough to set the contact areas glowing conspicuously. This reason has been assigned for the familiar loop of bright nebulosity near epsilon Cygni (Fig. 14.8).

In the part of a gas cloud that is too remote from a hot star to be much stimulated by it, the hydrogen is generally dark and

FIGURE 14.7
Two sharp interstellar hydrogen lines observed in the direction of the star HD 212571. The shift in these radio lines agrees with the sodium lines indicating that the hydrogen and sodium are in the same clouds. (The observations were obtained at the National Radio Astronomy Observatory by S. J. Goldstein.)

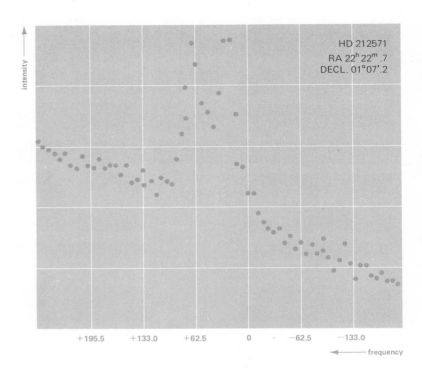

FIGURE 14.8
Hydrogen clouds and wreaths in Cygnus. In this mosaic of negative prints from the 48-inch Schmidt telescope, emission nebulae and stars appear black. Astronomers often work with negatives to avoid loss of detail in reproduction. The familiar Loop nebula is seen at the bottom. (Composite photograph by J. L. Greenstein, California Institute of Technology.)

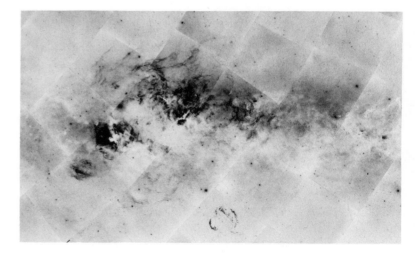

optically unobservable. Radiation by neutral atoms of cosmic hydrogen at a wavelength of 21 centimeters, previously predicted by H. C. van de Hulst, was first recorded with a radio telescope by H. I. Ewen and E. M. Purcell in 1951. This emission line has been employed effectively in tracing the spiral arms of the galaxy (see Section 17.6) to great distances from the sun. It would be difficult to stress the prediction and observation of the λ21-centimeter line too strongly. The last equivalent instrumental advance was the introduction of the photographic plate in the 1850s.

The emission by neutral hydrogen has also allowed us a detailed look at the center of the galaxy. It has been used to trace the extent of nearby galaxies as well. Where a bright source is located behind cooler neutral hydrogen the corresponding absorption line at λ21 centimeters is observed.

At λ18 centimeters we observe the rotational lines of the OH molecule. Again, depending upon circumstances, it may be observed in emission or in absorption.

14.7 Dimming of starlight by dust

Interstellar dust dims the light of stars behind it (Fig. 14.9). The dust also reddens the light of those stars, because dimming in the shorter wavelengths is greater than in the longer wavelengths of the light. A corresponding effect is observed in the reddening of the light of the setting sun by particles of our atmosphere. Careful observations show that the absorption (A) by the interstellar material is inversely proportional to the wavelength of the observation.

$$A \propto \frac{1}{\lambda}$$

A typical absorption curve as a function of wavelength is shown in Fig. 14.10. The wavelength dependence is just that expected for particles having sizes of 10^{-5} centimeter or slightly larger. If the particles were atoms or molecules the absorption should be inversely proportional to the fourth power of the wavelength

$$A \propto \frac{1}{\lambda^4}$$

which is shown by the light blue line in Fig. 14.10. We also know that if the particles were several times the wavelength in size we would have a neutral absorption, that is, all starlight is dimmed by the same amount in all wavelengths. Thus our observations tell us that the absorption is due to dust. We can not rule out the possibility that larger particles are dimming all light accordingly, or that gas is not present. Indeed, we shall see that the gas

FIGURE 14.9
*Photographic absorption by dust in two
regions of Aurigae. Star counts reveal one
dust cloud in region F and two in region G.*

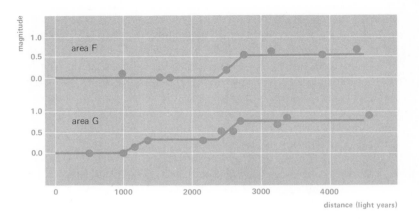

accounts for 100 times more of the interstellar mass than does the dust.

The **color excess** of a star is the difference of magnitude by which the observed color, blue minus visual $(B - V)$, exceeds the accepted value for a star of its spectral type; it is a measure of the reddening of the star by the dust. When the color excess is multiplied by an appropriate factor, we have the **photographic absorption**, that is, the degree to which the star is dimmed by the dust when photographed with a blue-sensitive plate, measured in magnitudes.

The distance of a reddened star determined by photometric means requires correction for absorption, which magnifies the actual distance. The corrected distance, r, is calculated by the formula:

$$5 \log r = m - M + 5 - K$$

where m is the observed photographic magnitude, M is the corresponding absolute magnitude for a star of this particular spectral type, and K is the photographic absorption.

Since dimming and reddening of starlight is caused by the interstellar dust, these quantities must vary depending upon where one looks in the Galaxy. The determination of dimming and reddening is not very difficult in principle, but in practice it is a most perplexing task. Thus, any method promising to yield distances, magnitudes, or colors independent of dimming and reddening is studied to the utmost.

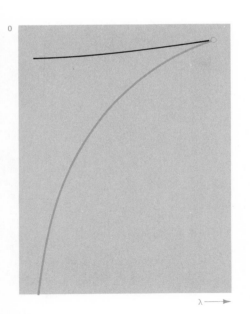

FIGURE 14.10
*Interstellar reddening as observed (yellow
line) approximates a $1/\lambda$ law. If micron-size
particles were responsible, the curve would
follow a $1/\lambda^4$ law (blue line).*

14.8 The dust grains

The dimming of stars by intervening cosmic dust is attributed to particles of the order of a hundred thousandth of a centimeter in diameter. Dust grains of this size would scatter the light in inverse proportion to the wavelength, which is not far from the observed relation for the reddened stars. Considerably larger particles would obstruct but not redden the starlight. The origin of the dust grains is not clearly understood; whether they build up from atoms of the interstellar gas or are particles transferred to the medium from the upper atmospheres of stars, or both, is conjectural.

Starlight becomes **polarized** in its passage through clouds of interstellar dust, as the photoelectric studies of J. S. Hall and W. A. Hiltner revealed independently. A possible explanation is that the polarization is produced by dust grains shaped like needles and rotating on their short axes, resulting in a concerted effect by setting themselves along magnetic lines of force of the Galaxy; a magnetic field of 10^{-5} gauss is indicated. This and other implications that our Galaxy has a general magnetic field have added other items to the growing list of problems in celestial hydromagnetics, not the least of which is the Zeeman splitting observed in the OH lines—if that is what it really is. Local galactic magnetic fields of some 10^{-2} gauss are implied by the OH lines and this is some 100 times larger than fields previously observed. Other radio observations, such as pulse delay times as a function of wavelength from pulsars, indicate field strengths of 10^{-5} gauss in agreement with the earlier results.

14.9 Interstellar molecules

We have seen that the interstellar material seems to bunch into nebulae. Within these clouds, shielded from the high-energy radiation of the ultraviolet and X-ray photons that would destroy them, interstellar molecules could form. The first interstellar molecule, CH, was discovered by T. Dunham in 1937. A few additional molecules were detected in the visual region of the spectrum, CN and CH^+ (singly ionized CH), but if our earlier reasoning is correct we cannot detect the more complex molecules, which are also more fragile, simply because they must be hidden deep in the interior of the great interstellar clouds.

Fortunately, radiation at radio wavelengths passes through the interstellar clouds, hence the clouds can be probed for emission from molecules. One other factor works in favor of detection by radio techniques. We recall that the strength of any atomic spectral line is proportional to the number of atoms involved, and this is also true for molecular lines. Thus, if the abundances of molecules are in a constant ratio, the very densest clouds will

TABLE 14.1
Interstellar molecules

Year	Molecule	Name
1937	CH	methyl idyne radical
1942	CN	cyanogen radical
1942	CH⁺	
1963	OH	hydroxyl
1973	SO	sulfur monoxide
1970	CO	carbon monoxide
1971	CS	carbon monosulfide
1971	SiO	silicon monoxide
1970	H_2	molecular hydrogen
1972	HD	hydrogen deuteride
1968	NH_3	ammonia
1968	H_2O	water
1970	HCN	hydrogen cyanide
1971	OCS	carbonyl sulfide
1972	H_2S	hydrogen sulfide
1972	DCN	deuterated hydrogen cyanide
1971	HNC	hydrogen isocyanide
1969	H_2CO	formaldehyde
1972	H_2CS	thioformaldehyde
1971	HNCO	isocyanic acid
1970	HC_3N	cyanoacetylene
1971	HCOOH	formic acid
1972	CH_2NH	methyleneamine
1970	CH_3OH	methyl alcohol
1971	$HCONH_2$	formamide
1971	CH_3CN	methyl cyanide
1972	$HCOCH_3$	acetaldehyde
1971	CH_3C_2H	methyl acetylene
1970	X	X-ogen

SOURCE L. Snyder, 1973.

have the highest concentrations rendering them more readily detectable.

In 1963 emission from the OH molecule was discovered at λ18 centimeters. This was followed in 1968 by the discovery of ammonia (NH_3) and water vapor (H_2O). These discoveries served to show that complex molecules do exist in the interstellar medium. Then, in early 1969 came the remarkable discovery at λ6.2 centimeters of formaldehyde (H_2CO), a polyatomic **organic molecule**. An organic molecule is one where hydrogen is attached to carbon; the term originates from the fact that such molecules are found in living organisms. This discovery has been followed by the detection of some 21 additional molecules, all of which involve hydrogen, carbon, nitrogen, oxygen, and sulfur, and most of which are organic. It is tempting to think that a carbon chemistry is possible if not common throughout the Milky Way. This thought has great consequences on our ideas of the origin and evolution of life. We will return to this in Chapter 19.

Table 14.1 shows the tremendous growth of the number of interstellar molecules discovered in the past few years and their increasing complexity. It demonstrates that the science of chemistry contributes to astronomy, and it has implications for the students of biochemistry as well. The last entry, X, is called X-ogen by Buhl and Snyder who discovered the line. The X-ogen line is clearly present and does not correspond to any line predictable by theory or observed in the laboratory. However, experience from the past cautions us to think that this line is probably caused by a molecule in an unusual energy level or a molecule that does not normally exist under terrestrial conditions.

The physics and chemistry of the interstellar molecules and medium is only now being unraveled. It will be several years, perhaps decades, before we will really begin to understand what we are observing. Many of the emissions observed are from levels that are not normally found on earth or in the laboratory, and in this sense they are similar to the forbidden lines that we have discussed for atoms. Molecules have several lines close together, and, as often as not, the relative strengths of these lines are not those we predict on the basis of calculations or laboratory measurements. We do not know the exciting mechanism that causes these anomalies, but work is being done in this area and answers will be available soon. We do not even know what causes abundance anomalies. The strength of the HD molecule ("heavy" hydrogen) compared to H_2 (normal hydrogen) gives a ratio of hydrogen to deuterium, an isotope of hydrogen, of 300; this is several orders of magnitude smaller than we would predict. This result, which was obtained by the spacecraft OAO-C (Orbiting Astronomical Observatory-C), is quite unexpected and typical of the subject of interstellar molecules.

The complex molecules are found in interstellar clouds almost everywhere in the Milky Way. By interstellar standards the molecules are quite abundant. Occasionally, the clouds are releasing large amounts of energy, and they are sometimes variable in their energy output. At least one cloud, with water in great abundance, has solar system dimensions.

Our conceptual picture of an interstellar cloud containing organic molecules is that in the densest innermost regions are found the most complex molecules along with all the other molecules making up the cloud. The molecules become less and less complex as we move outward from the center to regions where photons can penetrate and dissociate them. In the outermost regions only the tightly bound CO molecules can exist, and at the boundary even this molecule is photodissociated. In a cloud that is not very dense, perhaps the most complex molecule that can exist is water.

QUESTIONS

1 Define an emission and reflection nebula?
2 How did I. S. Bowen explain the forbidden lines in emission nebulae?
3 Why is the nebular light around Antares not visible in the blue light?
4 Why do we not see dark nebulae beyond 5000 light years?
5 Why can we say that the planetary nebula phase is brief in the life of a star and why do we think that the central star is an old star?
6 A planetary nebula increases its radius by 0.021 second of arc in 10 years. The radial velocity of the nebula is observed to be 21 kilometers per second and the diameter is measured to be 21 seconds of arc. How far away is the object, and how old is it?
7 What two facts about interstellar lines would serve to confirm that they are interstellar?
8 Why do we think the interstellar material is generally located in clouds rather than uniformly distributed?
9 Define a Strömgren sphere.
10 What is our present picture of an interstellar cloud containing complex molecules?

FURTHER READINGS

ABETTI, GIORGIO, AND MARGHERITA HACK, *Nebulae and Galaxies* (V. Barocas, Transl.), Faber and Faber, London, 1964.

ALLER, LAWRENCE H., *Gaseous Nebulae*, Chapman and Hall, London, 1956.

BLAAUW, ADRIAAN, AND MAARTEN SCHMIDT, EDS., *Galactic Structure*, University of Chicago Press, Chicago, 1965.

DAUVILLIER, A., *Cosmic Dust*, Philosophical Library, New York, 1964.

DUFAY, JEAN, *Galactic Nebulae and Interstellar Matter* (A. J. Pomerans, Transl.), Philosophical Library, New York, 1957.

SASLAW, W. C. AND K. C. JACOBS, EDS., *The Emerging Universe*, University Press of Virginia, Charlottesville, 1972, Chap. 4.

WOOD, F. B., ED., *Photoelectric Astronomy for Amateurs*, Macmillan, New York, 1963, Chaps. 5–7.

A rubbing of the great bronze star chart of Suchow. The chart, made to instruct the young prince of Suchow province in astronomy, includes the supernova of 1054.

The Lives of the Stars

15

15

In studying stellar evolution we try to understand how stars are born and how they spend their lives until they cease to shine. The idea that stars condense from nebulae was pioneered by Kant and Laplace in the eighteenth century and has persisted to the present time. Stars of large mass evolve faster than lighter stars and, during the course of their evolution, return material to the interstellar medium. Finally all stars end their careers as black dwarfs, white dwarfs, neutron stars, or black holes.

New stars are continually forming from condensations in the interstellar gas and dust, according to one present, popular theory of stellar evolution. The primitive stars are heated by contraction until they begin to shine. Thereafter, they go on shrinking until their central temperatures are hot enough to permit the building up of atoms of heavier chemical elements from lighter ones. Atomic synthesis, the formation of helium from hydrogen, replaces contraction as the main source of stellar energy. The star then settles down to a long life on the main sequence.

15.1 The birth of stars

The account of stellar evolution begins conveniently with structureless interstellar gas. When a denser cloud develops in the gas and is not considerably heated in the process, the cloud is likely to condense further under its own gravity and to break into smaller clouds. The fragmented cloud would be the beginning of a cluster or an association of stars (see Chapter 16).

How does a dense cloud form? How does the cloud ever become dense enough so that the self-gravitation is greater than the forces that would disrupt the cloud? The disrupting forces are the random motions in the cloud resulting from the heat caused by the contraction of the cloud. The densities required to make the cloud self-gravitating are on the order of 10^{-19} gram per cubic centimeter, whereas the average dense interstellar cloud has a density of 10^{-21} gram per cubic centimeter or somewhat lower.

A possible answer may be provided by clouds containing complex molecules. These clouds are generally quite cold; when formaldehyde is present, we can see it in absorption against the general cosmic background (see Section 18.15). This proves that the temperature of the inner part of the cloud must be less than $3°K$. If the temperature of the interstellar medium is $50°K$ or $100°K$, then the hotter interstellar medium exerts an inward pressure on the cooler cloud raising its density. This temperature difference will not cause a sufficient pressure difference, but it is a step in the right direction.

Photographs of the Milky Way show small roundish "globules" of dark material against backgrounds of star-rich regions (Fig. 15.1) and of bright nebulosities, such as M 8 (see Fig. 14.3). These dark spots, which continue in a sequence of diminishing sizes to the limit of the 200-inch telescope, are viewed with interest as perhaps representing the first visible stages in the formation of stars from nebulae.

An objection to the idea that such protostars can be observed is that stars are likely to form in the interiors of cosmic clouds, where they would be concealed by dust in the clouds. Exceptions are considered possible: One is that the star must begin to shine in the infrared first; infrared photons may then escape the cloud. Another is that the most massive stars of a very young group have already become hot blue stars. In the first case we can observe the stars as infrared objects. In the second, the intense radiations of the hot stars may have dispersed enough of the dust around the group so that the redder members can be seen still forming within the cloud. This may be the case with the T Tauri stars.

The T Tauri stars were so named after the prototype by A. H. Joy, who in 1945 first recognized them in considerable numbers in a clouded area of Taurus. Now observed in other heavily obscured regions of the Milky Way as well, these yellow and red stars are irregularly variable in brightness. Their spectra are characterized by strong bright lines of various elements superposed on ordinary dark line patterns. In some cases the stars illuminate portions of the surrounding clouds, producing fan-shaped reflection nebulae that may be variable in brightness and appearance.

Young stars very recently formed in the clouds, the T Tauri stars are believed to vary in brightness and to have bright lines in their spectra because of the instability of extreme youth rather more than by their interaction with the dust grains of the clouds. A few such stars appear in the photographs in places where no stars were previously observed; these may have been revealed lately by rapid thinning of the dust in front of them.

15.2 Arrival at the main sequence

When the contracting stars of a cluster or an association have begun to shine, they are enormous distended clouds of gas and dust and are quite cool. They radiate principally in the infrared and radio regions of the spectrum. Such stars have been observed and are referred to as **infrared stars.** Interestingly enough they are always found in highly obscured regions and intense radiation from water vapor is observed. When they move into the visible region of the spectrum they lie rather high up in the H-R diagram. The gravitational energy released in the deep interior of the cloud is carried outward by convective currents. Hayashi has shown that this causes the cloud to collapse rapidly to the point of **radiative equilibrium.** Since the energy radiated per square centimeter remains essentially constant, but the surface area decreases rapidly due to contraction, the protostar moves almost vertically down the H-R diagram. Once it comes into radiative equilibrium it moves to the left until it reaches the main sequence (Fig. 15.2). The stars are brighter and attain higher surface temperatures as their masses are greater. Thus they array themselves along this diagonal band in order of mass, the most massive stars at the blue end and the least massive at the red end.

The time required for the development of the stars of a group also depends on the masses. The more massive blue stars may arrive at the main sequence only a few hundred thousand years after their birth; these hot stars have presumably blown away enough dust from around the group so that we are able to look into the cloud and see how the other stars are progressing. The least massive stars may require a few million years to arrive at the main sequence. Thus in some very young clusters the blue stars are already settled in the sequence, while the other stars are still approaching it.

This recently recognized effect is shown in Fig. 15.3 for the very young cluster NGC 2264 in Monoceros, near the central line of the Milky Way about 15° east of Betelgeuse and 700 parsecs distant from us. The points representing the blue stars appear on the main sequence, whereas those for the yellow and red stars depart abruptly to the right from the standard sequence at type A0 and tend to lie about 2 magnitudes above the curve. Many of

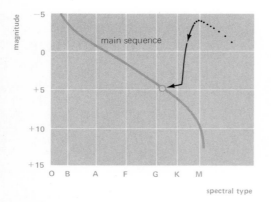

FIGURE 15.2
The Hayashi evolutionary track for a solar-type star. The star becomes visible as a very distended object (top of arrow), rapidly shrinks in size with little change in temperature until radiative equilibrium is established. At this point the star moves slowly to the left until it reaches the main sequence (yellow line).

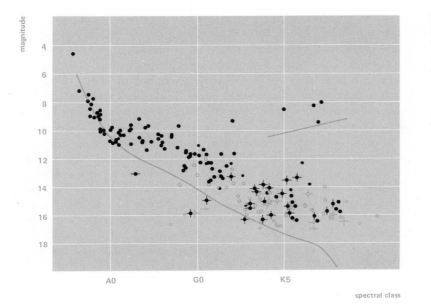

FIGURE 15.3
Color–apparent magnitude diagram of the young cluster NGC 2264. Most of the stars have not reached the main sequence. The dots represent photoelectric measures and the circles photographic measures. Vertical lines denote variable stars. Horizontal lines denote stars having Hα emission lines. The solid curve is the standard main sequence. (Diagram by M. F. Walker.)

the latter are variable stars of the T Tauri type, according to M. Walker.

An alternative explanation has been advanced that these stars actually lie along a line defining a second-generation main sequence. This might be the case if the interstellar medium from which the stars formed had been enriched with metals by an earlier generation of stars.

Some condensations may not have sufficient mass for radiative equilibrium to set in. In this case S. S. Kumar has shown that they will continue to contract into degenerate objects that he calls "black dwarfs." Presumably, if such condensations are gravitationally attached to a larger condensation, they become planets, but this part of the theory must still be confirmed.

15.3 The stable main-sequence stars

Contraction is halted when the stars reach the main sequence. This process, which was regarded in earlier theories as the main source of stellar energy, is inadequate to supply the stars during their long lives. Contraction suffices in the present account to heat the stars to central temperatures that are high enough to initiate the fusion of lighter chemical elements into heavier ones. Atomic syntheses then become the source of stellar energy; this process continues until a star's supply of atomic fuel is exhausted.

Main-sequence stars are stable for a time, neither contracting nor expanding. Therefore, they vary little in size, temperature, and brightness. Consider the sun as an example. At any level in its interior the sun is in mechanical equilibrium (hydrostatic equilibrium is the proper term). The weight of overlying gas is just supported by the upward push of the gas pressure at that level. Because the gas pressure depends on the temperature of the gas, among other factors, the temperature becomes known when the other necessary information is available. Obtained in this way, the sun's present central temperature is about 13 million degrees Kelvin. Another form of balance in the sun concerns its shining. The rate of radiation of energy from the sun's surface is 5 followed by 23 ciphers horsepower. The output of one ideal horse (however it was originally defined) is now equal to 7.46×10^9 ergs per second. In order to preserve the mechanical balance, this must also be the rate at which energy is provided in the hot core of the sun. As soon as the sun reached the main sequence, the central temperature became high enough to promote the fusion of hydrogen into helium with the release of energy at the required constant rate. This constant rate for a normal star is very constant indeed. There is fossil evidence on the earth that the sun's energy output has been constant to within 1 percent over a period of 3×10^9 years. There is reason to believe that this period has been even longer, at least 6×10^9 years, and will continue for another 5 to 10×10^9 years.

THE STARS IN MIDDLE AGE

The stars stay on the main sequence as long as they remain practically homogeneous in composition. The energy released by the fusing atoms keeps the stars shining steadily. As the building up of helium in their cores continues, the stars eventually lose stability; they then begin to expand and brighten, leaving the main sequence to become giant stars. The bluer stars withdraw sooner than the redder ones.

15.4 Fusion of hydrogen into helium

The relative weight of the nucleus of a hydrogen atom is 1.0080 and that of a helium atom is 4.003. The electrons normally associated with atoms need not be taken into account here, because these atoms are stripped of their electrons in the hot interiors of stars. When four hydrogen atoms unite to form one helium nucleus, 7 percent of the original mass is left over. The excess mass is converted into energy. By the relativity theory,

$$E = mc^2$$

the amount of energy, E (in ergs), released equals the mass, m (in grams), of the excess material multiplied by the square of the speed of light, c (in centimeters per second). Using this formula we can calculate that enough energy is thereby made available to keep the sun shining at its present rate if 700 million metric tons of hydrogen gas are converted each second to 695 million metric tons of helium in the sun's interior. This consumption is trivial compared with the vast supply of hydrogen in the sun.

Two possible processes by which hydrogen may be fused into helium in the sun and stars were explained in 1938 by H. A. Bethe. One of these, known as the **carbon cycle,** because carbon serves as a catalyst* to promote it, becomes effective in stars hotter than the sun. The second process, known as the **proton– proton reaction,** is considered more appropriate at the central temperatures of the sun and the main-sequence stars that are redder than the sun.

It is likely that the first **burning** process is that which transforms deuterium into helium. In this process a deuterium nucleus, which we have seen is quite abundant in space, interacts with a proton, or hydrogen nucleus, releasing energy in the form of a gamma ray and forming an isotope of helium (He^3). This process made possible by the temperature increase due to the collapse of the cloud then raises the temperature to a point where the proton–proton reaction can take place.

In the proton–proton reaction, in a form somewhat different from the original, six protons (or hydrogen nuclei) are required to form a helium nucleus. The reaction proceeds when two protons unite to form heavy hydrogen (an isotope of normal hydrogen called deuterium) releasing energy and a neutrino. Deuterium and a proton then react to form an isotope of helium releasing a gamma ray. When two helium isotopes combine, normal helium is formed and two free protons are released. The two unused protons go back into circulation. The excess mass in the four united protons is released as energy, which is carried up to the sun's surface to contribute to the sunlight. This reaction proceeds as the fourth power of the temperature and requires temperatures

---

* A catalyst is a substance that initiates a reaction. It remains unchanged after the reaction is complete.

of about 10^6 °K in order for it to take place. The reaction may be written symbolically in the following three steps:

(1) $_1H^1 + _1H^1 \rightarrow _1H^2 + e^+ + \nu$
                           (deuterium)

(2) $_1H^2 + _1H^1 \rightarrow _2He^3 + \gamma$

(3) $_2He^3 + _2He^3 \rightarrow _2He^4 + _1H^1 + _1H^1$

The notation is necessary but should not detract the reader from noting that a gamma-ray (γ) photon is released each time step 2 takes place and also that a neutrino (ν) is released when step 1 takes place. The neutrino is a nuclear particle having energy but no electronic charge and very little mass. The subscript indicates the electronic charge on the atom and the superscript indicates the total mass of the nucleus.

In stars somewhat more massive than the sun the carbon cycle dominates. In it carbon (C^{12}) combines with a proton forming an isotope of nitrogen (N^{13}) that decays into an isotope of carbon (C^{13}) releasing energy. This carbon isotope combines with a proton to form stable nitrogen (N^{14}) releasing additional energy. A proton then interacts with the nitrogen to form an istotope of oxygen (O^{15}) that immediately decays into another isotope of nitrogen (N^{15}). This nitrogen isotope then combines with a proton, but instead of forming oxygen, which we might expect, it forms a carbon nucleus plus helium. The end result is the conversion of hydrogen into helium, with the initial amount of carbon unchanged. This process proceeds with the twenty-first power of the temperature and operates at temperatures on the order of 10^7 °K. For reference we may write the carbon cycle symbolically in 6 steps:

(1) $_6C^{12} + _1H^1 \rightarrow _7N^{13} + \gamma$

(2) $_7N^{13} \rightarrow _6C^{13} + e^+ + \nu$

(3) $_6C^{13} + _1H^1 \rightarrow _7N^{14} + \gamma$

(4) $_7N^{14} + _1H^1 \rightarrow _8O^{15} + \gamma$

(5) $_8O^{15} \rightarrow _7N^{15} + e^+ + \nu$

(6) $_7N^{15} + _1H^1 \rightarrow _6C^{12} + _2He^4$

Again the important points for the reader are the release of energy through the high-energy, gamma-ray photons in steps 1, 3, and 4, and the release of neutrinos in steps 2 and 5.

Eventually the center of a star is helium. When the pressure becomes great enough, due to contraction, the helium **burns** to carbon by means of the **triple alpha process.** This process, first proposed by E. Salpeter, combines three helium nuclei (sometimes called alpha particles) directly into carbon. The triple alpha process proceeds as the nineteenth power of the temperature but requires a temperature on the order of 10^8 °K to operate.

This process may be written in two steps:

(1) $_2He^4 + _2He^4 \rightleftharpoons _4Be^8 + \gamma$

(2) $_4Be^8 + _2He^4 \rightarrow _6C^{12} + \gamma$

In step 1 the reverse arrow means that the beryllium decays back into two alpha particles spontaneously after a very short period of time. If, before this, another alpha particle interacts with the beryllium the second step takes place. Both steps yield high-energy, gamma-ray photons.

15.5 Evolution from the main sequence

While the hydrogen in the hot core of a star is increasingly converted to helium, the core contracts and grows hotter, and a new balance requires that the outer layers of the star expand. When the core of a star contains 12 percent of the total mass a change sets in. The star becomes brighter and redder, moving away from the main sequence. The evolution is more rapid for the blue stars, which consume their fuel at a faster rate, than for the redder stars. We will note this effect in the diagrams of galactic clusters in Chapter 16 and see how the main sequence evolves progressively from left to right as the clusters get older.

When the hydrogen in the core of the star is nearly exhausted, the interior has become hot enough so that its fusion extends to outer layers and goes on at a furious rate. Presently the star reaches the climax of its luminous career as a red giant. All its hydrogen fuel is then nearly consumed; if it is not considerably more massive than the sun, the star with its maximum central temperature of 100 million degrees Kelvin has not become hot enough for the synthesis of helium to heavier elements. Its only recourse is to contract, which it continues to do until the white dwarf stage is reached.

The diagram of Fig. 15.4 shows the probable evolutionary tracks of stars of the globular cluster M 3 since they withdrew from the main sequence more than 5 billion years ago. Stars originally a little bluer than the sun are now red giants, and those that were still bluer are presumably moving to the left in the diagram or have already declined to become white dwarf stars in the lower left corner. Meanwhile, the cluster stars that were similar to or redder than the sun have not yet risen far above the main sequence. The general evolution and relative sizes of the stars are shown in Fig. 15.5.

As an example of the evolution of a star of moderate mass, A. Sandage has predicted the future of the sun and its effect on the earth. He makes the reasonable assumption that the sun's evolutionary track in the temperature–luminosity diagram of Fig. 15.6 should closely resemble the color-magnitude diagram for the stars of comparable mass in the cluster M 67 (Fig. 16.5).

The sun today is approaching middle age; it has brightened only slightly in the 5 or 6 billion years since it reached the main sequence and will continue to brighten moderately for 6 billion years more. Life on the earth should go on in reasonable comfort to the end of that period, as far as this effect is concerned. Thereafter, the sun will begin to consume its remaining hydrogen so

FIGURE 15.4
Semiempirical evolution tracks of stars in the globular cluster M3. Values of log T_e (effective temperature) correspond to the following spectral types: 3.9 to A5, 3.8 to F5, 3.7 to K0 for the main sequence and G0 for giants, 3.6 to K5 for the main sequence and K0 for giants. (Diagram by A. Sandage.)

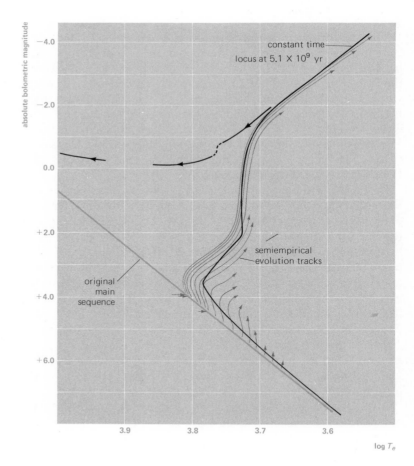

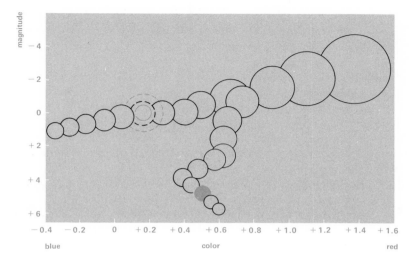

FIGURE 15.5
Relative sizes of stars on a logarithmic scale. The giant star in the upper right should actually be two hundred times larger than the sun which is shown shaded in the lower center of the diagram. The cluster variable stars are shown at maximum and minimum size along the horizontal branch.

rapidly that in the following 500 million years the temperature on the earth will rise 500° to 800°K. The oceans will boil away, lead will melt here, and life will cease to exist. At its maximum the sun will appear as a red globe in the sky 30 times its present diameter.

With little available fuel remaining, the sun will then decline rapidly, moving to the left and down in the diagram. Water on the earth will condense again to form oceans; these will soon freeze and carbon dioxide will freeze as well. The sun will end its visible life as a white dwarf star, according to this prediction, having a higher surface temperature for a longer time than that of the present sun but a diameter more nearly like that of the earth. The sunshine will be less than a thousandth as bright as it is now, and the temperature at the earth's surface will be more like the present low temperature at the surface of Neptune.

15.6 Evolution of more massive stars

All yellow and red main-sequence stars are expected to follow about the same track of evolution that we have traced for the sun. Stars considerably bluer and more massive than the sun will have shorter active lives. The most massive ones will exhaust their supplies of fuel in scarcely more than a million years. We will see how they have already broken away from the main sequence even in very young clusters such as the double cluster in Perseus.

The middle stages in the lives of blue main-sequence stars are likely to be more spectacular than in the case of the sun, because

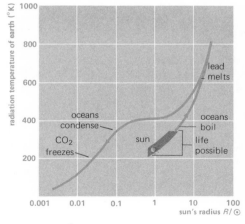

FIGURE 15.6
Predicted variation in the earth's radiation temperature as evolution alters the sun's radius. Temperature is given on the absolute scale. (Diagram by A. Sandage.)

they attain higher temperatures. They evolve upward to the right as we have already seen. They will then contract when helium burning sets in and move again to the left. Upon completion of this stage they again evolve to the upper right, until a new stage sets in such as carbon burning. In this way these stars build heavier elements and make several crossings of the H-R diagram at brighter and brighter levels (Fig. 15.7). Many such crossings can be traced, but this process cannot go on indefinitely.

At some point an instability sets in causing a catastrophic explosion, the violence of which depends upon the present mass of the star. Stars two or three times more massive than the sun probably shed their upper layers rather uniformly and may appear as a planetary nebula that slowly expands and dissipates. Several of these nebulae may be given off reducing the mass of the star to that of a white dwarf. Stars that are more massive, say 3 to 8 times the solar mass, are not so fortunate, particularly if they have a companion forming a binary system. When the instability occurs the star erupts explosively, providing a nova event that causes the star to shed its mass. This event may repeat at long intervals until the star reaches the mass of a white dwarf. The most massive stars have no recourse, they blow themselves to pieces in a supernova event. We discuss this further on page 336.

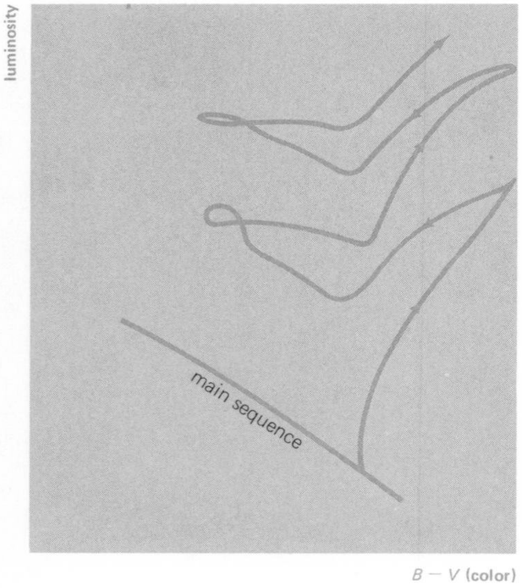

15.7 Synthesis of heavier elements

Two recent theories on the origin of chemical elements have been of special interest. In the first theory, G. Gamow and associates suppose that before the stars were born the elements were built up from neutrons in the course of half an hour in the very hot and dense early stages of the expanding universe. The second theory, described by W. A. Fowler and associates, supposes that helium is formed from hydrogen in evolving stars and that the process is extended to the forming of heavier elements in the more massive stars.

The latter theory supposes that the fusion of helium atoms into those of heavier elements begins in the core of the more massive star when it attains a temperature of 150 million degrees Kelvin. At this temperature three helium nuclei can be forced together to form a carbon nucleus, which in turn may capture helium nuclei to produce successively oxygen, neon, and perhaps magnesium.

When the supply of helium in the core of the star is nearly exhausted, the core contracts and grows hotter, extending the combustion of helium into the mantle. The outer layers expand and the star brightens until it becomes a supergiant. At a central

FIGURE 15.7
The evolutionary track of a massive star. The successive crossings of the H-R diagram is due to the "burning" of helium and heavier elements.

temperature of 5 billion degrees the building of elements should go as far as iron, the final synthesis that can release energy for the star's radiation. The synthesis of elements heavier than iron might then be accomplished, as in the Gamow theory, by successive captures of neutrons that were released abundantly in the preceding processes; it could possibly be extended to the heaviest known elements by use of the great energy provided by the explosions of supernovae. Explosions of great violence are known to produce the heaviest elements. As an example, traces of the very heavy element californium were identified in the residue of a thermonuclear explosion in the Bikini tests of 1952.

Thus, according to the theory, atoms of heavier chemical elements have been and will continue to be formed from lighter atoms in the evolving stars.

15.8 Return of material to cosmic clouds

In the course of their lifetimes the stars return to the interstellar clouds much of the material they originally acquired from the clouds. As examples, the supernovae may blow out more than the sun's mass of gas at a single explosion. The normal novae eject much less material on each occasion, but they explode much more frequently. The expanding planetary nebulae and the red supergiants are important contributors. On the average a star jettisons half its original mass during its lifetime. The other half remains when the white dwarf stage is reached. The very massive stars would of course have to return many times the sun's mass to the interstellar medium.

In the exchange of material between the clouds and stars, the clouds are becoming more complex chemically as the galaxy grows older. The stars of successive generations formed in the clouds contain a greater percentage of metallic gases compared with those of the lightest elements. All in all, as J. L. Greenstein remarks, a good case can be made for systematic change in the relative abundances of the elements.

It is convenient to suppose the first-generation stars were originally of pure hydrogen. A second-generation star might also contain helium, carbon, oxygen, and nitrogen, but no metals; however, no relic of either type is known. The sun would be at least a third-generation star, having all the elements heavier than hydrogen and helium as impurities. Although its age may be as much as 6 billion years, the sun is younger than the stars formed near the beginning of the metal-building process.

Actually the convenience of beginning with stars of pure hydrogen is just that—a convenience. Our understanding of physics is good enough that we can predict what such a star will look like at any time during its evolution based on its initial mass. We observe no such star, therefore the universe is older than the time of evolution of a pure hydrogen star (for a solar mass, about 10^{11} years) or the universe began with a certain amount of helium and other elements present or created shortly after the beginning. In order not to prejudice the reader we are using the terms creation and beginning to mean a starting point traceable by analytic and observational means. It is even conceivable that hydrogen is being created in just the right amounts to keep the relative abundances of the elements in the observable universe constant for all time. We return to these questions in Chapter 19.

THE DECLINING STARS

When the stars exhaust their available supplies of atomic fuel, they collapse and finally approach the end of their luminous careers as very dense white dwarf stars.

15.9 Gravitational collapse

There is at least one alternative to our discussion that must be taken into account. We have already seen that the more massive stars evolve faster than the less massive ones, and it is possible that in the more massive stars the synthesis already accomplished in the core before any catastrophic events occur can have a bearing upon the final outcome.

Chandrasekhar showed that for stars having a mass of 1.4 times the sun's mass or less, the evolution will be to the white dwarf stage as we have discussed for the sun, provided the nuclear processes have started. Since 1960, interest in the course of events for stars above this **Chandrasekhar limit** has increased, and an attempt to describe the detailed processes involved has been made.

When the nuclear fuel of a star is exhausted and the great majority of the interior is iron nuclei, the star contracts. The resulting compression raises the energy of the electrons and this pressure supports the remainder of the stellar envelope. However, the energy of the free electrons is now sufficient so that the iron can be converted into an isotope of manganese, which will not decay. With the electron pressure removed a catastrophic collapse sets in. The result of this collapse is to set up a shock front raising the temperature of the envelope to several billion degrees, thus inducing a nuclear explosion. This explosion literally blows away the outer envelope and the events of a supernova follow.

Up to the point of the nova event we have essentially refined ideas that have been generally accepted during recent years. Now these theories describe the events that take place in the remaining core. If the core does not contain too many heavy particles and if it was not compressed too greatly in the collapse mentioned, it then dampens out any remaining instabilities and becomes a white dwarf or perhaps a neutron star. If the collapse compressed the core beyond the critical point or if the number of heavy particles is too great, the core cannot come to equilibrium (a stable state) and it gravitationally collapses to zero volume and becomes a **black hole.** A black hole is an object whose surface gravity is so great that a photon cannot escape from it.

Neutron stars should be observable from outside the earth's atmosphere because they should radiate prolifically in the X-ray region of the spectrum. Also, they can explain the pulsar mechanism, as we shall discuss shortly.

We return to the observational problems and the question of time scale, in so far as age is concerned, in a later chapter.

In the events of a declining star, a series of collapses may occur with energy releases more on the order of regular novae until the star reaches the white dwarf stage. This part of such a decline may have been illustrated by the recent behavior of WZ Sagittae, which flared out as a nova-like star in 1913 and again in 1946. It is now a white dwarf, having about a hundredth of the luminosity of the sun.

15.10 White dwarf stars

White dwarf stars have diameters comparable with that of the earth, and masses averaging half the mass of the sun. They are globes of highly compressed gas that may be a million times as dense as the sun, aside from a less dense 100 kilometer envelope and a surrounding atmosphere. Although they comprise 3 percent of the stars in the galaxy, they are so faint that only a few hundred have been identified, many of these in the searches of W. J. Luyten.

Examples of white dwarfs are the faint companions of Sirius and Procyon. Another is a companion of the star 40 Eridani; D. M. Popper has observed that the lines in the spectrum of this star are shifted to the red by an amount that agrees well with the shift predicted by the theory of relativity for this very dense star. This red shift is proportional to the mass of the star divided by the radius of the star. Thus white dwarfs are just the stars we should use to look for this effect. Very recent work by Greenstein and Trimble fully confirms Popper's observations.

The spectra of more than 50 white dwarfs have been photographed with the Hale telescope by J. L. Greenstein and his colleagues. Some spectra show only helium, others only hydrogen, others only a few metallic lines, and still others no visible lines at all (Fig. 15.8). There is no relation between the spectrum patterns and the colors, as with normal stars.

The white dwarf stars, which are also yellow and even reddish, form a sequence below the main sequence in the color–magnitude diagram. They move very slowly to the right along their sequence as they cool, without any reduction in their diameters. M. Schwarzschild has said that a white dwarf requires 8 billion years to cool from the initial blue stage to the red stage and from there a longer interval before it ceases to shine. It may be that our Galaxy is not yet old enough to possess a single fully evolved star such as these.

15.11 Neutron stars

In addition to white dwarfs and black holes, an end point of stellar evolution during a supernova event is a **neutron star**. Such a star was first predicted by F. Zwicky in 1934 and the theory was more or less worked out during the period 1939 through 1947.

A neutron star must be very small and very dense, the diam-

FIGURE 15.8
Spectra of white dwarfs. Elements prominent in each spectrum are shown on the left. The wavelength scale is at the bottom. (Spectra by J. L. Greenstein, California Institute of Technology.)

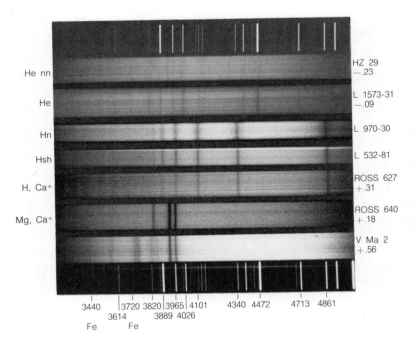

eters will be on the order of 10 kilometers and the densities on the order of 10^{14} grams per cubic centimeter. In this condition the neutrons align themselves and, if the star is rotating, this gives rise to a very strong magnetic field of approximately 10^{12} gauss. Electrons injected into this field will spiral and decelerate giving rise to synchrotron radiation in the plane of rotation of the star.

Since angular momentum is conserved, the star is rotating very rapidly, from a few thousandths of a second to several seconds for one rotation. The interaction of material with the neutron star should cause it to give off X-rays at the same period as its rotation. Also, the density at the surface of a neutron star is such that it forms a crust. When the star slows down the crust will collapse causing discrete discontinuities in the period of rotation. Such changes have been called **starquakes.**

If the theory is correct we can draw certain conclusions: Neutron stars should be found in supernova remnants; X-ray stars should be found where the neutron stars are; and, if we are close to the plane of rotation, we should see pulsed synchrotron radiation with the period of rotation of the star, that is, very short.

In the center of the Crab nebula there is a pulsar which is the type of object that fits our last requirement, the star was also the first X-ray star discovered, and the Crab nebula is obviously a supernova remnant. Other supernova remnants show pulsars that are also X-ray stars. To further strengthen the case, R. Kraft has pointed out that nova and supernova remnant stars seem to be binaries. A few pulsars and several X-ray stars are binaries, for example, HZ Herculis. Even more exciting, an X-ray star has been found in the Small Magellanic cloud that is obviously a binary system, but does not contain a pulsar like HZ Herculis as far as we can tell. Finally, discrete period changes should be observed and are.

In all, more than 130 X-ray sources have been found so far, perhaps 20 of which are identifiable with old novae or supernova remnants. Interestingly enough, half of all the X-ray sources seem to be associated with peculiar galaxies far removed from our own.

A point that we should not overlook is that neutron stars can be a source of gravitational radiation under very special conditions. The conditions are not so special as to be impossible, and J. Weber has reported pulses of gravitational energy at frequencies of occurrence that can be explained by a rotating neutron star.

While we are fairly certain that we have observed neutron stars it is not clear how one observes black holes. Perhaps it is possible for a degenerate star to orbit a black hole thus forming a binary system. This idea is being advanced to explain Cygnus X-1.

15.12 An interesting problem

Throughout this chapter we have drawn attention to the sun by using it as an example and explaining what the course of evolution is for the sun. Tacitly assumed is the fact that we take the sun to be a normal, well-behaved star evolving according to the processes described. We have seen that the various nuclear sources of energy release gamma rays and neutrinos. The highly energetic photons released in the same reactions are absorbed and reemitted many times before they escape as light from the sun. In fact the time for the escape of these photons is about 10^7 years. The neutrinos, however, escape directly from the sun and stars, thus offering us the opportunity to detect what happened in the solar interior only 8 minutes earlier (Fig. 15.9). In addition the various nuclear reactions produce neutrinos in different energy ranges so we can determine which of the reactions are taking place.

Neutrinos do not interact readily with matter so they are very difficult to detect. However, they can be detected and elaborate experiments have been undertaken to detect them. At first, the experiments were sensitive enough only to reach the expected detection level, and the response of the equipment was to just that level. Everyone was satisfied, but an experiment where the "noise" level is equal to the signal is not ideal so the experimental equipment was improved by a factor of five. The new experiment then showed that it was detecting neutrinos five times less frequent than predicted. Something was wrong.

Again the experiment was improved until it was 100 times more sensitive than the original experiment and again it detected neutrinos at this rate and no more. The neutrino flux from the sun is at least 100 times less than predicted, and this is cause for alarm.

There are several possible explanations. (1) The interior of the sun is not undergoing nuclear reactions, which means the sun is shining on energy released previously. (2) The nuclear reactions turn off and on, and at present we are in an off cycle. (3) Our knowledge of nuclear energy generation is wrong, and neutrinos are not produced as predicted. (4) Perhaps there is a layer of material in the sun that is not transparent to neutrinos? All of these possibilities tax our credibility. In any case it is clear that we do not know as much about the workings of a "normal" star

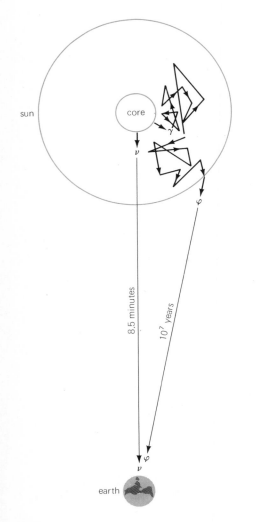

FIGURE 15.9
Neutrinos pass directly out of the sun and reach the earth only 8 minutes after they are released. The γ-ray photons are gradually degraded in a random walk to visible photons. This whole process takes some 10 million years.

as we thought we did and we are left with an exciting problem on our hands.

QUESTIONS

1 What must happen to a normal interstellar cloud in order for it to form stars?
2 Trace the early history of a protostar from the nebular state to the main sequence.
3 What are the three initial nuclear reactions that cause the release of thermonuclear energy?
4 What are infrared stars?
5 After a star leaves the main sequence, what thermonuclear reaction sets in to generate the required energy?
6 Trace the eventual history of the sun from the main sequence.
7 How does the evolution of a massive star differ from that of the sun?
8 How is the interstellar medium enriched in metals at the expense of hydrogen?
9 What are the three end products of a supernova event?
10 What are the arguments that pulsars are neutron stars?

FURTHER READINGS

BAADE, WALTER, in *Evolution of Stars and Galaxies*, C. Payne-Gaposchkin, Ed., Harvard University Press, Cambridge, Mass., 1963.

GAMOW, GEORGE, *Birth and Death of the Sun*, Viking, New York, 1946.

JOHNSON, MARTIN, *Astronomy of Stellar Energy and Decay*, Dover, New York, 1959.

MEADOWS, A. J., *Stellar Evolution*, Pergamon, New York, 1967.

PAGE, T. AND L. W. PAGE, *The Evolution of Stars*, Macmillan, New York, 1968.

PAYNE-GAPOSCHKIN, CECILIA, *Stars in the Making*, Harvard University Press, Cambridge, Mass., 1952.

SASLAW, W. C. AND K. C. JACOBS, EDS., *The Emerging Universe*, Univ. Press of Virginia, Charlottesville, 1972, Chap. 1.

TAYLER, R. J., *The Stars*, Springer-Verlag, New York, 1970.

A computer-printed star map generated for midnight 17 January 1974 as seen from New York City. Bright stars are shown as crosses. Orion is below and to the right of Saturn. Castor and Pollux are below and to the right of center. (Courtesy of IBM.)

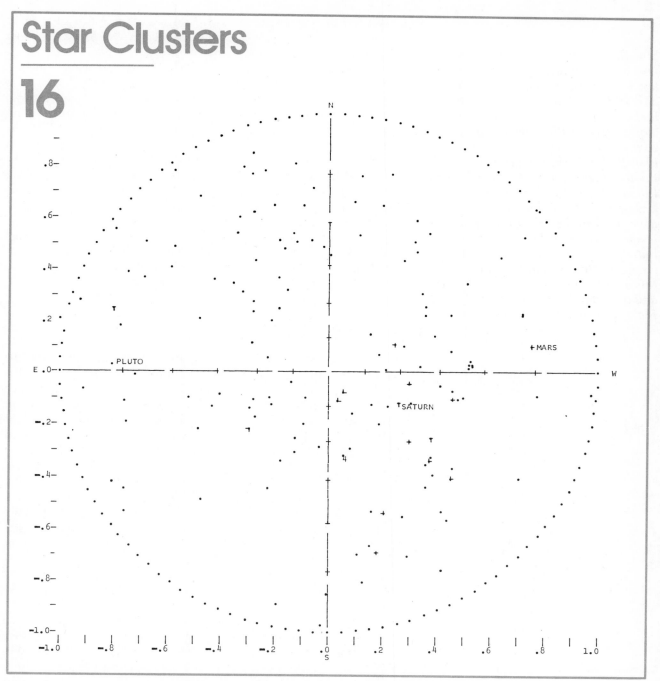

Star clusters are physically related groups of stars having their members less widely spaced than are the stars around them. The stars of a cluster had a common origin and are moving along together through the star fields, so that the cluster will maintain its identity for a very long time. Although the members of a cluster are of about the same age, the more massive stars have shorter lives and are further developed in their evolution than the less massive ones. Together, they give important information about the course of stellar evolution. Star clusters are of two types: galactic clusters and globular clusters. A related grouping of stars is called an association.

Clusters of stars associated with the Galaxy are called galactic, or open, clusters and globular clusters. They are found in other galaxies as well.

The brighter clusters are often called by special names, such as the Praesepe and Pleiades clusters. More generally, the clusters are designated by their running numbers in one of two catalogs where they are listed together with nebulae and galaxies. Thus the great cluster in Hercules is known as M 13 or NGC 6205. The first designation is by the number in the catalog of 103 bright objects, which C. Messier published in 1784 (a useful list of these objects and their positions in the sky is given in the periodical *Sky and Telescope* for March, 1954). Messier compiled this catalog to assist people searching for comets, since, in many cases, they could be mistaken for comets by an amateur astronomer. The second catalog designation of the Hercules cluster is by its number in J. Dreyer's *New General Catalogue* (1887), which, with its extensions in the later *Index Catalogue* (IC), lists over 13,000 objects.

GALACTIC CLUSTERS

Galactic clusters are so named because those in our Galaxy lie near its principal plane. They accordingly appear close to the Milky Way except a few of the nearest ones notably the Coma Berenices cluster, which is in the direction of the north pole of the Milky Way. They are also known as **open clusters,** because they are loosely assembled and are not greatly concentrated toward their centers.

16.1 Features of galactic clusters

The Pleiades in Taurus and the Hyades in the same constellation are familiar examples of galactic clusters. Their brighter stars are clearly visible to the unaided eye and those of the Coma cluster are faintly visible. The Praesepe cluster in Cancer, also known as the "Beehive," the double cluster in Perseus (Fig. 16.1), and some others are hazy spots to the unaided eye and are resolved into stars with binoculars. These are fine objects with small telescopes.

About 500 galactic clusters are recognized in our region of the galaxy. Their memberships generally range from two dozen to a few hundred, and even exceed 1000 stars in the rich Perseus clusters. The known galactic clusters are all within 6000 parsecs from the sun. More remote ones are too faint to be noticed in bright areas of the Milky Way or are concealed by dust in the dark areas. It is estimated that there are about 50,000 galactic clusters in the Milky Way; thus we see only 1 percent of the total number.

Galactic clusters may or may not have gas and dust associated with them. Those clusters with bright blue stars present always have gas and dust nearby. We will see that this is an effect of the age of the cluster.

16.2 The Pleiades cluster

This beautiful cluster meaning "the seven sisters" is visible to the naked eye as six stars in a hazy compact region (see Fig. 11.12 and color plate). The hazy appearance comes from the numerous faint stars of the cluster plus the nebula in which it is imbedded. Much speculation centers upon the question whether or not a seventh star was visible in antiquity when the cluster was named. If it was, it has either faded gradually until it was below detection by the naked eye, or one of the stars became a nova during a springtime period. There is no direct or indirect evidence for either supposition, and it is possible that the name was not meant to match the number of stars or that the poetry was written with another purpose in mind.

The Pleiades cluster has a membership of about 250 stars located in a tenuous cloudy region. The nebular material is visible around the brighter stars as reflection nebulae, as was first pointed out by V. M. Slipher in 1912. The brightest stars are of type B5 indicating that the cluster is rather young. The Pleiades does not have a separate giant population in the H-R diagram. Unlike many open clusters the Pleiades is sufficiently compact that it will retain its principal identity for the lifetime of its stars. This is because as a member is lost the cluster contracts and will do so on each successive loss until it is a tight cluster able to withstand dissolution forces from the outside. These forces arise in the tidal effects of the Milky Way and the shearing effect of Keplerian motion around the center of the Milky Way.

16.3 The Hyades cluster

The Hyades cluster is relatively so near us that the proper motions of its members offer a good example of the common motions of cluster stars. It comprises the stars of the V-shaped group itself, except the bright star Aldebaran which has an independent motion not shown in the diagram (Fig. 16.2), and also stars in the vicinity within an area having a diameter of 20°. The cluster of at least 150 stars has its center 130 light years from the sun. It is moving toward the east and is also receding from us, so that the

FIGURE 16.1
A fine pair of open clusters called h and χ Persei. These clusters contain many blue stars. They are among the youngest clusters (see Fig 16.5). (Copyright by Akademia der Wissenschaften der DDR. Taken at the Karl Schwarzschild Observatorium.)

FIGURE 16.2
Convergence of the Hyades cluster. The stars of the cluster are converging toward a point in the sky east of the present position of Betelgeuse. The lengths of the arrows show the proper motions in 50,000 years.

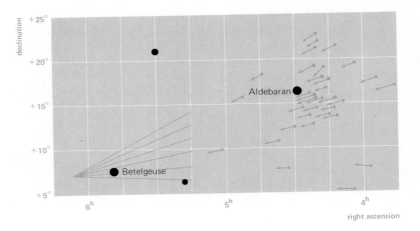

parallel paths of its stars are converging toward a point in the sky east of the present position of Betelgeuse in Orion. This motion toward the convergent point allows us to compute the distance to the cluster independently of the method of parallax as was first pointed out by L. Boss in 1908. Since there are many stars involved we have a valuable check on trigonometrically established calibrations. It is possible to show that this cluster passed nearest the sun 800,000 years ago at half its present distance from us.

The Hyades cluster being less populous than the Pleiades, but covering the same spatial volume, will probably suffer dissolution during the lifetimes of its stars. Some of its stars are grouped in **trapezium** type systems (3 to 7 stars) and eventually only these systems going their separate ways will give evidence of the once great cluster. The Hyades differs from the Pleiades in that its brightest stars are reddish K-type giants of about zero absolute magnitude. Also, nebular material is not very much in evidence in the Hyades, perhaps due to its age.

16.4 Convergent motion

It is often stated that if we know the tangential motion and the radial velocity of a star then we can find the star's space motion. This is not strictly true. What we should say is that if we know the radial velocity and the tangential velocity then we know the star's space motion. The difference is not trivial because to get the tangential velocity from the tangential motion, or proper motion, we must know the star's parallax, that is, its distance. In the case of a converging (or diverging) cluster we can convert

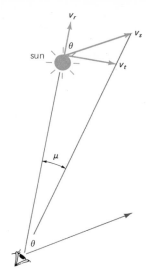

each star's proper motion into a velocity and hence determine its distance directly.

Suppose a group of stars is moving through space with motions parallel to each other. When viewed from a distant vantage point these motions will appear to converge at some distant point, as other parallel lines in space do. Irrespective of the sun's own motion, an imaginary line drawn from the sun to the convergent point is parallel to the motions of the moving group, and the angle θ formed by this line and a line from the sun to an individual star in the group is the angle between the direction of the star's space motion (v_s) and its radial velocity (v_r) with respect to the sun.

$$v_s \cos \theta = v_r$$

Simple trigonometry tells us that the tangential velocity (v_t) is determined once we know the angle (θ) and the radial velocity.

$$v_t = v_r \tan \theta$$

Knowing the tangential velocity in kilometers per second and the tangential motion in seconds of arc per year we can determine the distance.

$$\pi = \frac{4.74\,\mu}{v_r \tan \theta}$$

$$d_{\text{pc}} = \frac{v_r \tan \theta}{4.74\,\mu}$$

It is not actually necessary to calculate the tangential velocity.

This method is of great importance, because for the four or five nearest moving clusters we have three independent methods for determining their distances: the convergent method we have just discussed, trigonometric parallaxes, and spectroscopic parallaxes. The latter is often replaced with a curve fitting technique that fits the main sequence of stars in the cluster to the main sequence as defined by the nearby stars, as we will see on p. 348.

We cannot stress the point of independent methods of checking a measurement too strongly. In this case, it happens that the Hyades cluster is used to calibrate other more distant clusters containing both bright and variable stars that obey the period–luminosity relation. These stars are used in turn to calibrate other stars that then are used to calibrate other objects, and so on. Thus the distance calibrations in astronomy rest heavily (but not entirely) upon the Hyades cluster and any error compounds itself many times as we go farther into the cosmos.

16.5 The color–magnitude diagram

The color–magnitude diagram is an array of points representing the colors of stars in some standard system plotted with respect to their magnitudes. In the diagram of the Praesepe cluster (Fig. 16.3) the stars become redder from left to right—from blue stars such as Sirius at the left to yellow stars like the sun near the middle and to red stars at the right. Color has a known relation to spectral type, which also advances as the temperature diminishes; it is preferred to the spectral type in this and following diagrams, because color can be determined more easily and precisely. The stars become brighter from the bottom to the top of the diagram. In this case the apparent, or observed magnitude is plotted.

We prefer here to use color and magnitude as defined by the photoelectric system introduced by H. L. Johnson. Occasionally the less precise color index (C.I.) is used, which is defined as the difference between a star's photographic magnitude (m_{pg}) and its photovisual magnitude (m_{pv}).

The diagram of the Praesepe cluster can inform us of the distance and relative age of the cluster, as follows:

DISTANCE OF THE CLUSTER. Compare the Praesepe diagram with the standard spectrum—absolute magnitude diagram (Fig. 12.11). From the latter diagram and Table 12.4 we note that the sun's color is about +0.6, so that its absolute photographic magnitude, M, is $4.8 + 0.6 = 5.4$. From the cluster diagram, a main-sequence star having color of 0.6 has apparent photographic magnitude

FIGURE 16.3
A color–magnitude diagram of the Praesepe cluster. Note the evolved stars moving away from the main sequence at the upper left. (Adapted from a diagram by H. L. Johnson.)

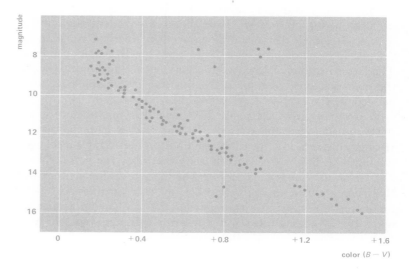

$m = 11.4$. By the formula, $\log d = (m - M + 5)/5$, $\log d = 2.18$. The distance, d, of the Praesepe cluster is accordingly about 150 parsecs.

AGE OF THE CLUSTER. This is also revealed by Fig. 16.3. Note that the top of the main sequence of the Praesepe cluster bends to the right and that a few of the brightest stars have moved still farther to the right. The point at which the stars begin to break from the normal main sequence is farther down the sequence as the cluster is older.

The distances of 18 galactic clusters are given in Table 16.1. The clusters were selected from a very extensive list by W. Becker in which he combined the best measurements of a number of observers including his own. After making allowance for the dimming and reddening of the cluster stars by intervening interstellar material, the main sequence of each cluster was then matched with that of a standard color–magnitude diagram. The difference, apparent minus absolute magnitude, at corresponding points in the two scales at the left of the diagram, gave all the information needed for calculating the distance of the cluster by the formula we have already noted. The difference, $m - M$, is called the **distance modulus** for obvious reasons.

TABLE 16.1

Distances of selected galactic clusters[a]

Cluster	Constellation	Distance	
		Parsecs	*Light years*
Hyades	Taurus	40	130
Coma	Coma Berenices	80	261
Pleiades	Taurus	125	408
Praesepe	Cancer	158	515
M 39	Cygnus	265	864
IC 4665	Ophiuchus	330	1080
M 34	Perseus	430	1400
NGC 1647	Taurus	550	1790
NGC 2264	Monoceros	715	2330
M 67	Cancer	830	2710
NGC 4755	Crux	1035	3380
M 36	Auriga	1270	4140
NGC 2362	Canis Major	1550	5060
NGC 6530	Sagittarius	1560	5090
M 11	Scutum	1700	5550
h Persei	Perseus	2150	7010
NGC 2244	Monoceros	2200	7180
χ Persei	Perseus	2460	8020

[a] Abstracted primarily from an extensive list by W. Becker.

$m - M$	Distance (pc)
0	10.00
1	15.85
2	25.12
3	39.81
4	63.10
5	100.0
6	158.5
7	251.2
8	398.1
—	—

In the margin we have given a convenient tabulation of distances in parsecs associated with various distance modulae through $m - M = 8$ magnitudes. Note that the distance for $m - M = 6$ is ten times that of $m - M = 1$ and $m - M = 7$ is ten times that of $m - M = 2$, etc. Therefore, $m - M = 9$ will be a distance of 631 parsecs, $m - M = 14$ is equivalent to 6310 parsecs, $m - M = 19$ is equivalent to 63,100 parsecs, and so on. For really large distances it is convenient to use the distance modulus.

We have already noted that the open or galactic clusters receive their name from the fact that they lie in or close to the Milky Way or galactic plane. If we were to plot their directions in the galactic plane and their distances, we would find that these clusters are not distributed at random. The galactic clusters seem to be arranged along lines in space (Fig. 16.4). We will see later that these lines agree quite well with the arms of the Milky Way galaxy as defined by other means.

16.6 Star clusters of different ages

The theory of stellar evolution described in Chapter 15 first traces the stars from their births in the nebulae to positions of temporary stability on the main sequence of the color–magnitude diagram. Here they are arrayed in order of color, from blue at the left to red at the right. In this order the stars of a cluster break from the sequence and move to the right to become giant stars. The greater the age of a cluster, the lower on the main sequence is the place where the stars begin to break from it.

A composite color–magnitude diagram for ten galactic clusters and one globular cluster is shown in Fig. 16.5. The colors, B minus V on the Johnson system, of the cluster stars were plotted against their absolute V magnitudes, and the trends of the points so plotted are represented here by broad lines. The ages of the clusters in years are read from A. Sandage's original scale at the right opposite the points where the curves begin to break from the main sequence. Above its point of departure a particular cluster has no stars remaining on the main sequence. Sandage and others have found reasons more recently for increasing the scale of ages but they regard these values as preliminary. There is some evidence that the age scale given is a factor of about 1.5 too small.

The cluster NGC 2362 is the youngest cluster represented in Fig. 16.5. The double cluster in Perseus is also in its youth. The Pleiades cluster is middle-aged, and the Hyades and Praesepe clusters are approaching old age. M 67, age 7 billion years, has had a longer life than most galactic clusters; it breaks from the main sequence where the stars are not much bluer than the sun. A still

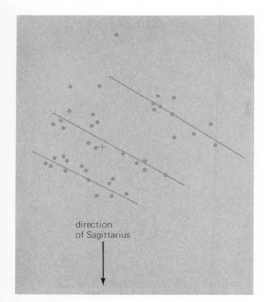

FIGURE 16.4
The distribution of galactic clusters taken from material by W. Becker. The young clusters are not distributed at random (see also Fig. 17.8). The direction to the center of the Milky Way is down.

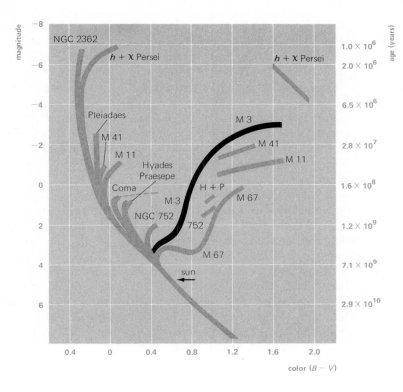

FIGURE 16.5
Color–magnitude diagrams for ten galactic clusters superposed plus the globular cluster M 3. Color is plotted against absolute visual magnitude. Age of stars evolving away from the main sequence can be read on the right. (Diagram by A. Sandage.)

older galactic cluster, NGC 188 (not shown in the figure), leaves the sequence at about the sun's position. Its stars are represented by a curve similar in form to that of M 67.

Note in Fig. 16.5 the roughly triangular area avoided by the curves of the galactic cluster stars. This vacant area, known as the **Hertzsprung gap,** is not completely explained. Also awaiting explanation is the evidence presented by Sandage and by Arp and Hodge, respectively, that field stars of the Milky Way and of the Magellanic clouds do not avoid a similar area of their color–magnitude diagrams.

16.7 Association of young stars

Several years ago V. A. Ambartsumian called attention to the fact that the early-type stars tend to occur in groups. He noted that the groups of O and B stars were not held together firmly enough by mutual gravitation to prevent their eventual dispersal by separate motions in different directions. However, he pointed out that the groupings could not be accidental and therefore the stars are to be associated with a common origin. He called these groupings **associations.** Evidently the associations are so young that they have not had time to disperse. Associations are designated by the predominant star type. Thus an association made up mainly of O stars is called an O association. Groups of T Tauri stars meeting Ambartsumian's definition of associations are called T Tauri associations.

An example is the zeta Persei association, named after its brightest member. At the distance of 520 light years from us, the 17 stars of the group are moving at the rate of 14 kilometers a second away from a center where they were presumably born only 1.3 million years ago. It came as a surprise that cosmic evolution could proceed so rapidly.

In one association of blue stars in Orion, at the distance of 1300 light years, the stars are withdrawing from their center generally at the rate of about 8 kilometers a second, so that they must have begun to separate 2.8 million years ago. Three "runaway stars" of the group, however, have speeds of 130 kilometers a second; these are 53 Arietis, AE Aurigae, and μ Columbae, which have already moved out of Orion into neighboring constellations, as is indicated by their names.

The type O9 star AE Aurigae has reached the vicinity of the nebula IC 450, which it is now illuminating. An unusual feature, described by G. H. Herbig, is that part of the nebula gives an emission spectrum, as would be expected from the presence of the very hot star, whereas another part gives a reflection spectrum and has more nearly the color of the star itself. It would seem that there has not been time enough since the star came near the nebula for the radiations of the star to complete the dispersal of the nebular dust.

GLOBULAR CLUSTERS

Globular clusters are spheroidal assemblages often of many tens of thousands of stars; they are much the larger and more compact of the two types of clusters.

16.8 The brighter globular clusters

About 120 globular clusters are recognized in the vicinity of our Galaxy and many more are likely to be hiding behind its dust clouds. Instead of crowding toward the Milky Way, as the galactic clusters do, they form a nearly spherical halo around the galaxy. They are somewhat more numerous toward the center of the Galaxy. Relatively scarce in space, not one of them has been seen within the distance of 20,000 light years (6100 parsecs) from us, where all our galactic clusters are observed. Their luminosity makes them visible afar in our Galaxy and around nearby galaxies.

The brightest globular clusters for us are omega Centauri, near the northern edge of the Milky Way in the south polar region, and 47 Tucanae, also in that region. They appear to the unaided eye as slightly blurred stars of the 4th magnitude and were given designations as stars before their true character was recognized. These two are the nearest globular clusters, at the distance of 22,000 light years, and are among the richest in stars.

M 13 in Hercules (Fig. 16.6) is faintly visible to the unaided eye, as is M 22 in Sagittarius. M 5 in Serpens, M 55 in Sagittarius, and M 3 in Canes Venatici can be glimpsed without a telescope in favorable conditions.

The Hercules cluster, M 13, at the distance of 7000 parsecs, is well known to observers in middle northern latitudes, where it passes nearly over head in the early evenings of summer. This cluster covers an area of the sky having two-thirds the moon's apparent diameter. Its linear diameter is 50 parsecs, or the distance of Spica from the sun.

More than 50,000 stars of the Hercules cluster are bright enough to be observed with present telescopes, although the stars in the

FIGURE 16.6
The beautiful globular cluster M 13 in Hercules. The stars are so dense in the center that the negative is "burned" out. (Official U.S. Navy photograph.)

central region are too crowded to be counted separately. The total membership is estimated as half a million stars, including the yellow and red stars of the main sequence, which are too faint to be observed. The slightly elliptical outline of the cluster, less than that of Jupiter's disk, suggests flattening at the poles by slow rotation; no other evidence of rotation has been detected.

The stars in the compact central region of the Hercules cluster have an average separation of 20,000 times the earth's distance from the sun, or about a twentieth of the spaces between the stars in the sun's vicinity. For anyone observing from there the night sky would have a splendor quite unfamiliar to us. Probably a hundred times as many stars as we see in our skies would be visible to the unaided eye, and the brightest ones would shine as brightly as the moon does for us. This, of course, would have its drawbacks for astronomers located there.

16.9 Variable stars in globular clusters

The presence in these clusters of many stars that are variable in brightness has been recognized for a long time. The majority are RR Lyrae variables, having periods around half a day; originally called cluster variables, they are now known to be even more abundant outside the clusters. More than 1600 variable stars were reported in 1961 by H. S. Hogg in the 80 clusters that had been searched for such objects. M 3 and omega Centauri are the richest in known variables. Variable stars of other kinds in the globular clusters include type II cepheids and Mira-type variables.

The RR Lyrae variables have about the same absolute magnitude ($+0.7$) and hence are excellent distance indicators to the distances to which they can be seen. The type II cepheids are brighter and follow the lower curve in the period–luminosity diagram (see Section 13.10); hence, they serve as distance indicators for even greater distances. Unfortunately many globular clusters are lacking or poor in these reliable types of variable stars and other methods must be used to find their distances. Usually the tenth brightest star in a globular cluster has the same brightness as the tenth brightest star in most other globular clusters so this is used for determining the distances of clusters lacking variable stars.

16.10 The system of globular clusters

In 1917 H. Shapley made the first important step toward the present understanding of our Galaxy by determining the arrangement of its globular clusters in space. His purpose was to answer a question of long standing: Do the stars go on out into space indefinitely, or do they form a system of limited extent? His idea was that the cluster system should have about the same dimensions and center as the system of stars.

Shapley measured the distances of the clusters using RR Lyrae stars as distance indicators. Having found the distances and of course knowing their directions from us, he then made a model of the cluster system, showing the sun's position in it. With later correction for the effect of intervening dust in magnifying the measured distances, it was determined that the cluster system primarily occupies a spherical volume of space 35,000 parsecs in diameter surrounding the flat disk of the galaxy proper. The center of the system is 10,000 parsecs, or 10 kiloparsecs, from the sun in the direction of Sagittarius and in the region of the sky where a third of the globular clusters are found (Fig. 16.7). This classic survey established the dimensions of our Galaxy and the eccentric position of the sun in it.

STELLAR POPULATIONS

The diagram that arrays the spectral types of stars in the sun's vicinity with respect to their magnitudes (Fig. 12.11) differs from the corresponding diagram for a normal globular cluster. Interest

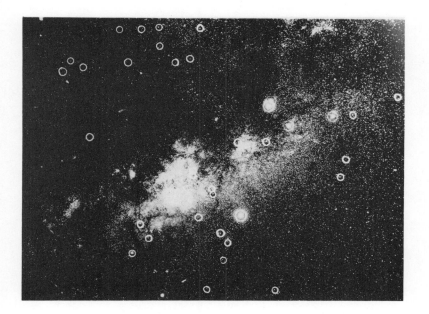

FIGURE 16.7
Distribution of globular clusters in the direction of Sagittarius. The clusters are circled for easier recognition. This one photograph contains one fourth of all of the Milky Way's globular clusters. Note their absence around the dark clouds due to obscuration. (Yerkes Observatory photograph.)

in this matter was increased by W. Baade's discovery with the 100-inch telescope in 1943 that a similar difference is found in the galaxy M 31 in Andromeda between stars in the spiral arms and in the central region or in the two elliptical companions of that galaxy. Baade called the two arrays populations I and II, respectively.

16.11 Two stellar populations

The **type I population** is represented by our region of the Galaxy, and was accordingly the first to be recognized. Population I frequents regions where gas and dust are abundant. Its brightest members are hot blue stars of the main sequence, which is intact (Fig. 16.8). Its giant members are mainly red stars around absolute magnitude zero. Open clusters are typical type I population.

The **type II population,** represented by a normal globular cluster, occurs in regions generally free from diffuse gas and dust. Its brightest stars are K-type giants of absolute visual magnitude −2.4, where the giant sequence begins. On the downward slope in the diagram this sequence divides into two branches, one of which runs horizontally to the left around magnitude zero. The second branch continues in about the original direction until it reaches the type I main sequence. To the left of the junction there are no original main-sequence stars remaining (Fig. 16.9).

Almost all of the globular clusters in our Galaxy are the type II population. In the Magellanic clouds (Section 18.6), however,

FIGURE 16.8
An idealized H-R diagram for population I stars. The position of the sun is indicated by the symbol ⊙.

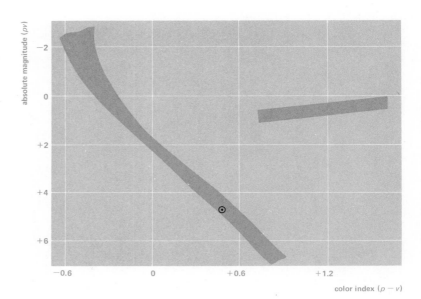

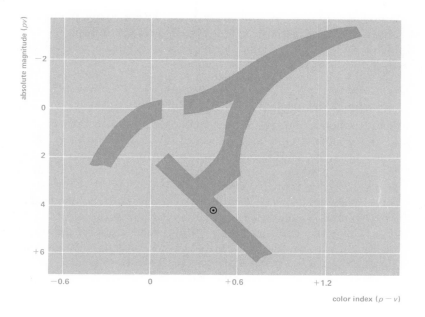

absolute magnitude (pv)

color index (p − v)

FIGURE 16.9
A generalized H-R diagram for M 3. The
gap in the horizontal branch is where the
cluster type variable fall. This diagram is a
typical population II diagram. The position
of the sun is indicated by the symbol ⊙.

there is about an equal division between populous red clusters and a formerly unfamiliar blue variety.

An extension of Baade's original type II diagram from photographs of the globular cluster M 3 with the 200-inch telescope is shown in Fig. 16.9. RR Lyrae stars were not included in this survey; a gap in the horizontal branch where they would appear suggests that this may be exclusively the domain of these variable stars. From the lower end of the vertical giant branch the remnant of the cluster's main sequence is shown extending down and to the right to yellow stars at the limit of faintness for the largest telescope.

Baade's types I and II represent extremes of young and old populations of stars. Intermediate types were needed to indicate the relative ages of stars in different parts of our Galaxy and of exterior galaxies. Indeed, it is now accepted that the two population types merely mark the extremes and that all intermediate types exist as well. The youngest stars correspond to Baade's type I population and are now referred to as **extreme population I.** Typical stars for this population are the cepheid variables, the T Tauri stars, and the blue supergiants. The next group contains the so-called **older population I** made up of the A stars, for example. In the middle of the two extremes is the **disk population** made up of planetary nebulae, ordinary novae, and the shorter-period RR Lyrae stars. Older yet is the **intermediate population II**, which is made up of the long-period variables and the so-called high-velocity stars. Finally, corresponding to Baade's population II is the **halo population II** composed of the globular clusters, subdwarfs, and longer-period RR Lyrae stars.

16.12 The origin and evolution of clusters

Because the two types of clusters are associated with the extremes of the population types, we should expect their histories to be quite different. The globular clusters are quite old as evidenced by the composition of their stars. Therefore they must be quite stable and were probably formed about the time that the Milky Way galaxy was formed, because we do not find any older stars. The globular clusters have highly elliptical orbits around the center of the galaxy and some of the orbits of these clusters have very high inclinations to the plane of the galaxy. Several of the globular clusters have orbits that actually take them through the central region of the galaxy where the star density is quite high. These few clusters must suffer disruptions and loss of stars with each passage, but the number of such passages is rather small, perhaps ten or so since the galaxy was formed. There is a question whether or not actual star collisions occur when a dense globular cluster passes through the dense central region. It is certainly possible, but the likelihood of its occurrence cannot yet be determined.

All evidence points to the fact that the extreme population type II stars are the oldest stars in the galaxy. Since the globular clusters are perfect examples, we can say that these clusters were formed from the same material from which the galaxy was formed. The atmospheres of the globular cluster stars then tell us what this material was; we find that 80 percent was hydrogen, a little more than 19 percent was helium, and there was just a trace of the other elements present. The population II stars are metal deficient, or perhaps we should say that the population I stars have an over abundance of metals.

The galactic clusters, on the other hand, show a great spread in ages. Some are almost as old as the globular clusters and others, such as NGC 2362, are less than 1 million years old. The only way these young stars could form is from the great interstellar clouds. Most young open clusters are imbedded in interstellar clouds (Fig. 16.10), which supports this point of view. The interstellar clouds have been formed from the metal-enriched material released by previous generations of stars. If this is so, then the stars in the young clusters should show higher abundances of the heavier elements, which they do.

FIGURE 16.10
A large-scale photograph of the relatively young Pleiades cluster. This cluster has no giant stars. Note the beautiful nebulosity. (Copyright by Akademia der Wissenschaften der DDR. Taken at the Karl Schwarzschild Observatorium.)

The open clusters, moving through fairly dense stellar regions, are subject to forces that tend to disrupt them. Their orbits of revolution around the center of the galaxy will serve to string out and disrupt a loose open cluster in about two or three orbits. The denser, tighter clusters can resist this shearing effect and only slowly dissolve to a minimum number of stars by ejection. This is the case with the Pleiades, which we discussed earlier.

QUESTIONS

1 Where does the name galactic cluster come from?
2 Why do we consider the Pleiades cluster to be a young cluster?
3 How do we determine the distance to a cluster showing convergent motion?
4 How do we use the color–magnitude diagram to get the distance modulus of a cluster? How would intervening material effect this determination?
5 If a cluster has a distance modulus of 17, what is its distance in parsecs?
6 How do we determine the ages of galactic clusters? Using the present scale, what is the age of the Hyades cluster?

7 Draw a H-R diagram and show where the Hertzsprung gap is located. Locate the region of the RR Lyrae variables in your diagram.

8 If a globular cluster does not contain variable stars that obey the period–luminosity relation, how do we determine its distance?

9 How did Shapley determine the general character of the galaxy?

10 What are the characteristics of the two population types?

FURTHER READINGS

BOK, BART J., AND PRISCILLA E. BOK, *The Milky Way*, 4th ed., Harvard University Press, Cambridge, Mass., 1973.

PAGE, T., AND L. W. PAGE, *The Evolution of Stars*, Macmillan, New York, 1968, Chaps. 2–4.

——— *Stars and Clouds of the Milky Way*, Macmillan, New York, 1968, Chap. 2.

SHAPLEY, HARLOW, *Galaxies*, 3rd ed., Harvard University Press, Cambridge, Mass., 1972.

TAYLER, R. J., *The Stars*, Springer-Verlag, New York, 1970, Chaps. 2, 6.

This is an effort by M. Easton in the late nineteenth century to depict the Milky Way as a spiral galaxy. Note that he has placed the center of the spiral structure in Cygnus, although, for unknown reasons, he has placed the sun at the center of the Galaxy.

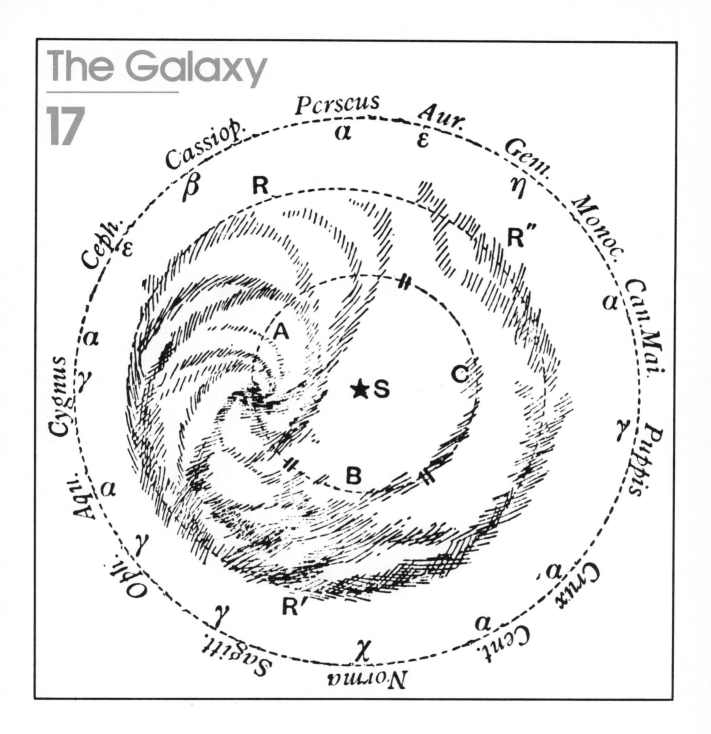

17

The galactic system, or system of the Milky Way, is so named because the luminous band of the Milky Way around the heavens is an impressive feature of the system in our view from inside it. This spiral stellar system, of which the sun is a member, is commonly known as the Galaxy as distinguished from the multitudes of other galaxies.

The Milky Way, long a sight admired by man, has only recently had its true nature devined. The luminous band is evidence of the disk of the Galaxy. The sun and solar system are located in this disk, and from this location we look out at the full expanse of the universe beyond.

17.1 The Milky Way in summer and winter

The full splendor of the Milky Way is reserved for one who observes on a clear moonless night from a place removed from artificial lights. The view with the unaided eye or with a very wide-angle camera is best for the general features. Photographs show the details more clearly than the eye alone can detect. An excellent collection of photographs with a 5-inch Ross camera is available in the *Atlas of the Northern Milky Way*. More recent and more penetrating photographs north of declination $-27°$ are those of the National Geographic Society–Palomar Observatory Sky Survey. These are pairs of negative prints in blue and red light made with the 48-inch Schmidt telescope (Section 5.5). Another excellent atlas is the *Lick Observatory Photographic Atlas of the Sky*.

The Milky Way is produced by the combined light of great numbers of stars that are not separately visible without the telescope. Its central line is nearly a great circle of the celestial sphere, so highly inclined to the celestial equator that it takes quite different positions in our skies in the early evenings of the different seasons.

At nightfall in the late summer in middle northern latitudes the Milky Way arches overhead from the northeast to the southwest horizon. It extends upward through Perseus, Cassiopeia, and Cepheus to the region of the Northern Cross. Here in Cygnus the Great Rift formed by cosmic dust clouds divides the Milky Way apparently into two parallel streams, which go on southward through Aquila into Sagittarius and Scorpius. This spectacular part of the Milky Way (Fig. 17.1) contains the bright star clouds of Scutum and Sagittarius; the latter is near the direction of the center of the Galaxy.

In the early evenings of late winter in our latitudes the Milky Way again arches overhead, now from northwest to southeast. It passes through Cepheus, Cassiopeia, Perseus, and Auriga, which is near the zenith in the early evenings of February. Here it is narrowed by a succession of relatively nearby dust clouds, which angle from northern Cassiopeia through Auriga to southern

FIGURE 17.1
The Milky Way from Scutum to Scorpius. The great Sagittarius star clouds are near the center of the picture. (Photograph from the Hale Observatories.)

Taurus. This is the direction away from the center of the Galaxy. The Milky Way then continues on past Gemini, Orion, and Canis Major, where it becomes broader and less noticeably obscured by dust.

The part of the Milky Way in the vicinity of the south celestial pole is either out of sight or else never rises high enough for favorable view in the United States. This region from Centaurus to Carina (Fig. 17.2) contains some fine star clouds and the dark Coalsack near the Southern Cross.

Except for the stars of Orion, the brightest stars of the winter

FIGURE 17.2
The Milky Way in the region of the Southern Cross. The obvious large dark cloud is the famous Coalsack. (Harvard College Observatory photograph.)

sky are noticeably displaced to the west of the Milky Way. This band of stars, called Gould's belt, contains the Pleiades and Hyades clusters and is a local manifestation of the structure of the Galaxy.

17.2 Galactic longitude and latitude

In descriptions of the galaxy it is convenient to denote positions in the heavens with reference to the Milky Way. For this purpose an additional system of circles of the celestial sphere is defined as follows.

The north and south **galactic poles** are the two opposite points that are farthest from the central line of the Milky Way. By international agreement they are respectively in right ascension 12^h49^m, declination $+27°.4$ (1950), in Coma Berenices, and 0^h49^m, $-27°.4$, south of beta Ceti. Halfway between these poles is the **galactic equator**, a great circle inclined $63°$ to the celestial equator; it crosses the equator northward in Aquila and southward at the opposite point east of Orion. The galactic equator passes nearest the north celestial pole in Cassiopeia and nearest the south celestial pole in the region of the Southern Cross. Thus the earth's equator is inclined $63°$ to the principal plane of the flattened Galaxy.

Galactic longitude was formerly measured in degrees along the galactic equator from its intersection with the celestial equator in Aquila, near R.A. 18^h40^m. By decision of the International Astronomical Union in 1958, the zero of galactic longitude is now changed to the direction of the galactic center (see Section 17.4), on the revised galactic equator in R.A. $17^h42^m.4$, Decl. $-28°55'$

(1950), in Sagittarius. As before, the longitude is measured through 360° in the counterclockwise direction as viewed from the north galactic pole; the new value equals the former one plus about 32°.

Galactic latitude is measured from 0° at the galactic equator to 90° at its poles and is positive toward the north galactic pole. Its present reckoning differs from the former one because of a recent slight revision of the position assigned to the galactic equator.

Possible confusion will be avoided in the following sections by specifying when the galactic coordinates are given in the old systems, otherwise the new system is being used. A possible new revision of the zero points has been discussed since 1967 but for the time being astronomers have decided not to revise the system.

STRUCTURE OF THE GALAXY

The Galaxy is an assemblage of the order of 100 billion stars. Its spheroidal central region is surrounded by a flat disk of stars, in which spiral arms of gas, dust, and stars are embedded. The center of the Galaxy is about 10,000 parsecs from the sun in the direction of Sagittarius. The disk rotates around an axis joining the galactic poles. Surrounding the disk is a more slowly rotating and more nearly spherical halo, containing globular clusters, scattered stars, and tenuous gas.

17.3 The disk of the Galaxy

The stars around us are concentrated toward the Milky Way. Stars visible to the unaided eye are 3 or 4 times as numerous around the galactic equator as they are in similar areas near its poles; the corresponding increase for large telescopes exceeds 40-fold, despite the greater obscuration by cosmic dust in the lower latitudes. This shows that the majority of the stars of our Galaxy are assembled in a relatively thin disk along the plane of the galactic equator. When we look in the direction of this equator, we are looking the longest way out through the disk and therefore at many more stars. The flat disk is estimated as 50,000 parsecs in diameter; its thickness is at least 3500 parsecs at the center and 1500 parsecs at the sun's distance from the center. Thus the disk is very thin, being only 2 percent of its diameter (see Fig. 17.3).

FIGURE 17.3
A schematic representation of the thickness of the disk of the Galaxy to its diameter. This emphasizes the extreme thinness of the disk.

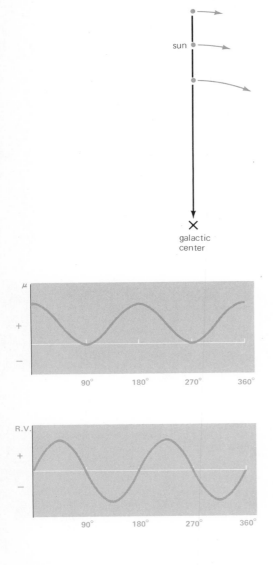

sun

× galactic center

μ

+

−

90°   180°   270°   360°

R.V.

+

−

90°   180°   270°   360°

17.4 The galactic center

The position of the center of the Galaxy was first announced by H. Shapley in 1917. He had determined the position of the center of the system of globular clusters that form a nearly spherical halo around the disk of the galaxy and had reasoned correctly that this must be the position of the galactic center itself. The direction originally assigned to the center in the former coordinate system was in galactic longitude 325°, latitude 0°, in Sagittarius. Shapley pointed out that the correctness of the position is indicated by the greater brightness and complexity of the Milky Way in this region.

As defined by international agreement in 1958, the direction of the galactic center is remarkably close to the original one. It is situated in Sagittarius at R.A. $17^h42^m.4$, Decl. $-28°55'$ (1950), or in former galactic longitude 327°.7, latitude $-1°.4$, and is taken as the new zero of galactic longitude. The position of the galactic center in space is about 10,000 parsecs from the sun. Thus the sun is situated in the disk of the Galaxy about halfway from the center to the edge in the direction of the constellation Auriga.

17.5 Differential galactic rotation

If the concept that the Milky Way is like other galaxies is correct and if Shapley's picture of the sun's position is correct, then a galaxy in rotation should show certain systematic effects. The sun is sufficiently far from the center of the Galaxy that it and the stars nearby are following circular orbits and Keplerian motion. Stars nearer the center are revolving about the center at greater speeds than those farther away. Oort reasoned that if this is so, the proper motions will show one effect and the radial velocities another.

Proper motions in the direction of the center of the Galaxy should be positive in reference to the celestial sphere. In the direction of the revolution of the sun (the apex, galactic latitude $l = 90°$) the proper motions should be zero. In the direction of the anticenter ($l = 180°$) the proper motions should be positive and again, in the direction of the anapex ($l = 270°$) the motions should be zero. The observed motions, derived from the averages of many nearby stars are plotted in Fig. 17.4.

The radial velocities show a different pattern. Consider Fig. 17.5. Stars in the direction of the center of the galaxy should exhibit no radial velocities, those in a direction 45° from the center will show a positive or recessional velocity, those toward the apex zero velocity, those at 135° from the center show

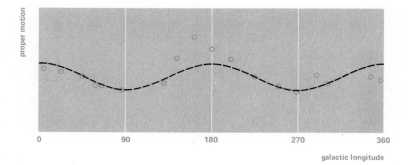

velocities of approach or negative velocities. In the direction of the anticenter the radial velocities will again be zero and so on around the plane of the Milky Way. The observed velocities are shown in Fig. 17.6.

The observed curves of Figs. 17.4 and 17.6 are in such good agreement with the predictions that we must conclude the Milky Way is in rotation as hypothesized. Furthermore, the observed curves can be described by simple geometrical relations that have multiplying constants dependent upon the mass of the Galaxy and the distance to the center of the Galaxy. These constants appropriately are called the **Oort constants.**

The sun's circular velocity in the galactic rotation, as computed by M. Schmidt, is 250 kilometers per second toward galactic longitude 90° in Cygnus. The sun's distance from the center of the galaxy is taken here to be 10,000 parsecs for convenience. The period of the rotation at this distance is 240 million years. The velocity of the galactic rotation diminishes with increasing distance from the center beyond the sun's distance; at the distance of 20,000 parsecs from the center the velocity is reduced to 194 kilometers per second. Thus in the outer parts of the Galaxy the rotation is not far from the planetary type.

The direction of the galactic rotation is clockwise as viewed from the north galactic pole. When we look toward the center in Sagittarius, the direction of the rotation between us and the center is toward the left.

17.6 The spiral structure

After the existence of galaxies as independent systems of stars was conclusively demonstrated by E. Hubble's detection of cepheid variables in the Andromeda galaxy in 1924, the frequently stated opinion that our own Galaxy might be a spiral remained without firm observational support for more than a quarter of a

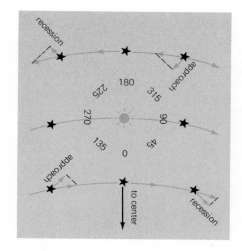

FIGURE 17.5
Effect of rotation of the galactic system on the radial velocities of stars. Stars nearer the center than the sun's distance are going around faster and are passing the sun. Stars farther from the center are moving more slowly and are falling behind the sun. Stars around longitudes 45° and 225° appear to be receding from the sun (positive radial velocities) while stars around 135° and 315° appear to be approaching the sun (negative radial velocities).

FIGURE 17.6
The average radial velocities of many stars are plotted as a function of their longitude. The resulting curve shows remarkable agreement with the predicted curve.

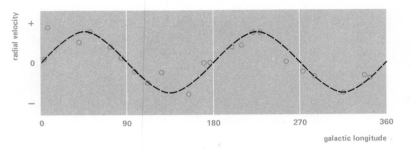

century. Hubble pointed out that galaxies divided rather nicely into three general classes—elliptical, spiral, and irregular. The spiral galaxies were characterized by lanes of gas and dust and bright stars. Even from our locations inside the Milky Way we can deduce that our Galaxy is a spiral because we see a flat Milky Way and plenty of gas and dust.

In 1949 W. Becker devised what he called a **synthetic galaxy;** this is as close a model as we could hypothesize to prove that our Galaxy is a spiral. Becker's procedure was to take a well-defined spiral galaxy (M 51), place the sun approximately two-thirds of the distance from the center to the edge, and sketch what the Milky Way would look like. One particular location reproduced the general features of the Milky Way as we actually see it quite well. This elegant proof left astronomers with an uneasy feeling, but it was the best that could be done. Fortunately this condition did not last long.

W. W. Morgan, S. Sharpless, and D. E. Osterbrock noted that emission nebulae were especially abundant in the arms of the Andromeda galaxy, a neighboring galaxy. They reasoned that the emission nebulae of our Galaxy should show the structure of a galaxy similar to that of Andromeda. They plotted the distances and directions of the visible nebulae of the Milky Way on the galactic plane (Fig. 17.7) and found clear indication of two arms, one passing near the sun referred to as the **Orion arm** and another more distant referred to as the **Perseus arm.** A third arm was indicated in the direction of the galactic center. Evidence for the third arm is now conclusive, and it is referred to as the **Sagittarius arm.** These arms coincide with the distribution of the bright O and B stars, the young open clusters, dust, and hydrogen gas.

The Orion arm contains the Orion nebula, the North American nebula in Cygnus, the Great Rift, and the Coalsack. The Perseus arm, more distant from the galactic center, passes some 2.2 kiloparsecs from the sun and contains the clusters h and χ Persei. The

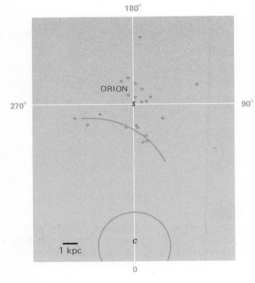

FIGURE 17.7
Parts of the spiral arms of the Galaxy traced by directions and distances of emission nebulae. The position of the sun is indicated. The galactic center is at C. (After a diagram by S. Sharpless.)

Sagittarius arm passes about 1.5 kiloparsecs from the sun and nearer the center of the Galaxy. The sun is located on the inner side of the Orion arm so the spacing between arms is about 2 kiloparsecs.

The most effective optical technique and the one used by Morgan and his colleagues is to photograph the Milky Way through a narrow filter that isolates the Hα line. The H II regions, by definition, will show strong Hα emission and hence stand out. Obscuration by dust limited distances to 3 kiloparsecs. Such distances are small compared to the scale of the galaxy and a more powerful technique is required. In the very year that Morgan, Sharpless, and Osterbrock announced their results, H. Ewen and E. Purcell detected the λ 21-centimeter line of hydrogen predicted by van de Hulst.

As we have noted in Chapter 13, the λ 21-centimeter radiation passes through the dust without difficulty and, furthermore, it will pass through intervening hydrogen provided there is a slight difference in the velocities along the line of sight. We have already noted that the spiral arms of galaxies are defined by gas and dust; much of the gas must be hydrogen since it makes up 70 to 80 percent of the mass of the universe. Differential galactic rotation provides the velocity differences we require. It is almost as if the radio technique was made to order.

To explain the technique let us assume a simple model of the Galaxy, one where all material—stars, gas, and dust—follow circular orbits around the center of the Galaxy. Now let us look through the Orion arm at the Perseus arm in a direction about 135° from the galactic center. First, as in Fig. 17.8, we will see the hydrogen in the Orion arm, it will fill the field of view of our antenna because it is so close and it will appear to be approaching us at a very slight velocity, perhaps 8 kilometers per second, again because it is so close so that its radius from the galactic center differs only slightly from that of the sun. Then we will see the hydrogen from the Perseus arm which does not fill our antenna beam but we will see more atoms and they will appear to be approaching us at a higher velocity, perhaps 45 kilometers per second, because it is located considerably farther from the galactic center than the sun. Our model, then, predicts two peaks in an intensity–velocity diagram. What do we see? Surprisingly enough, exactly what we predict and even more (Fig. 17.9). We see an additional tail that may indicate an arm beyond the Perseus arm. Now if we change our antenna direction slowly we can follow the peaks and hence trace out the spiral structure.

There is a guidepost that prevents the picture from becoming

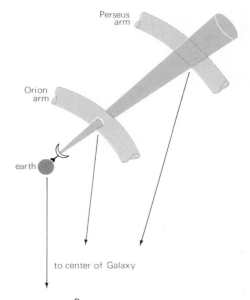

FIGURE 17.8
A radio telescope looking through two arms in the plane of the Galaxy. The beam "sees" only part of the near arm so the near arm's signal is less intense than that received from the far arm. Both arms are observed because of their different radial velocities.

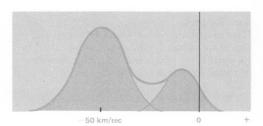

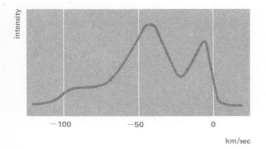

FIGURE 17.9
Observed hydrogen line profile in the direction of longitude 135°. Two arms are apparent. Note that the velocities are negative in agreement with Fig. 17.5. (After B. Burke et al., Carnegie Institution of Washington.)

FIGURE 17.10
The maximum radial velocity in a given direction occurs at the point where the line of sight is perpendicular to a line passing through the center of the Galaxy.

chaotic. As we follow an arm around the plane of the Galaxy toward the center it will at one point have its greatest radial velocity. At this point the motion of the arm is perpendicular to the center of the Galaxy (Fig. 17.10). If we know the distance from the sun to the center of the Galaxy we can then derive the distance from the center of the Galaxy to this tangent point. The various tangent points must give a reasonable picture or the scale (the distance to the center of the Galaxy) is wrong. A schematic radial picture of the Galaxy is shown in Fig. 17.11.

Since the H II regions are excellent optical tracers of spiral arms, it is a pity that they are visible only to distances on the order of 3000 parsecs. Because of the low densities in H II regions, N. Kardashev suggested that very high-level electron capture and a resulting cascade to the ground level may occur. This is called recombination and such lines are observed resulting from the transition of an electron from level 110 to 109 (recall that the Hα line arises when the electron goes from level 3 to level 2). This line presents the opportunity to observe H II regions at distances far removed from the sun.

Fortunately optical techniques have now been extended to 6000 parsecs and the scale is being determined with increasing confidence. The radio and optical values must agree and it appears that the distance from the sun to the center of the Galaxy is 8.7 kiloparsecs. For ease of computation and simple calculations we will continue to use 10 kiloparsecs.

17.7 The nature and origin of the spiral arms

What are the spiral arms and where do they come from? Following B. J. Bok and P. Bok we define the spiral features in the Milky Way and other spiral galaxies as long connected streamers of hydrogen gas and young stars. We have seen that the bright hot young stars, characteristic of population I are found primarily in the spiral arms and that the hydrogen gas can be followed to great distances in a spiral pattern. Other objects help reveal the spiral features, but as we move toward older and older stars, that is, more intermediate population types, the correlation is less strong. Thus, if we study the distribution of the ordinary type G giants (luminosity class III of the MK system, Section 12.21) we find no evidence of a spiral pattern at all.

We do not understand the origin of the arms, but there is one theory that has made some progress toward explaining them. Its principal proponents are C. C. Lin and W. W. Roberts. Lin assumes that the gravity field in the plane of the galaxy has a spiral field impressed upon it. The gas and stars are on circular orbits. When they reach the spiral they slow down, so there is a piling up of stars and material along the pattern. Here the density is high, and hence the term **density wave** is attached to the theory. A large dense cloud can move into the spiral, suffer compression and collapse to form stars.

In this theory the spiral pattern is a semipermanent feature of a spiral galaxy while the gas and stars that form it are constantly changing. An analogous phenomenon is familiar to traffic engineers. If an obstruction is placed on a high-density, high-speed freeway, the traffic quickly backs up and forms a compressed region of cars. Long after the original obstruction is removed the compression wave is evident although the population forming the compression wave has changed many times.

The beauty of the density wave theory is that it makes exact predictions that can be tested by observation. B. J. Bok stresses this point and so far has found several remarkable confirmations.

17.8 The radio view of the Galaxy

Many interesting features of the galaxy are found with radio techniques, although occasional fleeting glimpses by other techniques may be obtained. By analogy with other spiral galaxies our Galaxy has a spherical central region surrounded by a great halo and a very thin disk (see Fig. 18.2). The central region is about 5,000 parsecs in radius. There is a great ring of H II regions where the disk meets the central region. The spiral features begin there and spiral out to perhaps 25,000 parsecs from the center where a great ring of neutral hydrogen is to be found. The disk itself is only about 1500 parsecs thick.

This vast region has been surveyed at various radio wavelengths; one of these surveys is shown in Fig. 17.12. Several features are always present: The plane of the Galaxy is well defined and there is a general continuum centered on the Galaxy extending far out of the plane. At certain wavelengths, in particular those of neutral hydrogen, the continuum out of the plane of the Galaxy shows some interesting structures. A large feature in the general direction

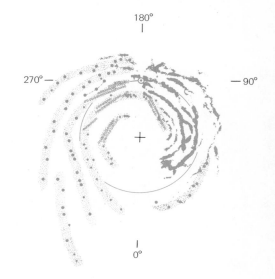

FIGURE 17.11
Spiral structure of the Milky Way galaxy as deduced from observations made in Holland and Australia.

FIGURE 17.12
A hydrogen survey at λ21 centimeters by G. Westerhout from Cygnus to beyond the center of the Galaxy. Some interesting features are marked. The longitude is marked in the old system. The center of the Galaxy is at 327.7° here and is, by definition, 0° in the new system.

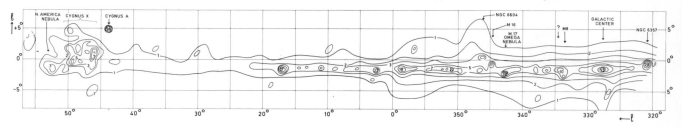

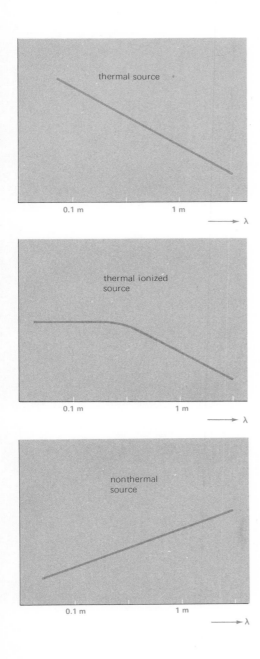

thermal source

0.1 m   1 m

     λ

thermal ionized
source

0.1 m   1 m

     λ

nonthermal
source

0.1 m   1 m

     λ

of the center of the Galaxy is associated with hydrogen moving up out of the Galaxy. This is referred to as a **spur** and appears at wavelengths as long as 2 meters. Looking toward the galactic poles we often see weak concentrations of hydrogen moving down toward the galactic plane with velocities greater than 100 kilometers per second. These are interpreted as clouds falling into the Galaxy, perhaps from intergalactic space, and are referred to as **high-velocity clouds.**

Every radio survey shows discrete sources or concentrations. Some are individual clouds, others are objects of an unusual nature, some are even stars. These sources can be studied at many wavelengths, analogous to photographing stars in different colors, and the intensity (I) plotted against wavelength (λ). If a source is thermal in origin it will be a straight line sloping upward as the wavelength decreases. If the source is thermal but has a large amount of ionized gas present, the intensity line will rise but then flatten out. If the source is nonthermal the intensity–wavelength line will slope upward toward longer wavelengths. The nonthermal source radiation is generally attributed to synchrotron radiation. Some sources, for example, the active sun, show an intensity plot that is a combination of a thermal and nonthermal source.

As the resolution of radio telescopes improved, many very large concentrations of radio radiation, such as Sagittarius A (the first radio source found in the region of Sagittarius), began to show structure and eventually proved to be numerous small sources. These particular sources are concentrated in the direction of the center of the Galaxy and are located there. There are at least five discrete sources comprising what we have been referring to as Sagittarius A. This name is now attached to the very strong continuum source located in the very center of the Galaxy. It is referred to as a continuum source because it is radiating at all wavelengths and in this case the origin of the continuous radiation is nonthermal. This source because of its location, is occasionally occulted by the moon so we can study it in detail. It is about 3 minutes of arc in diameter corresponding to about 10 parsecs at the distance of the galactic center. In the very center of this source is an exceptionally strong point source with a diameter less than 1 parsec. This point source is also a very strong source in infrared wavelengths. Sagittarius A, that is the continuous, nonthermal source, shows no structure as large as 1 parsec, the limiting resolution of the lunar occultation technique and equipment used to date.

The sources surrounding Sagittarius A are thermal sources emitting the hydrogen line at λ 21 centimeters and OH lines at λ 18 centimeters. The neutral hydrogen emission at λ 21 centimeters and the recombination lines (particularly hydrogen level 110 to 109) look exactly like that from H II regions. Our general description of the Galaxy describes the central region as pure population II, but here is clear evidence for the presence of gas even in population II regions.

Absorption due to hydrogen and some molecules can be seen against the Sagittarius A sources. The hydrogen is seen to be rotating very rapidly even within a few hundred parsecs of the center of the Galaxy and expanding outward in all directions as well.

Unfortunately we cannot see the center of the Galaxy visually. Scans in the infrared show the great bulge characteristically seen in the radio maps, but even the infrared photons do not penetrate the great clouds of gas and dust located between the sun and the galactic center. We are certain however, from our radio observations cited above, that the galactic nucleus is a very active and energetic region similar perhaps to many other galaxies.

17.9 Discrete radio sources

All radio surveys show discrete emission (and absorption) areas other than the galactic center and the galactic plane. Some of these regions are quite large in angular extent and even show detailed structure. Others are essentially point sources. These discrete sources are distributed all over the sky.

Many of the extended sources can be associated with emission nebulae (H II regions) and show the characteristic ionized thermal spectrum. Objects such as the Orion nebula, the Rosette nebula (see color plate), and many others are so identified. Other extended sources can be associated with emission nebulae but show nonthermal spectra. The most intense source of this type is Cassiopeia A, followed by Puppis A, and Taurus A. There are some traces of optical nebulosity in Cassiopeia A and Taurus A is centered directly on the Crab nebula. There is no optical counterpart to Puppis A.

Most of the nonthermal galactic radio sources can be associated with supernova remnants. The Crab nebula is the most striking example, but there are others including Cassiopeia A and the Cygnus loop. Another ring type nonthermal radio source is found in the direction of Tycho's supernova, and there can be no doubt that it is the remnant of that supernova.

Most of the discrete radio sources are very small excepting, of course, the sun. The sun dominates the radio sky in the daytime. Other stars have been observed by radio techniques. UV Ceti, the prototype of flare stars, shows bursts of radio noise coincident with flares that are observed optically. Some M-type supergiants have been observed in the water vapor lines and the eclipsing binary beta Lyra has been observed to have a nonthermal radio spectrum, which can be associated with the ionized gas streams between the two stars.

The great majority of the small discrete radio sources cannot be associated with stars. They are below the resolution of our instruments and at first were called radio stars. However, it soon became clear that many of these point sources could be identified with other galaxies, especially peculiar galaxies. Almost all of these sources have nonthermal radio spectra and show high-excitation emission-line spectra in the optical region.

The overall picture of the Galaxy is that of a large somewhat squashed ball having a radius of 25 kiloparsecs called the halo. The halo increases in stellar density toward the center of the Galaxy where the star density becomes more than one star per cubic parsec. Scattered symmetrically about the halo are the globular clusters, themselves spherical concentrations of stars. There is almost no gas or dust in the halo. All of the stars of the halo are old stars. Embedded in the halo are distributions of stars showing more and more flattening as the stars become younger and younger until finally we get to the very youngest stars which are found only in the plane of the Galaxy. The youngest stars are not spread over the entire plane but are concentrated in the spiral arms along with the gas and dust.

To illustrate the picture more completely we shall select the K-type giants of luminosity class III. Stars of this type in the halo have very weak metal lines in their spectra whereas similar stars in the young clusters of the spiral arms show stronger metal lines. To a first approximation, the strength of the metal lines is an indication of the age of the star.

Looking perpendicular to the plane of the Galaxy we find the density of all of the ordinary K giants falls gradually to the limits of our ability to detect and identify them. If now we divide these stars according to the strength of their metal lines, into three classes—strong, moderate, and weak—we find all three classes in about equal numbers in the plane of the Galaxy. At a distance of 1000 parsecs above the plane the strong line K giants make up only 2 percent of this type of star while the moderate and weak line stars are about equal in number. Strong-line stars do not exist 5000 parsecs above the plane of the Galaxy. The moderate-line stars are very few in number and the weak-line stars make up the overwhelming proportion of the K giants this far out of the plane.

Thus the distribution of stars hints of an evolutionary process within the Galaxy.

The description of the halo is characteristic of a general class of galaxies called elliptical galaxies. The description of the Galaxy is characteristic of another general class of galaxies called spiral galaxies. We now turn our attention to other galaxies.

QUESTIONS

1 What simple observation is a reasonable indication that the solar system is located in a flat system of stars?
2 Why do we say the galactic center is in the direction of Sagittarius?
3 How is the present system of galactic coordinates defined? In what constellation is the north galactic pole?
4 How did Shapley determine the direction and distance to the galactic center?
5 Two assumptions are made to predict the observed results of differential galactic rotation. What are they? Can you explain the observed curves in another way?
6 Why do we say the Milky Way is a spiral galaxy?
7 What is the definition of spiral arms?
8 What are good indicators or tracers of spiral arms?
9 How does the radio spectrum of a thermal source differ from that of a nonthermal source?
10 What are the various types of radio sources? Give some examples.

FURTHER READINGS

BOK, BART J., AND PRISCILLA F. BOK, *The Milky Way*, 4th ed., Harvard University Press, Cambridge, Mass., 1974.

PAGE, T., AND L. W. PAGE, *Stars and Clouds of the Milky Way*, Macmillan, New York, 1968.

PAWSEY, J. L., AND R. N. BRACEWELL, *Radio Astronomy*, Oxford University Press, New York, 1955.

SHKLOVSKY, I. S., *Cosmic Radio Waves*, Harvard University Press, Cambridge, Mass., 1960 (English transl.).

STEINBERG, J. L., AND J. LEQUEUX, *Radio Astronomy*, McGraw-Hill, New York, 1963 (R. N. Bracewell, transl.).

WHITNEY, CHARLES A., *The Discovery of Our Galaxy*, Knopf, New York, 1968.

A quasi-stellar object as seen on the storage oscilloscope after one day's observing by the Westerbork supersynthesis telescope. The contours can be converted to intensity points and a "picture" is rendered.

Cosmology

18

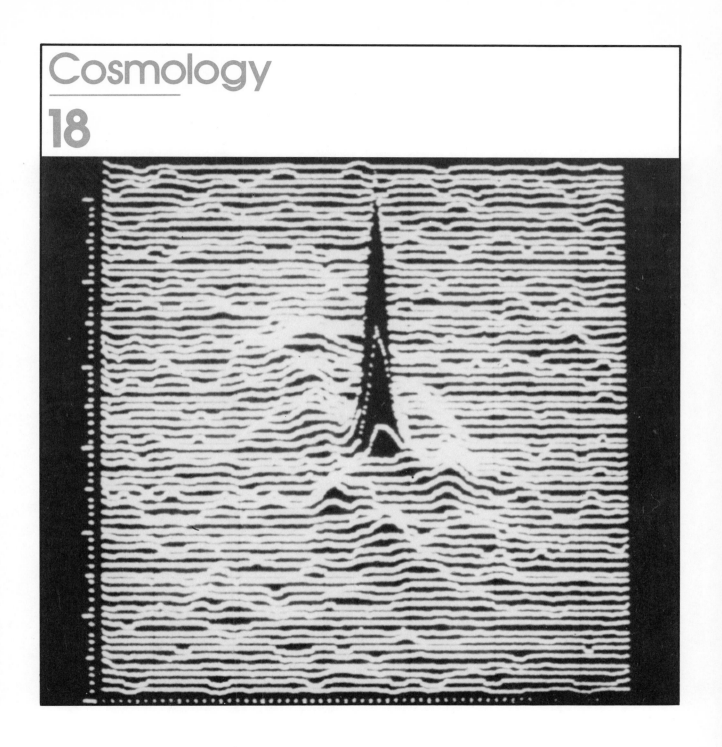

18

Cosmology is the study of the Universe. We begin this study by looking at the building blocks of the Universe —the galaxies—in some detail. The galaxies are structurally of three main types following an original classification by Hubble: spiral, elliptical, and irregular galaxies. Galaxies are assembled in large clusters and also in smaller groups such as the Local Group in which the Milky Way galaxy is located. The redshifts of their spectral lines are increasingly positive as the galaxies are more remote. Certain quasi-stellar objects (quasars) have the largest redshifts yet observed and are difficult to interpret.

STRUCTURAL FEATURES OF GALAXIES

Galaxies may be classified conveniently into three classes that correlate with stellar population types. The classes are called elliptical, spiral, and irregular, and they are composed principally of population type II stars, a mixture of types I and II stars, and population I stars, respectively.

18.1 The great debate

The serious study of galaxies began with V. M. Slipher at Lowell Observatory in late 1912 when, in a letter to his former teacher J. Miller, he presented a table of radial velocities of nine extragalactic nebulae (at that time the term extragalactic meant avoiding the galactic plane) obtained with the Lowell 24-inch refractor. Slipher pointed out that seven objects showed increasing recessional velocities that correlated with their apparent magnitudes. Furthermore, the velocities were astounding, the largest one being 1800 kilometers per second, far larger than that of any star. Miller counseled Slipher to be cautious and to add a few more observations before releasing this material. Finally in 1914 Slipher presented his observations on 13 extragalactic nebulae to the American Astronomical Society and received a standing ovation. The ovation was principally for the perseverance shown, a characteristic to this day of observational cosmologists, and the very large radial velocities observed. Some of Slipher's spectrograms covered 30 hours and required four different nights to complete the exposure. Little noticed at the time were Slipher's statements that the spectra were all characteristic stellar-type spectra and that one of the objects, the Sombrero nebula (NGC 4594), showed rotation. He concluded that the extragalactic nebulae of the type showing composite stellar spectra were separate island universes of stars, and all but the nearest ones were receding from us.

At about the same time A. van Maanen using the greatest telescope in the world measured the proper motion of the arms of the spiral galaxy called M 101. Clearly any object showing proper motion cannot be very far away, and so a contrast of views prevailed. The differing viewpoints became so sharp that a so-called great debate was arranged by the U.S. National Academy of Science where H. Shapley and H. Curtis presented the opposing points of view. The great debate took two different aspects, the first discussed on the first day concerned the nature of the Milky Way, and the second discussed on the second day concerned the nature of the extragalactic nebulae. In the first part of the discussion Shapley was right, clearly the Milky Way was centered at a point at a great distance from the sun. In the second part, Curtis leaning heavily upon Slipher's observations claimed the extra-galactic nebulae were island universes of stars. Shapley leaning upon van Maanen's observations claimed the contrary. In the second part of the discussion Curtis was right.

The stage was set for modern cosmology. It may be that it was Slipher's paper that precipitated Samuel Clemens' remark concerning science (and cosmology in particular): "One gets such wholesale returns of conjecture out of such a trifling investment of fact." However, the investment was not trifling and never was a person so right in his conjecture despite his small number of observations. It is in the nature of cosmology that one must try to make sense from a scarcity of observations.

In 1885 a supernova was observed in the Andromeda galaxy that aroused the curiosity of some astronomers. However, it wasn't until 1924 when Hubble observed cepheid-type variable stars in M 31 (the Messier Catalogue designation for the Andromeda galaxy) that astronomers were convinced that the island universe concept was correct. Hubble using Shapley's calibration of the period–luminosity relation assigned a distance to M 31 of 1.5 million light years (460,000 parsecs), which was later revised by W. Baade to 2.2 million light years or 675,000 parsecs using his newly revised calibration of the period–luminosity relation. Hubble and M. Humason fully confirmed Slipher's work and Hubble was able to classify galaxies into three principle types: spiral, elliptical, and irregular. Hubble refined this classification. The elliptical galaxies were classified as E0 for spherical systems through E1, E2, and so on to E7 depending on the degree of ellipticity. The spiral galaxies are classified as S0, Sa, Sb, or Sc depending upon how large the nucleus was and how tightly wound the arms were, with Sc galaxies having the smallest nuclei and the least wound arms. Irregular galaxies were not classified by definition. Hubble added a parallel classification to that of the spiral galaxies because a significant number of spiral galaxies showed a bar across their nuclei. These barred spiral galaxies he designated by SB0, SBa, SBb, and SBc.

Many astronomers have tried to attach evolutionary significance to the Hubble classification scheme (Fig. 18.1); Hubble himself never did. M. S. Roberts points out that, based upon our knowledge of the lifetimes of stars and clusters of stars and the age of the Universe (Chapter 19), one type of galaxy could not evolve into another type even if we are wrong in our estimate of the age of the Universe by a factor of 2 or more. Except for cataclysmic events in the nuclei of certain galaxies, the galaxies are

FIGURE 18.1
The sequence of types of galaxies as originally drawn by E. Hubble. The irregular galaxies were added at the right later.

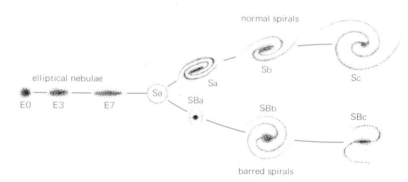

FIGURE 18.2
Well-developed Sb spiral galaxy called the Sombrero galaxy (NGC 4594). The disk is extremely thin, the apparent thickening at the outer reaches being a perspective effect. (Photograph from the Hale Observatories.)

not now very much different from what they were 10^{10} years ago nor will they be much different 10^{10} years in the future.

18.2 Spiral galaxies

Spiral galaxies are structurally of two types: normal and barred spirals. **Normal spirals** have lens-shaped central regions. From opposite sides of this central region two arms emerge and at once begin to coil around the centers in the same sense and the same plane. They are divided into three classes. **Class Sa** spirals have large central regions and thin, closely coiled arms; examples are NGC 4594 in Virgo (Fig. 18.2) and M 81 in Fig. 18.3. In **class Sb** the centers are smaller and the arms are larger and wider open; examples are M 31 and probably our own Galaxy. In **class Sc** the centers are smallest, and the arms are largest and most loosely coiled; an example is M 33 in Triangulum (Fig. 18.4).

Barred spirals have their two coils starting abruptly from the ends of a bright bar, which projects from opposite sides of the central region. They are classified in the series **SBa, SBb,** and **SBc,** paralleling the series of normal spirals. Through this series the central region diminishes while the arms build up and unwind. A third and less familiar type of spiral galaxy has its arms beginning tangentially from opposite sides of a bright ring around the center.

Spiral galaxies are presented to us in a variety of orientations. In the flat view they appear nearly circular. M 51 (Fig. 18.5) is an example; it has an irregular companion projected near the end of one arm. When they are moderately inclined to the plane of the sky, as in the cases of M 31 and M 81, another giant galaxy, these galaxies appear elliptical. In the side view they are reduced to bright streaks around thicker central regions. Characteristic of the spirals seen on edge is a dark band that sometimes seems to cut them in two; an example is NGC 4594, the Sombrero galaxy, shown in Fig. 18.2. The dust in their arms obscures their equatorial regions, just as the dust clouds of the Milky Way obstruct our view of what lies beyond.

One notable feature of spiral galaxies is the extreme thinness of the absorbing disk in comparison to the diameter of the disk. NGC 4594 shows this clearly, the apparent thickening of this layer toward the edges can be seen as due to a turning down of the left edge and a turning up of the right edge. This is known as the hat brim effect often seen in spiral galaxies. We have noted already the thinness of the disk of the Galaxy and should point out that G. Westerhout has called attention to the hat brim effect in the Galaxy's layer of neutral hydrogen. We will see that the Andromeda galaxy shows the same effect.

The nearest spiral galaxies have been studied at λ 21 centimeters and show two distinctive rings, one at about 20 percent of the diameter and the other just beyond the normal visible limits of the arms. The arms are located between these two rings and can be well represented as logarithmic spirals.

Any study of deep-sky photographs shows many spiral galaxies. At first they seem to be the preponderent type of galaxy, but this is a selection effect. First, the obvious arms attract our attention, and second, the spirals as a group are generally large galaxies having diameters ranging from 25 kiloparsecs to 60 kiloparsecs. Even so, present counts indicate that spiral galaxies make up about 65 percent of the galaxies. A typical, though large, spiral galaxy is M 31, which is near enough so that we can study it in detail.

The spiral galaxy M 31 in Andromeda is the brightest exterior spiral system and is one of the two spirals that are nearest us. Marked in Map 4 with its original designation as the "Great Nebula" in Andromeda, it is visible to the unaided eye as a hazy spot about as long and half as wide as the moon's apparent diameter. Only the central region appears to the eye alone and,

FIGURE 18.3
The fine Sb spiral galaxy in Ursa Major referred to as M 81 (NGC 3031). Note the peculiar galaxy M 82 nearby (see also Figs. 18.10 and 18.15). (Copyright by Akademia der Wissenschaften der DDR. Taken at the Karl Schwarzschild Observatorium.)

An Sc type spiral seen almost face on. Note the very small nucleus and the rather ill-defined arms. (Lick Observatory photograph.)

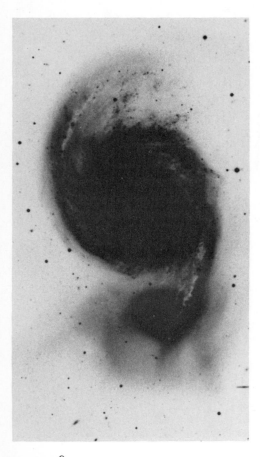

Galaxy M 51 and its interacting companion. This photograph, taken by S. van den Bergh was deliberately overexposed to bring out the interconnecting filaments between the two galaxies. (Photograph from the Hale Observatories.)

for the most part, to the eye at the telescope. Fainter surrounding features come out in the photographs, where the object is shown in its true character as a flat, double-armed spiral inclined 13° from the edgewise presentation. The flat disk is slightly turned up on one end of the major axis and down on the other as can be seen in Fig. 18.6.

In the photograph in Fig. 18.6 the tilted nearly circular spiral appears as an oval 3° long. Separate stars in the arms were first observed in 1924 by E. Hubble in his photographs with the 100-inch Mount Wilson telescope. The cepheid variable stars among them served to determine the distance of the spiral and to show it as another galaxy far beyond our own Milky Way.

The lens-shaped central region of the Andromeda spiral is surrounded by a flat disk of stars. Embedded in the disk are the spiral arms, which contain all the features observed in the Milky Way around us. Young stars, gas, and dust are localized in the arms. Patches of the gas appear as emission nebulae, made luminous by the radiations of blue stars in the vicinities. Based on these radiations W. Baade extended the tracing of the arms beyond their more conspicuous parts, increasing the diameter of the disk to 4.5°. At a distance of 675,000 parsecs from us, the diameter of the Andromeda spiral is about 55,000 parsecs or slightly larger than the diameter often assigned to the disk of our own Galaxy. Several hundred globular clusters of stars surround the disk, as they do in the halo around the center of our galaxy.

Before leaving spiral galaxies we should consider some interesting observations of M 51. M 51 is a relatively nearby spiral with a rather large companion galaxy partially hidden by the end of one of its arms. It is only slightly tilted to us and presents the classic appearance of a spiral galaxy. It is almost the same size as our Galaxy and shows all of the features that we see in the Milky Way—H I regions, H II regions, novae, supernovae, gas, dust, and an inner and an outer ring of hydrogen.

Recently M 51 has been studied in the radio continuum by D. S. Mathewson and his colleagues using the supersynthesis telescope at Westerbork. Their results show a long continuous ridge following the spiral pattern. This leaves us with the strong

FIGURE 18.6
The great Andromeda galaxy (M 31 = NGC 222) and two of its elliptical companions, NGC 205 and NGC 221. Andromeda galaxy is a member of the Local Group and is slightly larger than the Milky Way galaxy. (Photograph from the Hale Observatories.)

suggestion that the spiral arms are long connected streamers as defined by the Boks and that the density wave theory of Lin has credence.

18.3 Elliptical galaxies

Elliptical galaxies are so named because they appear with the telescope as elliptical disks having various degrees of flattening. They are designated by the letter E followed by a number that is 10 times the value of the ellipticity of the disk. The series runs from the circular class E0 to the most flattened E7, where the object resembles a convex lens viewed on edge. Examples are M 32, the nearly circular class E2 companion of the Andromeda spiral, and the second companion, NGC 205, of class E5 (see Fig. 18.7).

These galaxies are systems of stars that are generally dust-free. Almost all the gas and dust available for star building seems to have been exhausted. Individual red giants, the brightest stars of this older population, are visible in photographs of the nearer elliptical galaxies when they are taken through red filters with the largest telescopes.

This is the general picture of elliptical galaxies, but there is a great variety of elliptical galaxies and some do show gas and dust. These form a small fraction of all elliptical galaxies—less than

FIGURE 18.7
The highly elliptical galaxy NGC 3115. There is no evidence of gas or dust in this galaxy. (Photograph from the Hale Observatories.)

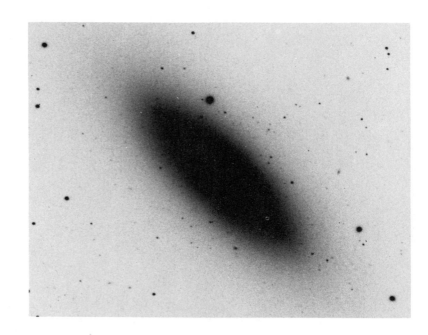

1 percent. The elliptical galaxies range from the largest and brightest galaxies, designated by gE, in the Universe with diameters of 85 kiloparsecs to the smallest galaxies, designated by dE, in the Universe with diameters as small as 200 parsecs. The giant elliptical galaxies are generally very strong radio sources. The giant elliptical galaxies are visible to the limits of our optical techniques, although a type of object (the quasi-stellar objects) may be visible at even greater distances and some unidentifiable radio sources may be at the limit of the observable Universe. Elliptical galaxies make up about 30 percent of the observable galaxies.

18.4 Irregular galaxies

Hubble's third great class of galaxies, the irregular galaxies is just as descriptive as the other two. These galaxies present a picture of an unorganized accumulation of stars, gas, and dust. They show all of the typical population I objects with gas and dust comprising for 50 percent of their mass. Despite this, many irregular galaxies contain globular clusters and red giants typical of population II. Irregular galaxies have a considerable size range. Some are as large as the large spiral galaxies, whereas others are comparable to the smallest elliptical galaxies. Five percent of all galaxies are irregular galaxies, although many galaxies may be classified as irregular when in fact they are peculiar. Typical irregular galaxies are the Milky Way's nearest neighbors, the Magellanic clouds.

The Magellanic clouds, named in honor of the navigator Ferdinand Magellan, are the two satellites of our Galaxy. Plainly visible to the unaided eye, they are too close to the south celestial pole (Map 6) to be viewed north of the tropical zone. As they appear generally in the photographs, the clouds have apparent diameters of 12° and 8°. At the distances of 50 kiloparsecs and 55 kiloparsecs, their linear diameters would be 9.8 kiloparsecs and 7.7 kiloparsecs, respectively. The neutral hydrogen of the clouds extends farther than do the stars and forms a link between the two objects. The clouds are classified usually as irregular galaxies, abbreviated I and are ragged assemblages that lack rotational symmetry. Yet the rotations of both clouds have been suspected. G. de Vaucouleurs presents strong arguments for a bar in the center of the Large Magellanic cloud (Fig. 18.8).

Among the initial achievements of the Australian 210-foot telescope are the first detailed radio pictures of the Magellanic clouds. The pictures consist of intensity contours recorded with parts of the continuous emission spectrum and also with the λ21-centimeter line of neutral hydrogen. Superposed on the low-

FIGURE 18.8
The Large Magellanic cloud, an irregular galaxy. Some astronomers believe the central portion is in the form of a bar. (Lick Observatory photograph.)

level backgrounds of the clouds are contour patterns of more intense discrete sources. About 100 of these are found to be thermal sources, of which more than half are identical in positions with optically cataloged emission nebulae. The Small Magellanic cloud has a double-peaked radio profile; it seems to contain two bodies of gas that differ in radial velocity by 30 to 40 kilometers per second.

Both clouds have variable stars, novae, planetary nebulae, etc. The small cloud even has an X-ray source that is a binary star. In general, the southern hemisphere of the celestial sphere has been lying fallow despite the fact that the center of the Galaxy is located there as are the Galaxy's two nearest companions. B. J. Bok reminds us that the Large Magellanic cloud (LMC) and the Small Magellanic cloud (SMC) are galaxies and may be studied using 50-centimeter (20-inch) telescopes with the same precision with which the Andromeda galaxy is studied using the 200-inch reflector.

Photographs in some of the Hubble type are shown in 18.9.

18.5 Other types of galaxies

Many galaxies just like stars do not fit into the Hubble classes nor, for that matter into more refined systems of classification. Many galaxies show strange but symmetrical forms suggesting that a superviolent event, usually in the nucleus, has occurred that is disrupting the galaxy. These galaxies are referred to as **peculiar galaxies** and often show the characteristic indicators of highly energetic events such as we have found in supernovae and their remnants.

M 82 is such a peculiar galaxy (Fig. 18.10). It shows filaments running out of its center perpendicular to its plane at velocities as great as 10,000 kilometers per second and even larger. These filaments emit polarized light, first observed by A. Elvius and J. Hall, typical of a synchrotron source. This galaxy is a strong nonthermal radio source and is an X-ray source and infrared source as well. Its spectrum shows numerous high-excitation forbidden lines. All of this sounds like a replay of our description of the supernova event and so we can assume that we are dealing with a highly energetic and violent phenomena. However, this cannot be even a super-supernova and the student should not be misled; it is an even more violent event. All indications are that M 82 was a normal galaxy, probably a spiral, that has now been torn asunder by an event in its nucleus of such violent proportions that it is disrupting the entire galaxy.

Another type of galaxy is the **radio galaxy.** Actually all galaxies emit energy in the radio region so we limit the name to the very strong radio emitters. Most of these galaxies are the giant elliptical galaxies, peculiar galaxies, N-type galaxies, and unidentifiable sources that have spectra typical of radio galaxies and that we presume are galaxies, perhaps at the edge of the Universe. Occasionally radio galaxies are variable in their energy output, not just in the radio region but in the infrared, visual, and X-ray regions. The time for these variations is quite short.

N galaxies are galaxies having bright star-like nuclei and strong emission in the radio spectrum. They often show faint, wispy halos. **Seyfert galaxies,** first discussed by C. Seyfert and named for him, are similar to N galaxies. They have very bright star-like nuclei that show high-excitation emission lines, which are very broad. Some, but not all, Seyfert galaxies are radio emitters. The Seyfert galaxies show traces of spiral structure. Seyfert galaxies may be an early stage of the N galaxies. To complete this picture we mention the **quasi-stellar objects** that show many of the fea-

FIGURE 18.9 [NEXT PAGE]
Types of spiral galaxies. The regular spiral sequence is shown on the left. The corresponding sequence of barred spirals is shown on the right. (Photographs from the Hale Observatories.)

NGC 1201   Type S0

NGC 2841   Type Sb

NGC 2811   Type Sa

NGC 3031 M81   Type Sb

NGC 488   Type Sab

NGC 628 M74   Type Sc

NGC 2859   Type SB0

NGC 2523   Type SBb(r)

NGC 175   Type SBab(s)

NGC 1073   Type SBc(sr)

NGC 1300   Type SBb(s)

NGC 2525   Type SBc(s)

FIGURE 18.10
M 82, an irregular galaxy. The central region of this galaxy has undergone a catastrophic event that is tearing it apart. (Photograph from the Hale Observatories.)

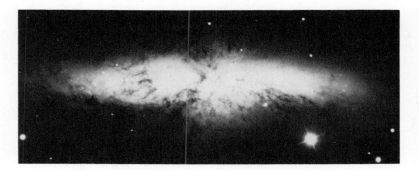

tures of Seyfert and N galaxies, but which are essentially star-like. Most of them are not strong radio sources, but those that are, are also infrared sources and X-ray sources. We return to these objects later in this chapter.

Dwarf galaxies are divided into two classes, dwarf elliptical galaxies and dwarf irregular galaxies. They are really the faint, low-mass galaxies extending the two major classes already discussed. The dwarf elliptical galaxies resemble globular clusters except that the largest globular cluster has a diameter of about 100 parsecs, whereas the smallest dwarf elliptical galaxy has a diameter of 200 parsecs. The majority of the known members of the Local Group of galaxies are dwarf elliptical galaxies (see Section 18.7). These galaxies are essentially free of gas and dust. Exhaustive studies by P. Hodge have failed to detect even a trace of interstellar material in these galaxies. They appear to be examples of pure population II objects.

Dwarf irregular galaxies are probably very numerous but even at 1 megaparsec they are extremely difficult to detect. Two of the Local Group (see page 393) are dwarf irregular galaxies, NGC 6822 and IC 1613. These two galaxies are much smaller than the Magellanic clouds but contain typical population I type objects such as cepheids, gas, dust, and H II regions. G. Reaves has shown that dwarf irregular galaxies make up a significant fraction of all galaxies and are at least as numerous as the spirals. Dwarf galaxies of both types are, in general, always associated with large galaxies or clusters of galaxies.

DISTRIBUTION OF GALAXIES

Galaxies appear to be evenly distributed over the entire sky, but they tend to group together in clusters. The Milky Way galaxy is a member of a small cluster of galaxies.

18.6 The arrangement of galaxies in the sky

The apparent distribution of the nearer galaxies over the face of the sky is shown in Fig. 18.11. This diagram resulted from a survey of the brighter galaxies by H. Shapley and A. Ames at Harvard Observatory. Few such objects are visible within 10° from the galactic equator, which is represented by the heavy curve in the figure; here they are concealed generally behind the dust of the Milky Way. Their numbers increase toward the galactic poles, where the least amount of dust intervenes. The increase would be fairly symmetrical in the two celestial hemispheres except for the conspicuous Virgo cluster at the left. Because of their apparent avoidance of the Milky Way the galaxies were called "extragalactic nebulae" before their true significance was recognized.

The more remote galaxies are similarly arranged in the sky, as was shown by Hubble's more penetrating survey with the Mount Wilson telescopes. If the obscuring dust were not present, the galaxies would be about equally numerous in all regions of the celestial sphere.

Figure 18.12 is an old diagram given in equatorial coordinates. If one removes the large concentration of galaxies at 12ʰ25ᵐ and +10°, the distribution in the two halves of the sky is about the same. There may be a slight excess in the north galactic polar region. This is attributable to observational effects, the north galactic pole lies in the northern hemisphere where the observational work was much more thorough than the southern hemisphere at the time of the drawing.

The distribution of galaxies in space can be considered when their distances as well as their directions from us are known. The

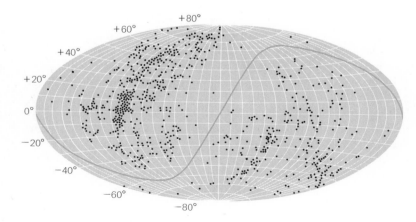

FIGURE 18.11
The distribution of galaxies as originally plotted by Shapley and Ames. The heavy skewed line is the galactic equator. The heavy concentration in the northern galactic region is due to the Virgo cluster of galaxies. (Harvard College Observatory diagram.)

FIGURE 18.12
A supernova event in NGC 4725. The photograph on the left was taken on 10 May 1940 and the one on the right (next page) was taken on 2 January 1941. (Photographs from the Hale Observatories.)

distances were derived in Hubble's pioneer studies in three successive steps: (1) by use of cepheid variable stars that he found in M 31, M 33, and five other relatively nearby galaxies; (2) by observing the apparent magnitudes of what seemed to be the most luminous stars in more remote galaxies, where the cepheids were too faint to be seen; (3) by measuring the apparent brightness of still more distant galaxies, assuming that galaxies of the same class have the same absolute brightness.

$$m - M = 5 \log d - 5$$

Later investigations with the 200-inch telescope showed the need for revising the early scale of distances.

One reason for revision was Baade's discovery (1952), that the classical cepheids are, on the average, 1.5 magnitudes more luminous than they were previously supposed to be. On this account and with the availability of the more precise photoelectric standards for the apparent magnitudes, the former values of the distances of most galaxies required multiplication by the factor 2.

More recently Sandage pointed out that what seemed to be the brightest stars in intermediate galaxies with the smaller telescopes are in fact small patches of emission nebulae. He estimated that these may appear 1.8 magnitudes brighter than the brightest stars in a particular galaxy and that the distances originally assigned to most galaxies might need to be multiplied by a factor as great as 10.

Present methods follow those used by Hubble and Baade. Cepheid variables are observed in nearby galaxies and the distances obtained on the assumption that the period–luminosity relation holds. This is checked by observing individual giant and supergiant stars as well as RR Lyrae stars. Whenever novae occur they are calibrated and cross checked. Globular clusters are visible around many galaxies and H II regions are visible in spiral galaxies. As we move to increasingly distant galaxies we depend more and more on the less reliable of these indicators until finally we assume that the tenth brightest galaxy in any given cluster of galaxies has very nearly the same brightness as the tenth brightest galaxy in any other cluster of galaxies. Fortunately a spectral feature appears that allows us to go to more distant galaxies; it is referred to as the **redshift** and we will discuss this feature later in Section 18.13.

Novae flare out in the galaxies, just as they do in our own (see Section 13.12). They are of two general types, normal novae and supernovae, with respect to the order of luminosity they attain.

Normal novae resemble those found in our Galaxy in their greater abundance and their lower luminosities at maxima. Surveys of many of these novae in the spiral M 31, reported by Hubble in 1929 and by H. C. Arp in 1956, reveal their characteristics more clearly with the elimination of the distance factor. Arp concludes that about 26 normal novae flare out annually in the Andromeda spiral; one-fourth of these are likely to be concealed by the dust clouds of this galaxy or in its brighter regions. The absolute magnitudes at the maxima ranged from -6.2 to -8.5.

Supernovae attain much higher luminosities and are less frequent. These spectacular outbursts of the more massive stars occur in a single galaxy only once in several centuries. They are most likely to be detected in repeated photographs of clusters of galaxies. Supernovae are divided into two groups. In **group I** they rise to absolute magnitude -15.5, or more than 100 million times as luminous as the sun. The extreme width of the lines in their spectra indicate the violence of the explosions. These supernovae are deficient in hydrogen. Examples are the supernova of 1959 in the spiral galaxy NGC 7331 (Fig. 18.12) and the outburst associated with the Crab nebula in our own Galaxy. Supernovae of **group II** attain about absolute magnitude -13.5 at maximum brightness. They probably constitute the more numerous group, but being fainter are less readily detected.

18.7 The Local Group

Our Galaxy is a member of a group of at least 23 galaxies that occupy an ellipsoidal volume of space almost 1 million parsecs in its longest dimension. Our Galaxy is near one end of this diameter and the Andromeda spiral is near the other end. These, M 33 in Triangulum, and Maffei 2 are the normal spiral members. The largest member may prove to be Maffei 1 which appears to be a giant elliptical galaxy. The less regular members are the Magellanic clouds and two smaller ones. The remaining 14 are elliptical galaxies, of which 10 are less populous and much fainter than the others. The first known examples of these dwarf ellipticals were discovered at Harvard Observatory in the constellations Sculptor and Fornax; the other four, two in Leo and one apiece in Draco and Ursa Minor, were found more recently in photographs with the Palomar 48-inch Schmidt telescope.

The members of the Local Group of galaxies are listed in Table 18.1, which contains their types, distances from us, and their apparent angular and linear diameters. The distances are taken primarily from the work of Sandage and S. van den Bergh. All of the data are subject to revision in the near future.

TABLE 18.1
The Local Group

	Type	Distance (mpc)	Apparent diameter	Diameter (kpc)
Milky Way	Sb	—	—	50.0
LMC	I	0.05	12°	9.8
SMC	I	0.06	8°	7.7
Ursa Minor System	dE	0.09	55′	1.2
Draco System	dE	0.1	48′	1.2
Sculptor System	dE	0.1	45′	1.2
Fornax System	dE	0.2	50′	3.1
Leo II System	dE	0.4	10′	0.9
NGC 6822	I	0.4	20′	2.8
NGC 185	E	0.5	14′	2.1
NGC 147	E	0.5	14′	2.1
Leo I System	dE	0.6	10′	1.2
IC 1613	I	0.7	17′	3.1
M 31	Sb	0.7	5°	55.0
M 32	E2	0.7	12′	2.1
NGC 205	E5	0.7	16′	3.1
And I	dE0	0.7	0′.5	0.5
And II	dE0	0.7	0′.7	0.7
And III	dE0	0.7	0′.9	0.9
And IV	dE	0.7	0′.2	0.2
M 33	Sc	0.7	62′	15.4
Maffei 1	gE	1.0	2°	35.0
Maffei 2	Sa	2.7	23′	18.0

The Local Group, considered a small irregular cluster of galaxies, shows the preponderence of small low-luminosity galaxies; hardly one of the 16 smallest members would be visible at a distance of 1 megaparsec. It would appear that more members are associated with M 31 than the Milky Way, but this could be the result of an observational effect.

18.8 Clusters of galaxies

Clusters of galaxies are believed to be the rule rather than the exception. They range in population from a very few to several thousand galaxies. As many as 10,000 rich clusters are listed in F. Zwicky's new catalog of clusters north of declination −30°, which are recognized in yellow-sensitive photographs with the 48-inch Schmidt telescope of Palomar Observatory. A **rich cluster** is here defined as containing more than 50 members having photographic brightness within 3 magnitudes of the brightest galaxy in the cluster.

Individual clusters are designated in the catalog by the equatorial coordinates of their centers; an example is Cl 1215.6 + 3025, where the first number is the right ascension in hours and minutes, and the second is the declination in degrees and minutes. Some of the more prominent clusters, such as the Virgo cluster and Perseus cluster (Fig. 18.13), are often designated by the names of the constellations in which they appear.

A **compact cluster** in the catalog has a single concentration of galaxies that appear close together in the photographs. An example is the Coma Berenices cluster; it has a membership of 9000 galaxies and a preponderance of S0 galaxies, especially in its densest central region. A **medium compact cluster** has either a single concentration where the galaxies are separated by several of their diameters, or else a number of pronounced concentrations. The Virgo cluster is an example; it has many spiral and giant elliptical galaxies. An **open cluster** has no outstanding concentration. An example is the Ursa Major cluster, which includes the great spirals M 81 and M 101.

G. Abell, in contrast, prefers to classify clusters of galaxies as **regular** and **irregular.** Regular clusters show spherical symmetry, they have a strong tendency toward a central condensation, their brightest member is a giant elliptical galaxy, and the appearance of large spiral galaxies is rather rare. Regular clusters range in membership from the great Coma cluster with 10,000 members to clusters having 1000 member galaxies.

Irregular clusters are amorphous in appearance and have a large spiral galaxy as their brightest member. Their composition ranges from the Virgo cluster and its 1000 members through the Local Group with its 23 members to groupings as small as 10 members.

Many astronomers consider clusters of galaxies as the fundamental features of the Universe. D. N. Limber, after a careful analysis, concludes that most clusters of galaxies are stable systems. He does not dismiss the possibility that clusters may have positive energy and hence be subject to disruption, as has been suggested by V. A. Ambartsumian, but he considers it unlikely.

18.9 Galaxies as radio sources

Galaxies differ in the character and strength of their radio emissions. The radiations of the majority are mainly of the thermal type (see Section 5.7); they are weaker in longer than in shorter radio wavelengths. Examples are our Galaxy and M 31. The radiations of some other galaxies are nonthermal; they are generally of the synchrotron type, which is stronger in the longer wave-

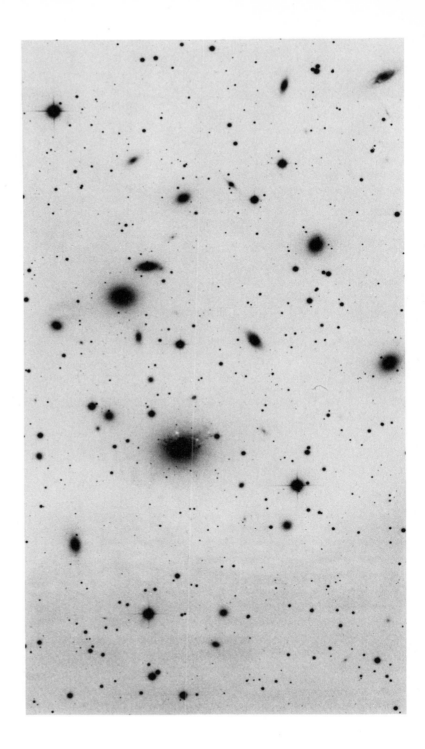

FIGURE 18.13
*A great cluster of galaxies in Perseus. Note
the evidence for gravitational interactions.
(Photograph from the Hale Observatories.)*

lengths and attains remarkable strength in some of these sources. A number of these sources have characteristics that are so much alike they may represent a class that occurs rather frequently. Examples are Cygnus A and Centaurus A.

The optical counterpart of the radio source Cygnus A (Fig. 18.14) is a pair of overlapping bright spots centered 2″ apart and surrounded by a larger dim halo. The appearance was interpreted at first as two colliding galaxies that had interpenetrated until their centers were only a few light years apart. The collision hypothesis for this and similar galaxies was favorably received by most astronomers until 1960, when it began to seem improbable that such strong emission could be produced by collisions. Moreover, the radio emission of Cygnus A comes mainly from two areas centered about 80″ on either side of the optical pair.

In 1960, I. S. Shklovsky concluded that Cygnus A is most likely a generically related double galaxy. A possible interpretation could be implied in a hypothesis of the instability of some clusters of galaxies, proposed in 1954 by Ambartsumian. Radio galaxies such as this one might be single originally, possessing enormous and unexplained energy that is tearing them apart. They might be splitting into pairs of galaxies with the production of intense radiation.

Several galaxies in early stages of explosion are already recognized. An example is the irregular galaxy M 82, a source of intense nonthermal radiation. A. R. Sandage's red photograph (Fig. 18.15) and C. R. Lynds' spectroscopic analysis have shown that enormous jets of material are streaming outward from above and below the main galaxy with velocities up to 10,000 kilometers per second. The explosion is estimated to have started 1.5 million

FIGURE 18.14
The optical counterpart of the radio source Cygnus A. This was first thought to be a pair of colliding galaxies, but is now thought to be a peculiar galaxy akin to M 82. (Photograph from the Hale Observatories.)

FIGURE 18.15
Peculiar galaxy M 82 taken in red light to show the high velocity filaments leaving the center of the galaxy. Compare this with Fig. 18.10. (Photograph from the Hale Observatories.)

years ago and to be expelling an amount of material 5 million times as massive as the sun.

18.10 Interactions between galaxies

We have noted several early conjectures that S0 galaxies might be spirals which lost the gas and dust from their arms in collisions and that radio sources such as Cygnus A might be powered by interpenetrating galaxies. Although collisions between galaxies are perhaps rare enough to be disregarded, effects of interactions between close pairs of galaxies are frequently observed. In some cases the structural features of the galaxies are distorted. In other cases bright filaments join the components of the pairs or appear in their vicinities, reminiscent of the gas streams around close binary stars. The original negatives of the pair NGC 4038 and 4039 (Fig. 18.16) show long filaments above and below the galaxies.

F. Zwicky has observed many pairs of interacting galaxies in his photographs with the 48-inch Schmidt telescope. In 1959 B. A. Vorontsov-Velyaminov published an illustrated catalog of 355 such pairs. The interacting pairs are believed to be revolving binary galaxies (Fig. 18.17).

Arp believes that interacting pairs of galaxies and even groups of interacting galaxies are the rule rather than the exception. Here we use interacting to mean a directly disturbing effect rather than a gravitationally stable system such as a cluster of galaxies. These interactions can often be interpreted as ejected members or parts of galaxies in agreement with Ambartsumian's proposal.

SPECTRA OF GALAXIES

The spectra of galaxies are composites, as would be expected for assemblages of stars of various spectral types. The lines are also widened and weakened by the different radial velocities of the individual stars. Doppler effects in the spectrum lines show the rotations of galaxies and redshifts of the lines, which increase as the distances of the galaxies from us are greater.

18.11 Rotations shown by spectra

The flattened forms of regular galaxies suggest that these galaxies are rotating. The character of the rotations may be determined from the spectra of spiral galaxies having their equators considerably inclined to the plane of the sky. When the slit of the spectroscope is placed along the major axis of the projected oval image

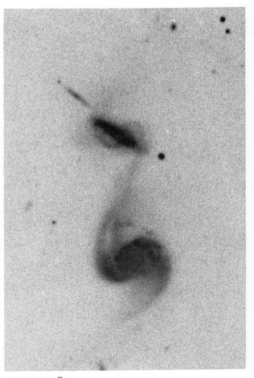

FIGURE 18.16
Two interacting galaxies, NGC 4038 and 4039. On the original negative the arms stretch more than ten times the diameter of the galaxies. (Photograph from the Hale Observatories.)

FIGURE 18.17
A pair of interacting galaxies referred to as Atlas 87. Note how the arms wrap around and the jet from the smaller member. (Photograph by H. Arp, Hale Observatories.)

of the spiral, the spectrum lines show Doppler displacements that depend on the direction of the rotation and its speed at different distances from the center of the spiral. V. M. Slipher was first to appreciate this technique and to apply it to the galaxies when he measured the rotation of the Sombrero galaxy in 1913.

As an example of such spectroscopic studies, Mayall and Aller observed that the rotation period of the spiral M 33 is the same within 18′ from its center. In the outer parts, between 18′ and 30′ from the center, the period increases with increasing distance, as it does in the outer regions of our own Galaxy (see Section 17.4). Thus the inner regions of spirals rotate as a rigid wheel whereas the outer regions are in Keplerian motion.

If we can correct our calculations for the projection effects of the galaxy we are measuring, we then know the velocity of the stars at a certain distance from the center of the galaxy. If this distance is sufficiently removed from the center we can assume that the star's orbit is determined by gravity, that is, the mass between it and the center is completely concentrated at the center. The above assumption means that we are dealing with Keplerian motion. Now if we assume a circular orbit we can obtain the mass of the galaxy.

This can be illustrated as follows: Suppose our spectrogram shows that stars located 10 kiloparsecs from the center have a velocity of 200 kilometers per second. Applying geometry first, we obtain the circumference of the orbit, and dividing it by the velocity we obtain the period, which we convert to years. Second, applying Kepler's harmonic relation, we find that the mass of the galaxy interior to this star is about 10^{11} solar masses. An approximation for the mass external to our point of measurement then gives the total mass of the galaxy in question, at least within a factor of 2.

A final question involves the direction of rotation of the spiral arms: Are the arms leading or trailing the rotation? Or, more simply, are the arms unwinding or are they winding up. Slipher's conclusion was that the arms are trailing and this conclusion was a point of controversy for several decades. His procedure was to determine the inclination of the spiral and then assign the closer arm by the arm having dust lanes close to and projected upon the nucleus of the galaxy under study. More refined techniques applied over the past 20 years have, in every case, verified Slipher's position and there is no doubt that the arms in spiral galaxies wind up. Thus the spiral M 33 (Fig. 18.4) would be expected to rotate in the clockwise direction.

18.12 New classification of galaxies

A modification of Hubble's structural classification of galaxies, devised by W. W. Morgan and explained by him in 1958, results from his studies of composite spectra as well as direct photographs of many galaxies. The principal feature of the new system is the idea that the stellar populations are older as the stars of a galaxy are more highly concentrated toward the center of the galaxy. At one extreme are the slightly concentrated irregular and Sc galaxies; their spectra contain strong hydrogen lines characteristic of young blue stars of Baade's population I, and are designated as group a. At the other extreme are the highly concentrated spiral and giant elliptical galaxies; their spectra show molecular bands characteristic of old yellow and red stars of population II, and are designated as group k.

Increasing central concentration of the galaxies is represented in Morgan's system by a succession of groups: a, af, f, fg, g, gk, and k, the lettering being in the same order as in the sequence of types of stellar spectra. The particular population group for a galaxy is followed by a capital letter denoting the form: S for normal spiral, B for barred spiral, E for elliptical, and so on. Finally, a number from 1 to 7 denotes the inclination to the plane of the sky, from flatwise, or nearly spherical, galaxies to edgewise presentation of flat objects. Thus the barred spiral NGC 1300, having rather small central concentration and young stellar population, is classified by Morgan as fB2. The normal spiral M 31, having high central concentration and an old population, is classed as kS5.

The population group assigned in each case is determined entirely by inspection of the central concentration of luminosity of the galaxy, but generally an equivalence with the spectral type is expected. Such correspondence may provide significant data in studies of the evolution of galaxies.

18.13 Redshifts

Most galaxies like stars become very interesting individually if studied carefully enough. Those that first attract our attention are apt to be abnormal. For example, the strong nonthermal radio source 3C8A was finally identified with NGC 1275, which when studied carefully showed high-excitation forbidden lines of oxygen, neon, carbon, etc. This was considered strange and peculiar until it was discovered that all Seyfert and N galaxies showed the same lines in the visual regions. They differed only in the radio and the infrared regions. Many apparently normal galaxies can also exhibit high-excitation spectra in their centers.

As galaxies of similar types become fainter and smaller and presumably more remote, a spectral feature becomes more and more obvious—the lines become more and more shifted to longer wavelengths, that is, to the red side of the spectrum. The redshift of the galaxies was discovered by V. M. Slipher in 1912 and is generally interpreted as a Doppler shift. We have already seen that this relation is given by the equation

$$\frac{\Delta\lambda}{\lambda} = \frac{v}{c}$$

where $\Delta\lambda$ is the shift in wavelength units from the normal wavelength λ, v is the velocity of the object, and c is the velocity of light. For velocities where v is much smaller than c, this simple relation holds to a high degree of accuracy. Thus a shift of 1 A in the titanium line at λ 4300 Å corresponds to a velocity of 70 kilometers per second.

For large shifts in the spectrum this simple formula no longer holds. If the shifts are due to large velocities, we must appeal to the theory of relativity where the Doppler shift is given by the relation

$$\frac{\lambda'}{\lambda} = \frac{1 - (v/c)\cos\theta}{(1 - v^2/c^2)^{1/2}}$$

where λ' is the measured wavelength, λ is the laboratory wavelength of the element, and θ is the angle between the line of sight to the observer and the direction of motion as seen from the source. We note in passing that this reduces to the simple Doppler relation in the first order. A little manipulation places this in the previous form and gives (for $\theta = 180°$)

$$\frac{\Delta\lambda}{\lambda} = \left(\frac{1 + v/c}{1 - v/c}\right)^{1/2} - 1$$

We can immediately see the difference between the two relations by calculating a few values as has been done in Table 18.2. Astronomers often use the abbreviation $Z = \Delta\lambda/\lambda$.

For velocities of 30,000 kilometers per second or less, the differences between the results from the two relations are insignificant (Table 18.2). However, for an object like the quasi-stellar object 3C9 (Fig. 18.18) with a measured $\Delta\lambda/\lambda$ ($= Z$) of 2, the differences are quite large. The first formula would require a velocity of twice the speed of light, which means that we would never be able to see the source we are seeing.

TABLE 18.2

Velocity (km/sec)	$\Delta\lambda/\lambda$ First order	$\Delta\lambda/\lambda$ Relativistic
30,000	0.1	0.105
60,000	0.2	0.225
90,000	0.3	0.363
120,000	0.4	0.528
150,000	0.5	0.732
180,000	0.6	1.000
210,000	0.7	1.381
240,000	0.8	2.000
270,000	0.9	3.359

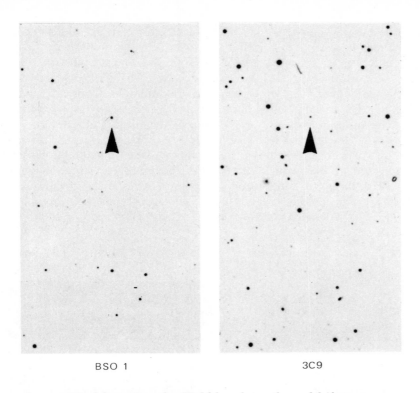

BSO 1   3C9

FIGURE 18.18
Two quasi-stellar objects. The one on the left is called a blue stellar object (BSO) because it is radio quiet but has a redshift equal to 3C9 which is shown on the right. Quasi-stellar object 3C9 has the same color as BSO 1, but is a strong radio source.

An empirical law given by Hubble relates the redshift or, more correctly, the velocity of recession to the distance of the object,

$$v = H \times r$$

Here v is the velocity of the source, H is the Hubble constant, and r is the distance. An approximate value of H is 50 kilometers per second per megaparsec, and r is given in units of millions of parsecs. For our example of 3C9 we obtain a value for its velocity of 240,000 kilometers per second from Table 18.2 and from Hubble's relation we find the distance to be 4.8 billion parsecs, or 16 billion light years. According to the Hubble relation the radius of the observable universe is 6 billion parsecs or 19 billion light years.

It is important to note the direct linkage between distance and time. When we observe a redshift that indicates an object is 19 billion light years away we are observing it as it was 19 billion years ago. Galaxies having velocities that are large with respect to the velocity of light are said to be at cosmological distances. We generally consider velocities of one-tenth the velocity of light and larger to be large with respect to the velocity of light. Studies of galaxies with large redshifts require great care in order to treat them in the same way that we do nearby galaxies. Their energy curves must be corrected for the redshift, for example.

Redshifts (Fig. 18.19) have been observed for many galaxies since Slipher's first observations. In addition, W. Baum has completed a uniform set of observations by means of 6-color photometry in order to obtain the bolometric magnitudes of galaxies. His diagram relating these magnitudes to redshifts (i.e., distances) for galaxies in 8 clusters is shown in Fig. 18.20. The points fall on a straight line, which agrees with the Hubble relation. Cluster 1410, the most remote of the 8 points, corresponds to the quasi-stellar object 3C295 and has a velocity of recession of about 132,000 kilometers per second. Possibly other points could be

FIGURE 18.19
Redshifts of galaxies in five different clusters at increasing distances. (Photographs from the Hale Observatories.)

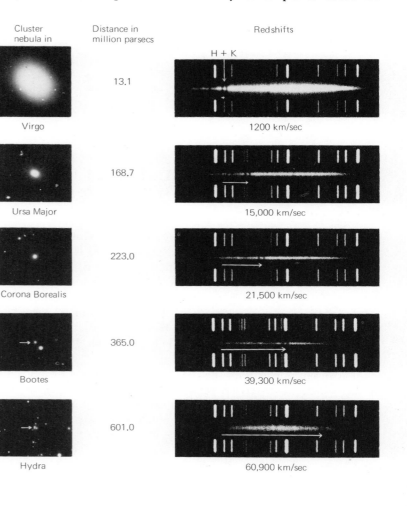

Cluster nebula in	Distance in million parsecs	Redshifts
Virgo	13.1	1200 km/sec
Ursa Major	168.7	15,000 km/sec
Corona Borealis	223.0	21,500 km/sec
Bootes	365.0	39,300 km/sec
Hydra	601.0	60,900 km/sec

placed on this diagram by including other quasi-stellar objects. However, we should realize that some of the redshifts may not be due to velocity, especially those of certain quasi-stellar objects. In such cases we will have to find a new explanation of their shifts. It would be a remarkable coincidence if Baum's magnitude–distance diagram held for the quasi-stellar objects and if subsequent findings proved that these objects are not at cosmological distances.

18.14 Quasi-stellar objects

In 1960 radio astronomers called attention to several radio sources with unusual characteristics. Despite increasing resolution the sources could not be resolved and their energy spectra seemed to have nonthermal components in the long-wavelength spectrum. Five of the objects were soon identified with optical sources. Spectrograms were obtained that displayed unidentifiable emission lines and a weak continuum with an excess of ultraviolet light. M. Schmidt fitted the emission lines of 3C273 uniquely to hydrogen if one assumes a redshift of the spectral lines corresponding to 0.16 times the velocity of light. Other emission lines were then identified with forbidden transitions of oxygen and neon. Another such object 3C48 was found to have a redshift corresponding to 0.3 times the velocity of light and the object 3C9 has been found to have a redshift corresponding to 0.8 times the velocity of light. These objects show emission lines of ionized carbon, silicon, etc., and occasionally absorption lines of the same elements as can be seen in Fig. 18.21.

The measured values of the redshifts of many quasi-stellar objects soon exceeded 1 and then even 2. The largest redshift observed up to late 1973 was for a quasi-stellar object called OQ 172 having $Z = 3.53$, that is, a redshift that is 91 percent of the velocity of light. Quasi-stellar objects having emission line redshifts greater than about 0.7 show absorption lines and, in general, the absorption lines are very sharp. The absorption lines also yield redshifts, some differing quite markedly from the emission line redshifts and always in the sense of smaller values. For example, the emission line redshift of one object, PHL 938 is 1.955, whereas one set of absorption lines has a redshift of 1.906 and another a redshift of 0.613. PHL 957 shows eight different redshifts lying between $Z = 1.26$ and 2.66. How can the same object have three or eight different redshifts? How can the absorption lines differing by 1.3 be so sharp? The physics of this problem is too detailed for discussion in this text and always leads to results open to interpretations that remain contradictory.

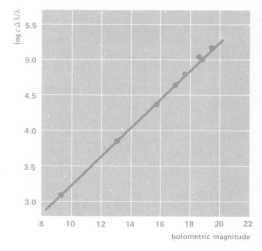

FIGURE 18.20
Velocity–distance relation for eight clusters of galaxies. From the relatively nearby Virgo cluster to the very remote Cluster 1410. For the mean of the galaxies observed in each cluster the logarithm of the rate of recession is plotted against the bolometric magnitude. (Diagram by W. A. Baum, Hale Observatories.)

FIGURE 18.21
The extremely redshifted quasi-stellar object 3C191 showing emission and absorption lines of highly ionized elements. The absorption lines are extremely sharp. (Kitt Peak National Observatory photograph.)

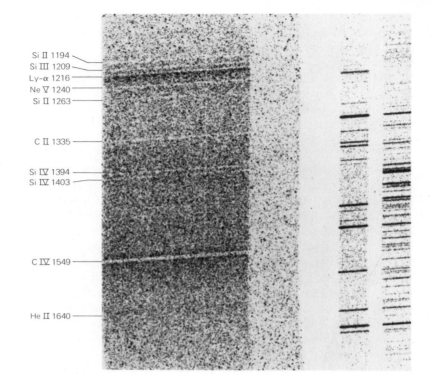

Here are unfamiliar objects, all of which are far away if we interpret the redshift in the customary cosmological way. Furthermore, they have very small dimensions but great radio and optical power output, as Table 18.3 shows. Our knowledge of spectroscopy requires that the forbidden transitions originate in a high-temperature, low-density condition. Greenstein has pointed out that the spectra are quite like those of planetary nebulae (see Section 14.4).

There might be other explanations of these objects. Arp has pointed out that quasi-stellar objects and peculiar galaxies occur together with a frequency greater than that of a chance coincidence; therefore certain quasi-stellar objects are associated with peculiar galaxies that we know are not at extremely great distances. Therefore the redshifts are not cosmological, that is, a result of Hubble's law. If this is so we must then find an explanation for the large redshifts of these objects.

Several hypotheses have been advanced. It has been argued

TABLE 18.3
Energy output of sources

Type of source	Radio power (ergs/sec)	Optical power (ergs/sec)
Normal galaxies	10^{38}	10^{44}
Radio galaxies	10^{43}	10^{44}
Quasars	10^{44}	10^{46}

that the redshifts are due to gravitational collapse of a large mass. However, J. Greenstein and M. Schmidt have shown that this cannot be the case from energy considerations. In addition, this mechanism does not reproduce the sharp emission lines, so for the moment the gravitational argument does not appear promising.

Arp, who has gone further than anyone else in trying to find an empirical explanation, has suggested that a catastrophic event caused two highly condensed objects to be expelled from a galaxy in opposite directions. Such a catastrophic event would explain the peculiar galaxy well enough as well as the paired symmetry that Arp finds in his studies. However, it would appear that some of these condensed objects should be expelled more or less toward us and that we should see almost as many blueshifts as redshifts among the quasi-stellar objects. So far we have not observed any blueshifts. Also, energy considerations show that any event powerful enough to eject such objects would undoubtedly rend the entire galaxy asunder. Along the same lines, J. Terrell has proposed that the source of the objects was the nucleus of the Galaxy. We see all the objects redshifted because the event took place some 10^7 years ago and hence they have long since passed the position of the sun in the Galaxy. If this were the case some of the objects might exhibit proper motions, which they do not.

Also suggested by Arp is the possibility that the quasi-stellar objects are plasma balls formed well outside a galaxy and are now falling into the nearest galaxy. Such a cause could hardly exhibit the symmetry Arp finds. We should also see such objects blueshifted as well and in about equal numbers which, as pointed out above, we do not. All of which indicates that we really do not have an explanation for the quasi-stellar objects that stands up on all counts.

In this discussion we have presumed that redshifts are caused by Doppler shifts. It may well be that there is another cause for the observed redshifts. Then Arp's final suggestion, "some as yet unknown cause," would be correct. Such a result would have momentous and monumental repercussions throughout astronomy and physics.

18.15 Other features of the Universe

So far we have only presented one large-scale feature and one characteristic feature of the Universe: clustering of galaxies and the correlation of redshift with distance. We have in fact presented a third, back in Chapter 1: *It gets dark at night*.

The fact that it gets dark at night leads to a paradox if one makes certain reasonable assumptions. This was first pointed out

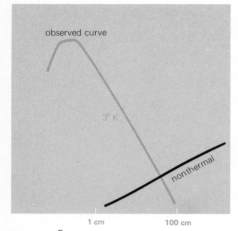

observed curve

3° K

nonthermal

1 cm   100 cm

FIGURE 18.22
*The 3°K isotropic radiation curve observed
in the microwave region of the spectrum.*

by H. Olbers in 1826. He reasoned that in an infinite universe the line of sight will eventually intercept a star, hence the celestial sphere should be uniformly bright, roughly equal to the brightness of the sun. It gets dark at night, however, and Olbers' assumption must be modified in some way to explain this. There are a number of explanations that will remove Olbers' paradox the most simple one being an expanding universe (Section 19.2).

A fourth general feature of the Universe has been stated repeatedly, 70 percent or more of the mass of the Universe is hydrogen. This percentage is remarkably constant as is the ratio of hydrogen atoms to helium atoms, which is slightly more than 9 to 1.

Recently, in 1965, A. Penzias and R. Wilson observing the sky at centimeter wavelengths noted an isotropic radiation field. Everywhere they pointed their telescope they obtained the same intensity of radiation. Careful fitting showed that their observations could be satisfied by a thermal radiation field having a temperature of 2.7°K (Fig. 18.22).

An isotropic radiation field has also been observed in the X-ray region with remarkable consistency on every X-ray rocket and satellite flight. This observation remains to be interpreted although it has been suggested that it arises from electrons ejected from galaxies striking a very hot (10^8 °K) **intergalactic** gas.

Years ago Hubble tried to count galaxies down to various brightness levels, the purpose being to see if the distribution of galaxies is uniform in space or whether significant departures from a uniform distribution occur. The task was far too time consuming and Sandage has shown that such studies using plates to the limit of the 200-inch telescope would not reach the distances required. Counts of radio sources, however, can be carried out almost automatically and many such counts have been made. Tentatively, there seem to be too many faint radio sources, meaning that there were more radio sources in the past then there are now.

QUESTIONS

1 Define cosmology.
2 What caused the necessity for the great debate?
3 Why is the Hubble classification of galaxies not an evolutionary classification?
4 Describe the principal features of a typical spiral galaxy.
5 What type of galaxies are the Large and Small Magellanic clouds?
6 Describe a typical Seyfert galaxy.

7 How de we obtain the distances to distant galaxies?
8 What are the Zwicky classes of clusters of galaxies? What are the Abell classes and their features?
9 Why are velocities that appear to be twice the velocity of light allowed in relativistic mechanics but not in Newtonian mechanics?
10 List the seven significant features about the Universe as a whole. Can you think of additional ones?

FURTHER READINGS

GAMOW, GEORGE, *The Creation of the Universe*, Viking, New York, 1952.

HODGE, P. W., *Galaxies and Cosmology*, McGraw-Hill, New York, 1966.

HUBBLE, EDWIN P., *The Realm of the Nebulae*, Yale Univ. Press, New Haven, 1936; paperbound, Dover, New York.

SASLAW, W. C. AND K. C. JACOBS, EDS., *The Emerging Universe*, Univ. Press of Virginia, Charlottesville, 1972.

SHAPLEY, HARLOW, *Galaxies*, 3rd ed., Harvard Univ. Press, Cambridge, Mass., 1972.

—— *The Inner Metagalaxy*, Yale Univ. Press, New Haven, 1957.

SWIHART, T. L., *Astrophysics and Stellar Astronomy*, Wiley, New York, 1968, Chap. IV.

The Scientific Endeavor, Centennial Celebration of the National Academy of Sciences, Rockefeller Institute Press, New York.

An engraving dating from about 1750 depicting the inhabitants of the moon. The notion of life on other worlds persists to the present day, but such ideas are no longer used as vehicles of satire as they were in this engraving.

Cosmogony

19

19

Cosmogony is the study of the origin and evolution of the Universe. Using the redshifts as a measure of distance we find an expanding universe. This, plus a knowledge of the distribution of various types of galaxies and objects in the Universe, allows an interpretation of the type of universe we live in and its age. To accomplish the necessary observations we must make use of instruments in space and intensify our ground-based efforts.

At the end of Chapter 18 we presented the general observations concerning the Universe as a whole. We deliberately left out the quasi-stellar objects because their true nature is not yet certain. We now look at the theories that attempt to explain the observations.

Any one wishing to discuss cosmogony in a rational way must begin from some certain well defined rule. This is similar to baseball where we decide in advance that a fair ball hit beyond a certain distance on the fly is declared a home run, counting as one run (although it may result in as many as four runs). Imagine the chaotic results if you could run around the bases until the original ball was retrieved? So it is in cosmogony, we begin with the rule that the cosmological principle must be satisfied. The **cosmological principle** states, simply, that the Universe must look the same at a given time to an observer located anywhere within the Universe. Some propose a much more stringent condition. This condition, called the **perfect cosmological principle**, requires that the Universe look the same to any observer regardless of time.

It goes almost without saying that we reject any theory that does not begin from some *reasonable* rule and produce testable predictions. Any rule that allows an infinite (or even a finite) number of exceptions must be rejected.

19.1 Some theories

The perfect cosmological principle really allows for only one theory, the so-called steady-state theory proposed by H. Bondi and T. Gold, which has been eloquently defended by F. Hoyle.

One of our most convincing observations is that the Universe is expanding. If it expands beyond the observable horizon then matter must be created to replace that which is going over the horizon. The rate of creation is very small and essentially untestable. The perfect cosmological principle requires that the Universe look the same everywhere, therefore, the number of galaxies of any given type near and far must be the same. Any significant deviation proves the theory wrong; the Universe on a large scale must be homogenous.

On the other hand, starting from the cosmological principle several universes are predicted, an essentially static universe, a universe starting from time zero (the creation or "a big bang") and moving to infinity, or a universe that expands to a certain size, contracts and then expands again, that is, a pulsating universe. The last two possibilities look the same in all ways except that the latter predicts that the expansion of the universe decelerates, stops, and then contracts.

These theories assume that the theory of general relativity holds, that the laws of physics apply everywhere, and that the universal constants of physics are valid. Indeed, these statements are implied by accepting the cosmological principle.

Other consistent theories can be developed that are based upon the large dimensionless numbers that appear in physics, as was first pointed out by P. Dirac. One of these numbers is the ratio of the electrical force to the gravitational force between the proton and electron, which has a value of 10^{40}. Another number is the ratio of the radius of the Universe to the radius of the electron, which also has the value 10^{40}, and so on. Dirac assumed that these ratios remained constant. Since the universe is expanding a consequence of this is that gravity, the mass of the proton, the charge on the electron, etc., must be changing. Dirac allowed the charge on the electron to vary with time. Later P. Jordan and more recently R. Dicke developed similar theories allowing the universal constant of gravity to vary with time. The possible variations on this theme are almost numberless.

Depending upon interpretation the Universe may be closed and unbounded and static, or closed and unbounded and dynamic (Figs. 19.1 and 19.2), or open and unbounded, and so on.

19.2 Tests

Now let us review the observations:

1 Galaxies exhibit redshifts, and the redshifts lead to the Hubble relation (Fig. 19.3).

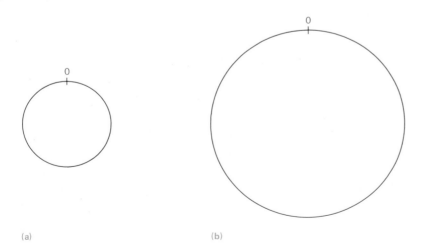

(a)   (b)

FIGURE 19.1
A circle is a one dimensional line connected upon itself by constant positive curvature. A circle describes an infinite closed universe. If the circle changes size it is descriptive of a dynamic universe, if not it is descriptive of a static universe.

FIGURE 19.2
A plane surface of positive curvature is a sphere and helps one visualize a two-dimensional infinite but closed (bounded) universe. This too can be thought of as a static or dynamic universe.

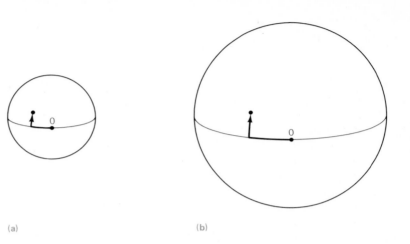

(a)   (b)

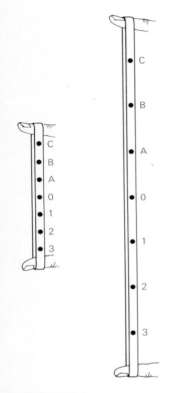

FIGURE 19.3
A schematic explanation of the Doppler redshift in galaxies. The points on a rubber-band (left) move away from each other uniformly as the rubberband is stretched (right). Point 1 has moved three times farther away from the arbitrary origin (O) during the time of stretching. The same is true for point 3; thus point 3 has had to move three times faster than point 1. This is true for any and all points with respect to all the others. Thus, any point on the rubberband will see itself as the origin and the points farthest from it will be moving the fastest.

2 Galaxies group into clusters.

3 It gets dark at night.

4 There is an isotropic background radiation consistant with a 2.7°K source.

5 There may be an isotropic X-ray background.

6 There seem to be more radio sources at cosmological distances than nearby.

These observations are used to test any theory concerning the origin and evolution of the Universe. They are not equally important since we don't know if point 2 (clustering of galaxies) is meaningful and we are not yet certain that point 5 (isotropic X-ray background) is even true. Point 1 tells us that the Universe is expanding and this alone rules out various static and Newtonian, that is, nonrelativistic, universes. It also explains Olber's paradox (observation 3).

Thus observations 1, 4, and 6 must tell us what type of universe we have. Observation 6 tells us that we are in an evolutionary universe. Earlier, 10 billion years or so ago, there were many more radio sources than there are at present.

In 1932 Alpher, Bethe, and Gamow hypothesized that all of the Universe was originally concentrated in a primeval "atom" that became unstable and exploded. This has become known as the "big-bang" theory. They predicted that the cooling primeval fireball should leave the universe permeated with radiation at a temperature of nearly 3°K. Expressed in another way, the 2.7°K isotropic radiation is the primeval fireball seen at a redshift of

$Z = 1000$. The discovery of the 2.7°K isotropic background radiation by Penzias and Wilson lends strong support to the "big-bang" origin.

The radio source counts may be more revealing. The number of sources, assuming uniform density, should be proportional to the volume investigated, which is proportional to the cube of the radius. The intensity of the radiation decreases as the square of the radius, that is, it is inversely proportional to the square of the radius. Hence in simple three-dimensional space, the number of sources should be inversely proportional to the three-halves power of the intensity. Symbolically

$$N \propto S^{-3/2}$$

where N is the number of sources and S is the intensity (in radio terminology the flux). Thus, if we plot the logarithm of N against the logarithm of S a straight line of slope $-3/2$ results, if we are in a simple constant universe. Any deviation is significant and indicates an evolving universe. When distances become large in an expanding universe our relation becomes

$$N \propto S^{-3/2 + \alpha}$$

where α is always positive, but depends upon the model of the universe selected. At very faint (weak) intensity levels the slope of the line appears to be -1.8, which substantiates the argument against a constant density of radio sources. There appear to be too many sources in the distant past.

In models of the Universe employing general relativity, a factor called the acceleration parameter, q_0, appears and is critical. Its value determines which model is correct. It is the quantity observable in the sense that it can be determined from observation, and considerable effort is going into its evaluation. If $q_0 = -1$, the Universe is a steady-state universe. If $0 < q_0 < \frac{1}{2}$, the Universe is hyperbolic, if $q_0 = \frac{1}{2}$ the Universe is Euclidian (obeys simple, straight-line geometry), and if $q_0 > \frac{1}{2}$, the Universe is spherical and decelerating. The results according to Sandage are shown in Fig. 19.4. Sandage's value for q_0 is $+0.95$.

The weight of the general character of the Universe seems to favor a big-bang or pulsating model. It is best to reserve judgment on this matter, however.

19.3 Other considerations

We have not made any use of the ubiquitous quasi-stellar sources. If they are cosmological they clearly indicate an evolutionary universe and rule out steady-state models. Most astronomers will

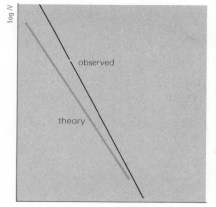

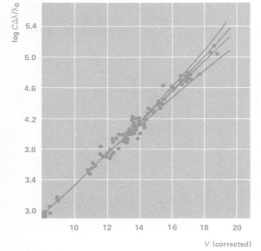

FIGURE 19.4
*Sandage's graph of corrected magnitude
against redshift for the brightest galaxy in
84 clusters. The lines are for different values
of the acceleration constant, q_0. The points
seem to favor a decelerating universe, which
leads to a pulsating model of the Universe.
(Reproduced from the Astrophysical Journal,
by permission.)*

not commit themselves, but Sandage in a series of very careful
papers concludes that the redshifts of the quasi-stellar objects are
cosmological in origin. The upper limit on their redshifts is 3.5,
which represents about 91 percent of the age of the Universe. Thus
the quasi-stellar objects were "turned on" well after the universe
was formed. They apparently represent a critical stage in the history of the Universe. Why this stage is not closer to a redshift of
1000 raises new questions for us to answer. On the grand scale
our observations have an enormous gap between redshifts of 3
and 1000. It is essential that we fill in this gap.

ORIGIN AND EVOLUTION

The cosmogonical theories, in general, assume a uniform, isotropic, expanding universe. They do not really allow for the
irregularities that become galaxies and clusters of galaxies, let
alone stars and life.

19.4 The origin and evolution of galaxies

Galaxies exist therefore they originated somehow. We do not
know how. W. C. Saslaw assumes an early instability, perhaps
during the formation of the elements a few minutes after the
Universe began. These instabilities remained and eventually
became detached clouds that condensed into galaxies or, more
likely, clusters of galaxies. Once galaxies exist we can then follow
with reasonable certainty their evolution, the formation and
evolution of stars, etc. Galaxies evolve very slowly.

The three basic types of galaxies must be related to their early
history (just as with stars), and the only conceivable difference
would be their initial rotation. If the rotation was small, an
elliptical galaxy results; if the rotation was intermediate, a spiral
galaxy results; and if the rotation of the cloud was high, an irregular galaxy formed.

In a galaxy stars will be formed, evolve, return material to the
galaxy, new stars will form, evolve, and so on. Near any one of
these myriads of stars, planets may form, and life may exist.

19.5 The conditions of life

We do not really propose discussing in great detail all of the
aspects of the origin of life. In this context we refer to life as a
branch of carbon chemistry and assume that we are interested in
"intelligent" life after establishing the chemistry.

We have seen how stars are formed. A cloud that forms stars always forms many stars, most in multiple systems but some as single stars. These single stars have a large revolving cloud in their plane of rotation with a cosmic composition that includes the interstellar molecules. Some planets can form from this cloud, just how and how many is a matter of opinion at the moment. The number can be anything equal to or greater than zero.

In the case of the solar system we are fairly certain the terrestial planets formed by accretion; the gaseous planets may be original condensations or may also have formed by accretion. We know that the earth is about 5×10^9 years old and that 3.6×10^9 years ago a great event occurred, perhaps the capture of the moon, which may or may not be important to life. Life on earth began after this event, perhaps 3×10^9 years ago, thus the evolution of life requires something on the order of 3×10^9 years.

This sequence of events can occur anywhere in the Universe. There is no reason to believe that the sun and its history is unique in any way.

Of course the conditions must be reasonably hospitable wherever life begins. If the initial step occurs by chance, then the conditions should be such that many chances take place. Temperature and pressure must not change too greatly nor too rapidly on a short time scale. For a given set of conditions those life forms which adapt most readily survive. The evolution of life forms will naturally converge toward the most efficient forms possible.

One of the conditions only recently appreciated is the role of the fundamental constants of physics. If these constants are varied only slightly our carbon chemistry will not work, the region of variation is highly restrictive. In a Dirac-type universe (Section 19.1), life would exist only during the phase when the physical constants were balanced just right.

We cannot pursue this problem as it is beyond the scope of the book. Some hint into the problem can be given, however. The fundamental constants are the coupling constants for electromagnetic interactions, gravitational interactions, the strong and weak nuclear interactions. Very small changes in these constants cause significant changes in the charge on the electron that greatly effect the chemistry of the amino acids (subunits of proteins), for example. If the charge becomes too small the amino acid bonds break, if the charge becomes too large the amino acids crystallize. Thus such a universe, where the constants are slowly varying, will allow for life based on a carbon chemistry only over the interval where the constants are close to their present values. A special condition such as this seems too fortuitous to be true.

19.6 Does life exist elsewhere?

This intriguing question has no answer at the present time. If you ask this question of any person his answer will merely reflect his personal bias. All astronomers can do is exhaustively determine whether the conditions for life exist elsewhere. A sufficient proof for life elsewhere in the Universe would be the reception of an understandable communication. This is not an easy criteria. Communication techniques change quite rapidly, thus the development of a civilization must reach the right stage at the right time and distance or communication will not occur.

On the other hand, failure to receive a communication does not prove that life does not exist.

What are the conditions for life? Briefly summarized:

1 The proper central star, nonvariable, neither too energetic nor too feeble—an F through K main sequence star will do.

2 A planet with water located at the proper distance, the size and mass being within 50 percent of that of the earth.

3 Time—at least 3×10^9 years since the solidification of the planet.

4 A universe where the fundamental constants are constant.

It is then the task of the chemist and biologist to tell us if life will begin.

One of the exciting challenges for astronomy is to seek out those places where life may exist.

ASTRONOMY FROM SPACE

The advancing capability of man to place large stabilized payloads in orbit around the earth has led astronomers to employ this useful tool to press forward the frontiers of astronomy. Space astronomy can be classified into three areas: (1) solar research, (2) planetary research, and (3) stellar research, where the term stellar is used in the very broadest sense.

19.7 Solar research

The sun is known to be quite variable in the ultraviolet and X-ray region of the spectrum. This can be deduced from the effects of these wavelengths upon the atmosphere. The origin of this radiation is from high-temperature processes that are little understood.

Hence, a detailed study of the sun in the high-energy region can be most fruitful.

A start in this study has been made with the various Russian, American, and European satellites, such as OSO (Orbiting Solar Observatory). Additional information can be obtained from radiation probes scattered around the sun in positions far distant from the earth. After all, the earth samples only a small portion of the radiation emanating from the sun. Interplanetary probes, such as the Mariner (Fig. 19.5) and Venera series are starting such studies.

19.8 Planetary research

Besides studying radiation from the sun and the solar magnetic field, interplanetary probes allow astronomers a chance to sample interplanetary space itself. The size and mass of particles can be directly ascertained. Dramatic advances can be expected in our knowledge of the material making up the interplanetary medium.

Even more spectacular is the opportunity for obtaining close-up information of the planets. The recent Mariner photographs of Mars are an example of the ability of space probes to assist our studies by eliminating certain hypotheses from further consideration. We can expect similar results, possibly more exciting results from probes to other planets. The probe that would yield the most information would be one to Jupiter and such a probe,

FIGURE 19.5
Artist's conception of Pioneer F as it flew by Jupiter. This successful spacecraft passed Jupiter on 2 December 1973 and was accelerated on a long journey out of the solar system never to return. Pioneer F is an example how modern space technology has enabled astronomy to make giant strides toward the understanding of the solar system. (Drawing courtesy of the National Aeronautics and Space Administration.)

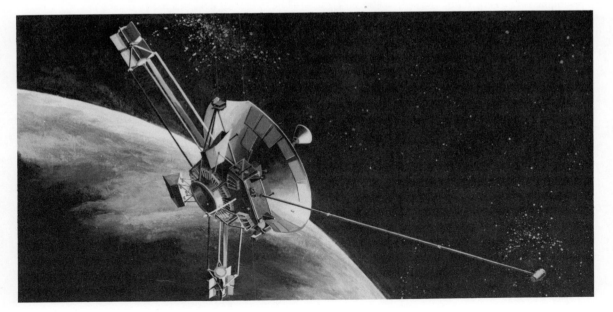

Pioneer F (called Pioneer 10) has completed the first part of its mission by taking several excellent pictures at various distances from Jupiter. Results from other experiments on the spacecraft are not yet available. Because of the long lead time for such probes the challenge of the observations and interpretations will be before astronomers for decades.

Perhaps with these new tools, we can find the answer to the origin of comets. Is there a tie between the remnant solar nebula where the comets originate and the interstellar medium nearby? Such information will perhaps bring us closer to information of the origin of the Solar System and perhaps even the origin of the Galaxy.

19.9 Stellar research

Study of the interstellar medium is already in progress with the Orbiting Astronomical Observatory (OAO) series (Fig. 19.6). This study and its bearing upon cosmology has already been mentioned. From measurements of the absorption of ultraviolet radia-

FIGURE 19.6
The highly successful spacecraft OAO-C in a prelaunch check out. This 4/5-meter telescope was dedicated to the study of the interstellar medium and was the most complex satellite ever launched. It reached and in most cases surpassed all of its goals. (National Aeronautics and Space Administration photograph.)

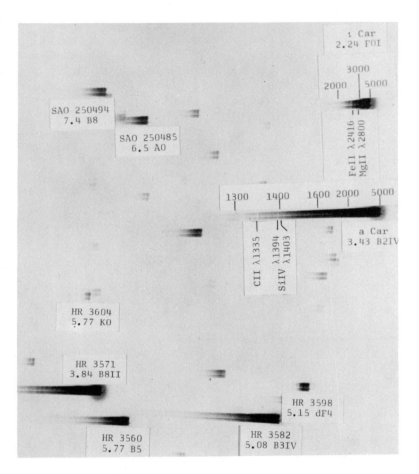

FIGURE 19.7
Ultraviolet stellar spectra photographed during the Skylab 1 mission (see color plates) using the University of Texas 15-centimeter aperture ultraviolet objective prism spectrograph. The field is in Carina and the wavelength region is identified on a Carina. (University of Texas, National Aeronautics and Space Administration photograph. Principal Investigator K. G. Henize, Astronaut's Office.)

tion we will be able to deduce the mass of interstellar hydrogen. Similar observations of high-temperature stars will have bearing upon the interaction of such stars on the interstellar medium surrounding them (Fig. 19.7).

The search for X-ray point sources is very important. Fortunately, recent X-ray rocket and satellite experiments have turned up some 130 galactic and extragalactic X-ray sources, a number of which are point sources. A search for γ-ray sources in similar experiments is underway but the results so far are less conclusive.

At the other end of the electromagnetic spectrum, fruitful areas of research in space are open. The atmosphere cuts off radiation above a wavelength of 30 meters. Strong sources of low-frequency radio radiation would be nonthermal in origin. We can already predict that the Crab nebula will be such a source. A detailed study of this nebula coupled with ground-based observations will shed a great deal of light on the energy sources operating in it.

There is no purpose served in rating the results from the various space probes, any such list would be subjective. The effect of space technology upon mankind is undeniable. One example is the revolution in communications between people. It is also revolutionizing astronomy.

Removing the spectacular aspects, the lunar sampling missions have provided answers to long-standing questions and, of course, raised new ones. The various Mariner missions to Mars, particularly Mariner 9 have completely changed many of our ideas concerning Mars. The highly successful X-ray satellite *UHURU* has opened a whole new area of study, answered a host of questions, and raised many more. And so on the story goes—Venus, the sun, and the interstellar medium have all been studied. Answers have been found and questions have been raised.

Further in the future we can anticipate the use of high-resolution, perhaps even diffraction-limited telescopes, some of which will be manned (Fig. 19.8). Such a telescope would add information to every area that we have studied so far. For example, it

FIGURE 19.8
Artist's conception of a diffraction limited large space telescope (3 meters) in orbit. The next step in space astronomy. (Drawing courtesy of the National Aeronautics and Space Administration.)

would resolve many galaxies into stars and allow a detailed study of their contents. The variable stars in these galaxies could be studied and hence more reliable distances obtained. A telescope with such capabilities would open up vistas undreamed of, even now. In fact, all space research can be expected to yield unexpected results and new problems for study.

ASTRONOMY FROM THE EARTH

In order for the full potential of the space astronomy effort to be realized, there must be a similar effort on the ground. Certain essential observations, especially those pertaining to tests of cosmological theories, require large telescopes and a considerable amount of time. To make progress more large- and intermediate-size telescopes are needed.

19.10 Optical research

As exciting as space astronomy is and will prove to be, ground-based astronomy is at least its equal. Space astronomy is relatively expensive, so we must use it wisely and apply it where other approaches are unavailable. We have already discussed one example where space astronomy can be applied to the specific task of looking at variables in galaxies unresolved from the ground. The integrated magnitudes for these space-resolved galaxies can be obtained cheaply from great telescopes on the earth and the statistical studies of fainter unresolved galaxies carried out. Thus, one should make maximum use of the two capabilities.

Another example is in the study of the masses of stars. A thorough study of the motion of an unresolved astrometric binary from the ground yields only the mass function. A few observations of the resolved system, coupled with the ground-based study, would quickly yield the separate masses.

In those areas where light-gathering power is the criteria, such as, spectroscopy and photometry, ground-based work will press forward with newer and many more large telescopes over the next few decades. The detailed study of stellar spectra, populations, and variation will go on at an ever-increasing pace. Even the study of **astrometry** (parallaxes, motions of stars, time and fundamental reference systems) promises new and interesting advances. The success of the U.S. Naval Observatory's 61-inch (Fig. 19.9) reflector in obtaining the parallaxes of stars down to the 16th magnitude argues well for this branch of study. The establishment of

FIGURE 19.9
A modern astrometric reflector dedicated to finding the distances of the nearby stars. (Official U.S. Navy photograph.)

an absolute reference frame by using external galaxies is another example of work best carried out from the surface of the earth.

The life cycles of stars can be used to predict the evolution of a galaxy. The low-mass red stars may live 2×10^{10} years and in the beginning of the life of a galaxy contribute very little to its brightness and color. The galaxy would be dominated by the light of its massive, bright, blue stars. As evolution proceeds fewer such stars are born. As these stars rapidly evolve and burn out, the brightness of a galaxy must diminish until, finally, it is dominated by the light of the faint red stars. Such ideas as this can be tested within the next 100 years. However, before the above ideas can be fully tested, we must understand how a star changes in the H-R diagram as it evolves. To do this there must be thorough observational studies of all types of stars in minute detail. This means large telescopes where coudé plates can be taken of faint stars (Fig. 19.10). Only now are we starting to make a beginning toward understanding the peculiar A stars and we have not even scratched the surface of the variable star problem. The observations must be obtained in order that the theoreticians will have a reasonable footing for their work.

Another most interesting problem of high priority is a thorough explanation of the quasi-stellar sources already discussed. The key to the explanation may well lie in the determination of their true angular size, a problem that may tax the imagination of astronomers for decades to come. Most observations needed for testing cosmological consequences can be carried out from the ground. Intermediate- and large-size telescopes are needed to carry this work forward. Just as important, of course, is the need for trained astronomers to use these instruments and interpret the results.

19.11 Radio research

This relatively young branch of astronomy is enjoying a healthy growth that will continue over several decades, and the search for weak sources will continue with more and more refined equipment. The study of known sources over a greater spectral range should intensify as techniques develop.

The desire for a more detailed study of the structure of sources is leading to the rapid refinement of radio interferometry. There is the hope now that a huge array of large steerable parabolas can be constructed to yield very high resolution (Fig. 19.11). The success of the large interferometer of the National Radio Astronomy Observatory at Green Bank makes this next step logical. An alternative approach is to tape record signals received by two dis-

FIGURE 19.10
The 4-meter Mayall reflector dedicated in 1973 and placed in regular operation during the same year. This great reflector with others being built will greatly enhance man's observational capabilities. (Kitt Peak National Observatory photograph.)

tant telescopes and to bring the tapes to a common laboratory for analysis. The tapes can be synchronized by inserting time marks from previously synchronized atomic clocks (see Section 4.9) as the recordings are being made. Such experiments by Canadian and National Radio Astronomy Observatory astronomers have been successful. These recent experiments have shown that many quasi-stellar objects have diameters of less than 0.01 second of arc. Similar experiments using intercontinental baselines are becoming a matter of routine.

19.12 Theoretical research

There are two factors that have shaped and accelerated theoretical investigations in astronomy, namely, the rapidly increasing num-

FIGURE 19.11
Artist's conception of the new telescope called the very large array (VLA). Each leg of the Y of this instrument is about 7 kilometers long. The instrument will be usable at very short wavelengths as an interferometer, a supersynthesis telescope, and as individual telescopes. The antennae are paraboloids 25 meters in diameter. (Drawing courtesy of the National Radio Astronomy Observatory.)

ber and quality of observations and the development of modern computers and techniques (Fig. 19.12). In the first factor we include the results of physicists in determining the lifetimes of energy states and the like. In the latter we include the calculation of important nuclear cross sections and other long, laborious calculations.

The carrying through of the calculations of the declining era of a star was impossible prior to the development of modern computer technology. We can look forward to even more exact computations of model stellar atmospheres, interiors, and evolutionary tracks.

19.13 Summation

The oldest of the sciences is presently experiencing a growth that is unparalleled in all its long history. No one denies that this growth was accelerated with the coming of the space age, but the results have merited the growth. The discovery, study, and interpretation of the quasi-stellar radio sources, not to mention pulsars, are good examples.

The excitement and promise of modern astronomy is breathtaking to behold. Perhaps the most recent comparable period in this science was during the early part of this century when V. M. Slipher discovered the receding galaxies and H. Shapley discovered the true extent of the Milky Way. These great steps had been preceded by the H-R diagram, the discovery of the period–luminosity relation, and fundamental work in atomic physics, both theoretical and in the laboratory.

We are now witnessing major advances in nuclear physics and we have before us the discovery of the quasi-stellar sources. We can now look forward to the new surprises and expanding knowledge that always accompany major developments.

FIGURE 19.12
A large astronomical computer facility. Computers are used to reduce data, guide telescopes, and calculate detailed models of stellar atmospheres, stellar interiors, and stellar encounters. Many formerly unsolvable problems or problems solved by approximation can now be solved rigorously by applying large computers.

QUESTIONS

1 What is the cosmological principle and why is it central to cosmogony?
2 Give four theories concerning the evolution to the universe.
3 What is the role of source counts in cosmogony? Can you think of a similar situation earlier in the history of astronomy?
4 Why is the parameter q_0 so important to the cosmologist?
5 Explain why Field wishes to fill in the redshift gap between the quasi-stellar objects and the 2.7°K isotropic radiation.
6 What are the basic conditions for life as we know it?
7 Give some arguments for believing life exists elsewhere in the universe. Give some opposing arguments.
8 Which type of space probe do you think has been most successful? Why?
9 How can a space telescope advance ground-based observations?
10 NASA has a large space telescope (3-meter aperture) under intensive study. Can you list five ways such a telescope will advance our knowledge?

FURTHER READINGS

BONDI, HERMANN, *Cosmology*, Cambridge University Press, New York, 1960.

GAMOW, GEORGE, *The Creation of the Universe*, Viking, New York, 1952.

GLASSTONE, SAMUEL, *Sourcebook on the Space Sciences*, Van Nostrand, New York, 1965.

HODGE, P. W., *Galaxies and Cosmology*, McGraw-Hill, New York, 1966.

MCVITTIE, G. C., *Fact and Theory in Cosmology*, Eyre and Spottiswoode, London, 1961.

SASLAW, W. C., AND K. C. JACOBS, EDS., *The Emerging Universe*, Univ. Press of Virginia, Charlottesville, 1972, Chaps. 5, 7–10.

The Scientific Endeavor, Centennial Celebration of the National Academy of Sciences, Rockefeller Institute Press, New York.

SNEATH, P. H. A., *Planets and Life*, Thames and Hudson, London, 1970.

SWIHART, T. L., *Astrophysics and Stellar Astronomy*, Wiley, New York, 1968, Chap. IV.

WHITROW, G. J., *What Is Time?*, Thames and Hudson, London, 1970, Chaps. 4–8.

General References

Current contributions to the literature of astronomy appear in periodicals, such as:

The Astronomical Journal. Published at the rate of 10 numbers a year for the American Astronomical Society by the American Institute of Physics, 335 East 45th Street, New York, New York, 10027.

Astronomy and Astrophysics. Published monthly for the European Southern Observatories by Springer-Verlag, 1 Berlin 33, Heidelberger Platz 3 (Germany).

The Astrophysical Journal. An International Review of Spectroscopy and Astronomical Physics. Published biweekly for the American Astronomical Society by the University of Chicago Press, 5801 Ellis Avenue, Chicago, Illinois, 60637.

Griffith Observer. A popular booklet published monthly by the Griffith Observatory, P.O. Box 27787, Los Angeles, California, 90027.

The Journal of the British Astronomical Association. Published 10 times a year, 303 Bath Road, Hounsley West, Middlesex, England.

The Journal of the Royal Astronomical Society of Canada. Published bimonthly. 252 College Street, Toronto, Ontario. The Society also publishes for each year *The Observer's Handbook.* A useful reference for astronomical data and events of the ensuing year.

Mercury. Published bimonthly by the Astronomical Society of the Pacific, North American Hall, California Academy of Sciences, Golden Gate Park, San Francisco, California, 94118. The Society also publishes monthly *Leaflets* written in popular style by various astronomers.

Monthly Notices of the Royal Astronomical Society. Published monthly, Burlington House, London W. 1.

The Observatory. A review of astronomy. Published monthly. The Editors, Royal Greenwich Observatory, Herstmonceux Castle, Hailsham, Sussex, England.

Scientific American. Contains frequent articles on astronomy and allied sciences. Published monthly by Scientific American, 415 Madison Avenue, New York. Easily readable semitechnical reviews.

Sky and Telescope. Published monthly. Sky Publishing Corporation. Harvard Observatory, Cambridge, Massachusetts. Easily readable technical and popular articles.

For students interested in astronomy as a career, a booklet entitled "A Career in Astronomy" is available from:

The Executive Officer
American Astronomical Society
211 Fitz Randolph Road
Princeton, New Jersey 08540

Glossary

A

Acceleration A change in velocity of a body.

Achromat A compound lens of two or more simple lenses corrected to remove chromatic aberration over a broad region of the visual spectrum.

Albedo The reflectivity of a body compared to the reflectivity of a perfect reflector of the same size, shape, orientation, and distance. Usually given as a percentage.

Almanac A yearbook containing astronomical tables of positions of celestial bodies, sunrise, sunset, moonrise and moonset, etc. The American Ephemeris and Nautical Almanac are the almanacs most used by astronomers in the United States.

Apastron That point in the true relative orbit farthest from the primary star.

Aperture Technically the clear diameter of a telescope, but often the diameter of the principal element or of the entrance pupil.

Apex Usually the solar apex, the direction toward which the sun is moving. The apex of a cluster's motion is either the convergent point or the divergent point.

Aphelion The farthest point from the center of the sun in a solar orbit.

Apogee The farthest point from the center of the earth in an earth orbit.

Apsides The two points where the major axis of an orbit intersects the orbit.

Asteroid A minor planet.

Astrometry The measurement of stellar and planetary positions.

Astronomical unit The mean radius of the earth's orbit.

Astrophysics The study of the physical properties of the sun, planets, and stars. It includes solar physics and planetary physics.

Atmosphere The low-density, gaseous sphere surrounding the earth and various planets. Also the gaseous layers lying above a star's photosphere.

B

Balmer limit The atomic transition from the second electron orbit of hydrogen to the continuum. The confluence of the Balmer lines near this point in a stellar spectrum.

Binary (binary star) A pair of stars orbiting each other by their mutual gravitational attraction.

Bipolar Having two poles. A magnetic field having both a plus pole and a minus pole is said to be a bipolar magnetic field.

Bolometer A device to measure infrared radiation.

Bolometric magnitude The magnitude assigned to a body when the total energy from far-ultraviolet to long radio waves is used.

C

Catalyst An agent that assists or speeds a reaction with no change in itself in the end product. Carbon is a catalyst in the carbon–nitrogen cycle.

Celestial mechanics That branch of astronomy that deals with the orbits and motions of planets and stars.

Centrifugal force The tendency of a body in motion to move away from a centrally restraining force.

Cepheids Intrinsic variable stars with periods between 1 and 60 days.

Chromosphere That part of a stellar atmosphere lying just above the photosphere. It is about 15,000 kilometers thick.

Circumference Usually the measured boundary of a plane figure.

Coelostat A stationary telescope looking at a flat mirror that is in a polar mount. The drive is arranged to keep the sun or a star field fixed in the telescope.

Collimator Optical lens (lenses) and/or mirror (mirrors) designed to render light in parallel rays. A telescope is a reverse collimator.

Color excess The difference between the measured color of a star and the normal color for a star of its type.

Color index The difference in magnitudes between intensities measured in two different colors. More precisely, the difference between visual and photographic magnitudes.

Conjunction Two celestial bodies on the same longitude. Inferior conjunction occurs when a planet or body passes between the earth and sun. Superior conjunction occurs when a planet is on the opposite side of the sun from the earth.

Constellation A historical grouping of the bright stars. Constellations are useful for indicating specific portions of the celestial sphere.

Convection In astronomy, the transfer of energy by the bodily transport of a hot plasma to a cooler region where it can cool by radiation.

Coronagraph A telescope so arranged as to block out the bright disk of the sun in order to photograph the corona.

Cosmogony Theories and hypotheses dealing with the origin of the Universe.

Cosmology The study and measurement of the Universe.

Coudé focus A fixed focus in a room in line with and south of the polar axis.

D

Declination The astronomical coordinate measuring the position of a body above (+ or N) and below (− or S) the celestial equator. It is always given in circular measure (e.g., degrees).

Density (1) The mass of a body divided by its volume (units are usually grams per cubic centimeter). (2) The number of objects per unit volume, such as the number of G stars per cubic parsec.

Distance modulus The apparent magnitude of a star minus its absolute magnitude, $m - M$. The term is derived from the distance relation: $m - M = 5 \log r - 5$.

Diurnal An event taking place each day.

E

Eccentricity The measure of the degree of flattening of an ellipse. The letter e is generally used in astronomy to denote this parameter.

Eclipse A celestial body passing between the observer and another body.

Ecliptic The path the sun seems to follow and, therefore, where solar eclipses tend to occur. Actually the plane of the earth's orbit projected to the celestial sphere.

Elongation The angular distance from the center of motion, generally a planet's angular distance from the sun.

Ephemeris A table of predicted positions. Generally a table of predicted positions as contained in the American Ephemeris and Nautical Almanac.

Ephemeris time The time interval of one second as based upon the earth's annual orbit in 1900. Clocks using this second are keeping ephemeris time.

Equinox A point where the sun's path (ecliptic) crosses the celestial equator. When the sun is moving north it occurs around 22 March and is called the vernal equinox; when the sun is going south it occurs around 22 September and is called the autumnal equinox.

Erg The amount of work done by a mass of 1 gram moving 1 centimeter.

Escape velocity The velocity needed to escape the gravitational attraction of a body. It is equal to the parabolic orbital velocity, $v_p = 2G(m_1 + m_2)/a$.

F

Fluorescence The emission of light at a longer wavelength than that of the absorbed radiation.

Focal length The distance between the objective and its focus when looking at parallel incoming light.

Focus The point where incoming radiation is imaged by an optical system.

Forbidden transitions Changes in energy levels not normally allowed of a body.

Fundamental stars A set of carefully chosen stars distributed over the whole sky whose positions are carefully measured and which serve as positional standards for other stars and objects.

Fusion (1) The nuclear process joining atoms together to form heavier elements and releasing energy. (2) The point of change from solid to liquid.

G

Galaxy (1) A great assemblage of billions of stars, gas, and dust gravitationally restrained. (2) When capitalized, the Milky Way galaxy is implied.

Gamma radiation That region of the spectrum shortward from a wavelength of 8 Å.

Geiger counter An electron tube that discharges each time a charged particle passes through it. It is used by astronomers to discover cosmic rays, X-rays, and gamma rays.

Geodesy The science dealing with measuring the size and shape of the earth.

Geomagnetic storm Disturbance of the earth's ionosphere and magnetic field by an unusual influx of charged particles from the sun.

Grating (1) Bars across the objective causing diffraction of the light. This would be called an objective grating. (2) A ruled grating, usually blazed, used to diffract the light for spectroscopic purposes.

Gravity The term given to the attractive force between two masses.

H

Heliostat A coelostat used to observe the sun.

High-velocity stars Stars with high radial velocities with respect to the sun. In reality their large radial velocities reflect the high velocity of the sun with respect to them.

Hydrocarbon Compounds of carbon and hydrogen; methane, CH_4, is important to the gaseous planets.

I

Igneous rocks Rocks formed from molten materials.

Inertia The property of matter to remain in its current dynamical state unless acted upon by an outside force.

Infrared radiation That region of the spectrum extending from a wavelength of 7500 Å to about 1 millimeter.

Insolation Solar energy received by the earth.

Interferometer Any instrument that recombines electromagnetic radiation with a known change in phase.

International Astronomical Union (IAU) The international organization of astronomers.

Ionosphere The shell of atmosphere lying between 70 kilometers and 320 kilometers above the surface of the earth. It contains a high density of ionized atoms and particles.

J

Julian Day (Julian date) Day numbers beginning at noon and starting at an arbitrary date B.C. Used especially by variable-star observers.

K

Kiloparsec 1000 parsecs, equivalent to 3200 light years.

Kinetic energy Energy due to mass and motion ($\frac{1}{2}mv^2$).

L

Leap year That year every four years when an extra day is added to the month of February to keep the civil calendar in step with the sun. Century years are not leap years.

Luminescence Visible glow of a material induced by invisible radiation.

M

Magnetometer An instrument for measuring a magnetic force.

Main sequence The concentrated band of stars in a luminosity–temperature (H-R) diagram.

Mass The property of a body that resists a change in motion. It is a unique property and, in the body's own reference frame, remains constant regardless of where the body is located.

Megaparsec 1 million parsecs, equivalent to 3,200,000 light years.

Microwave Short radio waves; their wavelengths lie between 1 centimeter and 1 meter.

Minor planet Any of thousands of small bodies in orbit around the sun with mean orbital radii between the distances of Mars and Jupiter.

Molecule Two or more atoms chemically united forming the smallest unit possessing the properties of a compound.

Monochromatic Literally, one color. Usually light of a sufficiently narrow wavelength that can be treated as having a single wavelength.

N

Nadir The direction opposite the zenith.

Nebula (1) A cloud of gas surrounding a star. (2) An interstellar gas cloud.

Node The intersection points of an orbital and an arbitrary plane, such as the plane of the sky.

O

Objective In astronomy, the main lens or mirror of a telescope.

Oblate A circle or sphere flattened at the poles.

Observatory The astronomer's laboratory where his telescopes are mounted.

Occultation A larger astronomical body passing between the observer and a smaller body.

Opacity The blocking of radiation.

Opposition A planet is in opposition when it, the earth, and the sun are in a line.

P

Parallax The apparent change in location of a body due to a change of position of the observer.

Parameter A measurable quantity characteristic of a system to be described.

Parsec A parallactic second, equivalent to 3.26 light years.

Penumbra (1) The partial shadow in a solar and lunar eclipse. (2) The gray outer regions of a sunspot.

Periastron The closest point to the center of the primary star in a stellar orbit.

Perigee The closest point to the center of the earth in an earth orbit.

Perihelion The closest point to the center of the sun in a solar orbit.

Photosphere That level where the atmosphere of the sun becomes opaque. The visible surface of the sun.

Planet A satellite of a star, in nearly circular orbit, which was never capable of shining by a self-sustained energy reaction.

Plasma An ionized gas.

Platonic year The time required for the earth to complete its precessional cycle—approximately 25,800 years.

Polarizer Any substance that polarizes light or radiation passing through it.

Precession (1) The slow drift of the poles of a spinning body, due to a force or set of forces acting to try to tip its poles. (2) In astronomy, when unqualified, it usually means the change in coordinates due to the precession of the earth's poles.

Primary mirror The main mirror of a reflecting telescope.

Prism A wedge-shaped piece of glass used to disperse light into its familiar rainbow colors in spectroscopy.

Proper motion The cross motion of a star as seen projected on the sky and always given in seconds of arc (usually per year but occasionally per century).

Protostar The central region of a condensed cloud that is about to become a self-luminous star.

Q

Quadrature A planetary elongation 90° east or west of the sun.

R

Radial velocity The line of sight velocity of an object given in kilometers per second.

Radiant Usually refers to the point from which a meteor shower appears to be coming.

Radiation (1) The processes of emitting energy. (2) Often used in terms of the energy emitted.

Radio astronomy That branch of astronomy dealing with the study of the cosmos at radio wavelengths.

Relativity Generally one of two theories by A. Einstein. The special theory treats time and distance as depending upon the motion of the object and the observer. The general theory relates the structure of space to gravitation.

Retrograde The apparent westward motion of a planet through a star field.

Right ascension The astronomical coordinate measured along the celestial equator eastward from the vernal equinox. It is usually given in units of time but occasionally in angular measure.

S

Satellite (1) Any body that orbits a planet. (2) Also, any inferior body that orbits a larger body.

Secondary mirror Usually the mirror at the front end of a reflector with the purpose of changing the direction of the light reflected to it by the primary mirror.

Sedimentary rocks Rocks formed by deposition, either by settling out of water or by precipitating out of a solution.

Seeing The atmospheric effects on an image as seen through a telescope.

Solstice Times when the sun appears to stand still in the annual north and south motion.

Spectrogram A photograph of a spectrum produced by a spectrograph.

Spectrograph An instrument that renders light parallel upon a dispersing agent (prism or grating) and then photographs the resulting spectrum.

Spectroheliograph A special instrument that photographs the sun in monochromatic (one-color) light.

Specular reflection Reflection, as from a mirror.

Spicule Small grass-like jets at the surface of the sun.

T

Telescope A somewhat misnamed instrument used by astronomers to collect radiation from celestial objects.

Telluric lines Lines in a spectrum arising from atoms and molecules in the earth's atmosphere.

Thermocouple A junction of two wires of different metals that changes voltage when the amount of heat falling upon it changes.

Totality Commonly used during total solar and lunar eclipses to denote the total phase of the eclipse.

Transit (1) A smaller astronomical body passing between the observer and a larger body. (2) A celestial body crossing the observer's meridian. (3) A telescope mounted in such a way that it can observe an object only when that object crosses the local meridian.

Transverse Lying across. Transverse vibrations are vibrations across the direction of motion or perpendicular to the line of sight.

Triple alpha process A nuclear process, first discussed by E. Salpeter, where three helium nuclei combine directly to form a carbon nucleus.

Turbulence Irregular, random motions in a fluid or gas.

U

Ultraviolet radiation That region of the spectrum extending from a wavelength of 100 Å to that of about 4000 Å.

Umbra (1) The black shadow in a solar or lunar eclipse. (2) The dark center of a sunspot.

Universal time The local time of the $0^h0^m0^s$ meridian (Greenwich, England).

Universe The totality of physical reality.

X

X-radiation That region of the spectrum extending from a wavelength of 8 Å to one of 100 Å.

Z

Zodiac The band centered on the ecliptic divided into 12 equal sections. Each section was assigned a constellation and sign by the ancients.

ENGLISH–METRIC CONVERSION UNITS

The principal advantage of the metric system over the English system is that the metric system is based upon powers of ten. Any powers-of-ten system is as good as any other, but the metric system has the advantage that it has been adopted by more people than any other single system.

1 inch = 2.54 centimeters
1 foot = 30.48 centimeters = 0.3048 meter
1 yard = 91.44 centimeters = 0.9144 meter
1 mile = 160930 centimeters = 1609.3 meters = 1.6093 kilometers

1 ounce = 28.3495 grams = 0.0283 kilogram
1 pound = 453.6 grams = 0.4536 kilogram

1 pint (fluid) = 47.32 centiliters = 0.4732 liter
1 quart (fluid) = 94.64 centiliters = 0.9464 liter

1 kilometer = 0.6214 miles
1 meter = 1.0936 yards
1 centimeter = 0.3937 inch
1 liter = 2.1134 pints (fluid)
1 gram = 0.0353 ounce

Brief table of decimal multiples

Decimal	Power notation	Prefix	Symbol
0.001	10^{-3}	milli-	m
0.01	10^{-2}	centi-	c
0.1	10^{-1}	deci-	d
1	10^0		
10	10	deca-	da
100	10^2	hecto-	h
1,000	10^3	kilo-	k
1,000,000	10^6	mega-	M
1,000,000,000	10^9	giga-	G

Powers-of-ten notation is extremely convenient, as can be ascertained from the last line of the table. The rules for multiplication and division in this notation are simple. For multiplication we have $10^a \times 10^b = 10^{a+b}$, thus $10^2 \times 10^5 = 10^7$. For division we have $10^c \div 10^d = 10^{c-d}$, thus $10^2 \div 10^5 = 10^{-3}$.

Greek alphabet

A	α	alpha	I	ι	iota	P	ρ	rho
B	β	beta	K	κ	kappa	Σ	σ	sigma
Γ	γ	gamma	Λ	λ	lambda	T	τ	tau
Δ	δ	delta	M	μ	mu	Υ	υ	upsilon
E	ε	epsilon	N	ν	nu	Φ	ϕ	phi
Z	ζ	zeta	Ξ	ξ	xi	X	χ	chi
H	η	eta	O	o	omicron	Ψ	ψ	psi
Θ	θ	theta	Π	π	pi	Ω	ω	omega

Great refracting telescopes (>65 cm)

Year	Optician	Observatory and location	Aperture (cm)	Focal length (cm)
1897	Alvan Clark	Yerkes Observatory, Williams Bay, Wisconsin	102	1935
1888	Alvan Clark	Lick Observatory, Mt. Hamilton, California	91	1760
1893	Henry Brothers	Observatorie de Paris, Meudon, France	83	1615
1899	Steinheil	Astrophysikalisches Observatory, Potsdam, Germany	80	1200
1886	Henry Brothers	Bischottsheim Observatory, University of Paris, at Nice, France	76	1600
1714	Brasher	Allegheny Observatory, Pittsburgh, Pennsylvania	76	1411
1894	Howard Grubb	Royal Greenwich Observatory, Herstmonceux, England	71	850
1878	Howard Grubb	Universitäts-Sternwarte, Vienna, Austria	67	1050
1925	Howard Grubb	Union Observatory, Johannesburg, South Africa	67	1070
1883	Alvan Clark	Leander McCormick Observatory, Charlottesville, Virginia	66	1000
1873	Alvan Clark	U. S. Naval Observatory, Washington, D. C.	66	990
1953*	McDowell	Mount Stromlo, Canberra, Australia	66	1100
1897	Howard Grubb	Royal Greenwich Observatory, Herstmonceux, England	66	680

* First used in Johannesburg, South Africa 1926

Great reflecting telescopes (> 3 meters)

Year	Observatory and location	Aperture (meters)
U.C.	Zelenchukskaya Astrophysical Observatory, U.S.S.R.	6.0
1948	Hale Observatory, Mt. Palomar, California	5.1
1973	Kitt Peak National Observatory, Kitt Peak, Arizona	4.0
U.C.	Cirro Tololo Inter-American Observatory, Cirro Tololo, Chile	4.0
U.C.	European Southern Observatories, La Silla, Chile	3.6
U.C.	French National Observatory (site to be decided)	3.5
1959	Lick Observatory, Mount Hamilton, California	3.0

Some radio telescope systems

Year	Type	Observatory and location
1963	1000-ft fixed spherical dish reflector, and 100-ft dish for interferometer	Arecibo Observatory, Arecibo, Puerto Rico (100-ft. at Los Canos)
1970	5 to 60-ft. paraboloids, interferometer, equatorial mounts	Radio Astronomy Institute, Stanford, California
1951	250-ft paraboloid, altazimuth mount	Jodrell Bank, England
1964	125-ft × 83 ft. elliptical bowl, altazimuth mount	Jodrell Bank, England
1970	328-ft. paraboloid, altazimuth mount	Max-Planck Institute for Radio Astronomy, Bonn, West Germany
1970	12 to 82-ft paraboloids, 1-mile baseline east–west	Westerbork, Netherlands
1959 1969	2 to 90-ft paraboloids, and a 130-ft paraboloid altazimuth mount, interferometer	California Institute of Technology, Owens Valley, California
1965	140-ft paraboloid, equatorial mount, and 300-ft paraboloid, transit	National Radio Astronomy Observatory, Green Bank, West Virginia
1966	210-ft paraboloid, altazimuth mount	Goldstone Tracking Station, Mojave Desert, California
1970	120-ft paraboloid, equatorial mount, and 600-ft × 400-ft parabolic cylinder, interferometer	Vermilion River Observatory, Illinois

A brief chronology of astronomy

ca. 3000 B.C.	the earliest known recorded observations are made in Babylonia
ca. 1400 B.C.	earliest known Chinese calendar
ca. 1000 B.C.	earliest recorded Chinese, Hindu observations
ca. 800 B.C.	earliest preserved sundial (Egyptian)
ca. 500 B.C.	Pythagorean school advances concept of celestial motions on concentric spheres
ca. 430 B.C.	Anexagoras explains eclipses and phases of the moon
ca. 400 B.C.	Philolaus speculates that the earth moves
ca. 400–300 B.C.	several cosmological systems involving moving concentric spheres proposed by Plato, Eudoxus, and others
ca. 350 B.C.	earliest known star catalog (Chinese)

ca. 250 B.C.	Aristarchus advances arguments favoring a heliocentric cosmology
ca. 200 B.C.	Eratosthenes measures earth's diameter
160–127 B.C.	Hipparchus develops trigonometry, analyzes generations of observational data, obtains highly accurate celestial observations
ca. A.D. 140	Ptolemy measures distance to moon; proposes geocentric cosmology involving epicycles
1054	Chinese observe supernova in Taurus
1543	Copernicus publishes *De Revolutionibus*
1572	Tycho Brahe observes supernova; immutability of celestial sphere cast in doubt
1546–1601	Tycho accurately measures motions of the planets
1608	Lippershey invents the telescope
1609	Kepler, using Tycho's measurements, shows planets move in ellipses.
1609	Galileo uses telescope to observe moons of Jupiter and crescent phase of Venus, thus lending support to Copernican hypothesis; Galileo conducts experiments in dynamics
1675	Romer measured the velocity of light
1686–1687	Newton's *Principia*: Newton combines the results of terrestrial and celestial natural philosophy to obtain the fundamental laws of motion and gravity
ca. 1690	Halley shows the big comets observed every 75 years are one and the same comet in elliptical orbit; he discovers proper motions of stars
ca. 1690	Huygens makes estimate for distance to stars, based upon assumption that the sun is merely a typical star
1727	Bradly observes aberration of starlight, conclusive proof of Copernican theory
ca. 1750	Wright proposes disk model for Milky Way
1755	Kant proposes nebulae are "island universes"; proposes solar system formed from rotating cloud of gas
1738–1822	W. Herschel constructs large telescopes; discovers Uranus; observes gaseous nebulae
1801	First asteroid discovered
1802	Solar spectrum first viewed
1838	first stellar parallax measured (by Bessel)
1840	Draper produces first astronomical photograph
1842	Doppler effect explained shift in visible spectrum
1843	the effect of motion on eight spectra is explained by Doppler
1845	Earl of Rosse discovers spiral structure of some "nebulae"

1846   Neptune discovered independently by Leverrier and Adams

1850–1900   Development of spectrum analysis; stellar spectra used for first time to obtain temperatures and compositions of stars

1877   Schiaparelli sees "canals" on Mars

1905   Einstein's special theory of relativity

1905–1920   Einstein develops his theory of gravitation; general relativity

1915   100-in. reflecting telescope constructed at Mount Wilson

1914   Slipher discovers that spiral nebulae are receding from us; Lemartre, DeSitter, and Eddington explain this phenomenon using general relativity

1924   Hubble measures distances to spirals and confirms the viewpoint that they are galaxies in their own right

1910–1930   Russell, Eddington, and others develop the theory of stellar structure

1920–1930   Shapley, Oort, Linblad investigate rotation of Milky Way galaxy

1931   Pluto discovered by C. Tombaugh

1930–1960   Nuclear physics develops; used to explain the energy source of the stars

1931   Jansky discovers extra-terrestrial radio radiation

1947–1960   astronomical instruments sent by rocket above earth's atmosphere

1949   Great 200-in. reflector went into routine operation

1951   Observation of hydrogen at λ21-cm

1957   Sputnik I orbits earth

1959   Russian space probe hits the moon

1961   Yuri Gagarin becomes first man in space

1963   discovery of quasi-stellar objects

1965   discovery of 3°K background radiation

1965   first close photographs of Mars by Mariner 4

1968   discovery of pulsars

1969   Apollo 11 lands first men on the moon

ELECTROMAGNETIC SPECTRUM

The electromagnetic spectrum extends from very long waves (low frequency) to very short waves (high frequency). The long waves are called electric waves and the short waves are called gamma rays. Between them lie the various radio and microwave regions through the infrared, visible, ultraviolet, and x-ray regions. The radio region is divided into very low frequency, low frequency, medium frequency, high frequency, very high frequency, and ultra high frequency where it overlaps the microwave region. The upper part of the microwave region is divided into super high frequency and extremely high frequency where it overlaps with the infrared region. Frequency may be converted to wavelength by the simple relation $\lambda = c/f$, where λ is the wavelength in the units of c, the velocity of light, and f, the frequency in Hertz (cycles per second). Thus the regions shown range from a wavelength of 300,000 kilometers down to 3×10^{-5} angstroms.

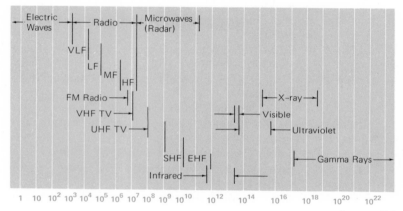

ABOUT THE AUTHORS

The late ROBERT H. BAKER is known to thousands of students for his two major textbooks, ASTRONOMY, now in its Ninth Edition, and AN INTRODUCTION TO ASTRONOMY, the present book in its Eighth Edition. Dr. Baker was a graduate of Amherst College, and he received his Ph.D. from the University of Pittsburgh in 1910. He served for a year as Assistant Professor of Astronomy at Brown University, and in 1911 he was appointed Professor of Astronomy at the University of Missouri, which post he held until 1922. He spent the year 1922–1923 as a Martin Kellogg Fellow at Lick Observatory. In 1923 Dr. Baker was appointed Professor of Astronomy at the University of Illinois, the position he held until his retirement in 1951. In 1931–1932 and again in 1938–39, he was a Research Associate of Harvard University while on sabbatical leave from the University of Illinois.

LAURENCE W. FREDRICK is Hamilton Professor of Astronomy and Chairman, Department of Astronomy, and Director, Leander McCormick Observatory, University of Virginia. He has been General Secretary of the American Astronomical Society since 1968. A graduate of Swarthmore College (B.A. 1952, M.A. 1954), he received his Ph.D. from the University of Pennsylvania in 1959. He is a member of the National Research Council, American Astronomical Society, International Astronomical Union, and the U. S. National Committee for the IAU.

From 1959 to 1963, Professor Frederick served as Astronomer at the Lowell Observatory. Previous to 1959 he was Research Associate at Flower and Cook Observatory and at Sproul Observatory. He has contributed to many professional journals, and his present program concerns the application of image intensifiers and other automated electronic devices to astronomy.